EXCHANGE IN OCEANIA

A graph theoretic analysis

PER HAGE
and
FRANK HARARY

CLARENDON PRESS · OXFORD
1991

Oxford University Press, Walton Street, Oxford OX2 6DP
Oxford New York Toronto
Delhi Bombay Calcutta Madras Karachi
Petaling Jaya Singapore Hong Kong Tokyo
Nairobi Dar es Salaam Cape Town
Melbourne Auckland
and associated companies in
Berlin Ibadan

Oxford is a trade mark of Oxford University Press

Published in the United States
by Oxford University Press, New York

British Library Cataloguing in Publication Data
Hage, Per 1935–
Exchange in Oceania : a graph theoretic analysis.
1. Social anthropology. Applications of structural models
I. Title II. Harary, Frank
306.0228
ISBN 0–19–827760–1

Library of Congress Cataloging in Publication Data
Hage, Per, 1935–
Exchange in Oceania : a graph theoretic analysis / Per Hage and Frank Harary.
p. cm
Includes bibliographical references and index.
1. Ethnology—Oceania—Mathematical models. 2. Ceremonial exchange—Oceania—Mathematical models. 3. Graph theory 4. Oceania—Social life and customs—Mathematical models.
I. Harary, Frank. II. Title.
GN663 . H34 1991 306′. 099—dc 20 90–45810
ISBN 0–19–827760–1

Typeset by Latimer Trend & Company Ltd, Plymouth
Printed in Great Britain by
Biddles Ltd, Guildford & King's Lynn

To J. A. Barnes

The whole tribal life is permeated by a constant give and take.

Bronislaw Malinowski, *Argonauts of the Western Pacific*

Mathematical reasoning consists in constructing a diagram according to a general precept, in observing certain relations between parts of that diagram not explicitly required by the precept, showing that these relations will hold for all such diagrams, and in formulating this conclusion in general terms. All valid reasoning is in fact thus diagrammatic. This, however, is far from being obviously true.

Charles Sanders Peirce, *Lessons from the History of Science*

ACKNOWLEDGEMENTS

PARTS of this book were conceived during the first author's visit to the Cyclades and Paris in the summer and fall of 1983. The book was outlined at a conference in Hoboken in the winter of 1984 when the second author was Research Professor of Computer Science and Electrical Engineering at the Stevens Institute of Technology, and written at the University of Utah and the University of Michigan in 1985. The revisions were completed in Las Cruces in the spring of 1986 when the second author was Distinguished Visiting Professor of Mathematics and of Computer Science at New Mexico State University. We wish to thank our hosts during our travels: Svetlana Alexeieff Rockwell in Sifnos and Paris, Frank Boesch in Hoboken, and Keith Phillips in Las Cruces. The first and second authors thank Andrea Morguloff-Hage and Lucia Muñoz Hayakawa respectively for their most enthusiastic support.

We owe a substantial debt of gratitude to many individuals for their contributions and assistance. John Barnes went over the entire first draft and made numerous precise, subtle and informed comments on the anthropology and on the application of the structural models. Fred Buckley and Gary Chartrand made an energetic and perceptive study of the graph theoretic material and clarified points in our presentation. Claude Lévi-Strauss and Robin Fox provided helpful and important comments on the anthropological content of several chapters, and encouraging comments on the entire work. Mac Marshall clarified various points of Oceanic ethnology and, together with William Alkire, William A. Lessa, and John Fischer, supplied crucial facts of Carolinian ethnography. Saul Riesenberg and Thomas Gladwin made available unpublished field notes which enabled us to complete the reconstruction of the overseas exchange network analysed in Chapter 3. Many colleagues commented on ethnographic or theoretical aspects of individual chapters: Alan Barnard, Aletta Biersack, Frederick Damon, Anthony Good, Ward Goodenough, Roger Green, Bernard Grofman, Charles Hockett, Geoffrey Irwin, Raymond Kelly, Patrick Kirch, Mervyn Meggitt, Douglas Oliver, Garth Rogers, Richard Salisbury, Erik Schwimmer, Andrew Strathern, John Terrell, and Roy Wagner. Robert Anderson played a vital role as the party responsible for the superb Oceania collection at the University of Utah Library. Brent James with Karen Stephenson did the computer work for Chapters 3, 4, and 5. As before in *Structural Models in Anthropology*, Ursula Hanly transformed a handwritten manuscript into a perfect typescript, drew the figures, and helped to maintain the smooth and uninterrupted flow of the work. To one and all we express our thanks.

It is a pleasure to dedicate this book to John Barnes, who was our gracious host at Cambridge University in 1980–1 when we launched our project in structural analysis. Among other things, he provided a pleasant, spacious office for us in the Social and Political Sciences Committee building while we held visiting fellowships at Robinson College and Churchill College respectively.

An outline of Chapter 3 was given as an Alfred P. Sloan Foundation Lecture, 'A Graph Theoretic Analysis of Voyaging, Exchange and Settlement in Micronesia' at the Population Studies Center, University of Michigan. The models in Chapters 2 and 5 were introduced in a paper, 'Melanesian Exchange Systems' presented at the First Hoboken Conference on Graph Theory at the Stevens Institute of Technology. A survey of the uses of graph theory in the analysis of exchange structures was presented in a talk, 'Some Genuine Graph Theoretic Models in Anthropology' given at the Computing Research Laboratory, New Mexico State University and appeared as an article in the *Journal of Graph Theory* special issue celebrating the 250th anniversary of the discovery of graph theory by Leonhard Euler in his 1736 paper, *Solutio problematis ad geometriam situs pertinentis*.

We make grateful acknowledgement to the following for permission to reproduce the figures and maps listed below:

The Royal Anthropological Institute for Fig. 1.5; P. V. Kirch and Cambridge University Press for Fig. 1.8 from *The Evolution of the Polynesian Chiefdoms* (1984); R. C. Green and Harvard University Press for Fig. 1.10 from 'Lapita' in J. D. Jennings (ed.), *The Prehistory of Polynesia* (1979); the Institut d'Ethnologie for Fig. 1.11 from M. Leenhardt, *Notes d'ethnologie néo-calédonienne* (1980); the University of Sydney for the map in Fig. 2.11 from F. L. S. Bell, 'Warfare among the Tanga', *Oceania*, vol. 5 (1935), and for Figs. 6.14–6.17 from R. F. Fortune, 'A Note on Some Forms of Kinship Structure', *Oceania*, vol. 4 (1933); W. H. Alkire and the *Canadian Journal of Anthropology* for Map 3.1 from 'Technical Knowledge and the Evolution of Political Systems in the Central and Western Caroline Islands of Micronesia', *CJA*, vol. 1 (1980); W. de Gruyter for Fig. 3.1 from H. Damm and E. Sarfert, *Inseln um Truk*, in Ergebnisse der Südsee-Expedition 1908–10 (1935); W. A. Lessa and the American Anthropological Association for Figs. 3.8 and 3.9 from 'Ulithi and the Outer Native World', *American Anthropologist*, vol. 52 (1950); G. J. Irwin for Fig. 4.9 from 'Pots and Entrepôts: A study of settlement, trade and the development of economic specialization in Papuan prehistory', *World Archaeology*, vol. 9 (1978); G. J. Irwin and *Mankind* for Fig. 4.10 from 'The Emergence of a Central Place in Coastal Papuan Prehistory: A theoretical approach', *Mankind*, vol. 9 (1974); Oxford University Press for the cover drawing and Fig. 6.4. from F. E. Williams, *Drama of Orokolo* (1940); W. H. Alkire and the University of Illinois Press for Fig. 7.1 from *Lamotrek Atoll and Inter-Island Socioeconomic Ties* (copyright 1965 by the Board of Trustees of the University of Illinois); D. Lewis and Curtis Brown (Aust.) Pty. Ltd. Sydney, for Fig. 7.3 from *The Voyaging Stars* (1978); K. O. L. Burridge and the University of Sydney for Fig. 7.10a from 'Marriage in Tangu', *Oceania*, vol. 29 (1958); and the Bishop Museum Press for Fig. 1.14 from P. H. Buck *Samoan Material Culture* (1930) and Fig. 8.12 from E. G. Burrows, *Ethnology of Futuna* (1936).

P.H.
F.H.

CONTENTS

1

Graph Theory and Exchange Structures

> The object of social structure studies is to understand social relations with the aid of models.
>
> Claude Lévi-Strauss, 'Social structure'

The eventual application of mathematics to the study of social structure in Oceania was clearly foreseen by W. H. R. Rivers, who, in the introduction to his monumental *History of Melanesian Society*, observed:

> It would seem as if the only really satisfactory plan would be to employ symbols for the different relationships and it is probable that the time will come when this will be done and many parts of the description of the social systems of savage tribes will resemble a work on mathematics in which the results will be expressed by symbols, in some cases even in the form of equations (1914: 10).

Rivers was speaking of kinship, but surely the fundamental requirement is a mathematical model suitable for the analysis of social relations in general.

Our purpose is to advance the structural analysis of exchange relations in Oceania through the application of graph theory and to provide a general foundation for further research on this subject. In our previous work, *Structural Models in Anthropology* (*SMA*), we introduced graph theory as a family of models for the study of social, symbolic, and cognitive relations. We now focus on the topic of exchange in a part of the world where its practice and forms are known for their exuberance and complexity. We apply and extend these graphical models and introduce new ones in the analysis of ethnographic data from Melanesia, Micronesia, and Polynesia. Our intention is not to give an encyclopaedic account or even a detailed survey of exchange forms in Oceania, but rather to demonstrate, with reference to diverse empirical cases, how graph theoretic models can contribute to the innovative as well as the rigorous analysis of these forms. While agreeing completely that the role of the structural anthropologist is only to discover and study 'structured or structurable islands' that bathe in an 'ocean of contingency' (Lévi-Strauss in Bucher 1985), we none the less wish to indicate that the islands are more numerous and varied than commonly imagined, that the ocean resembles the Pacific more than it does the Atlantic.

To set the stage for our presentation we shall proceed as follows in Chapter 1. We begin with a brief overview of research on exchange in Oceania and state in general terms the distinctive advantages of using graph theory as a source of structural models. We then introduce, in an informal way, the simplest kinds of graphs, those which we call primordial because they are the building blocks of all larger structures and the simplest expressions of elementary structures. We wish to emphasize right at the outset that the ultimate value of graph theory for anthropology will depend not just on the use of its pictorial representations, but also on the application of its theorems. Hence we next illustrate, under the heading of 'necessary relations', the kinds of conclusions which theorems permit one to draw in the analysis of empirical structures. We illustrate the different types of graphs and some of their uses, choosing examples drawn from the anthropological literature in order to demonstrate their naturalness and implicit appeal as models of exchange structures. In conjunction with this informal survey we preview the research to come in the following chapters. We conclude by briefly reviewing a few basic concepts from set theory which underlie our definition of a graph.

The presentation throughout this book is organized in a logical graph theoretic manner and gives the basic essential concepts for a general model of exchange. Without any doubt further research will add to these concepts as well as discover new empirical applications. The account, as in *SMA*, is self-contained. It presupposes no background in graph theory and should be readily accessible to the non-mathematical reader.

Prologue

A mathematical, graph theoretic analysis of Oceanic exchange systems is overdue and urgent. It should have commenced much earlier given the stimulating ethnographic beginning in Malinowski's (1922) monograph on the *kula* ring, and the brilliant potential theoretical foundation contained in Williams's (1932,

1934, 1936) and Fortune's (1933, 1935a) alliance theories of exogamy and their model building of specific forms of marriage exchange.[1] In *Argonauts of the Western Pacific*, Malinowski gave an immensely detailed account of an endlessly fascinating exchange system,[2] and, as Lévi-Strauss (1950) has observed, one which presents the Melanesians themselves as implicit authors of the modern theory of reciprocity. In *Papuans of the Trans-Fly*, Williams formulated a theory of exogamy which sounds distinctly modern because, like Lévi-Strauss (1969), he interpreted the exchange of women as the 'supreme gift' in a larger set of exchanges occurring at all levels of society:

Pursuing this argument I suggest that the exchange of girls in marriage falls into line with these other exchanges. The unmarried girl is, so to speak, the supreme gift. The insistence on reciprocity is the same, and the transaction serves the same purpose, that of binding the contracting groups together in a bond of mutual restraint and fellowship. We have seen that groups united by marriage acknowledge this bond; that they maintain it by reciprocal services, and that the norm of conduct between them is one of respect and goodwill.

On the face of it (and in this hypothesis I am deliberately eschewing the kinship or intra-group attitude towards exogamy) there is no reason why the men of the group should not marry their own girls, just as they might live entirely on the food which they themselves produce. But on the contrary they insist on giving and receiving, and the function—I hesitate to call it the motive, because it is not clearly felt—which underlies this giving and receiving is, I suggest, that of enabling the group to enter into relations with other groups. They can do so by other kinds of exchange; but the strongest bond they can form with outsiders is this of exchanging girls in marriage (Williams 1936: 168).[3]

[1] In emphasizing Williams's and Fortune's theoretical contributions, one should not lose sight of their major ethnographic achievements: Williams (1928, 1930, 1936, 1940, 1977) and Fortune (1932, 1935b). These works remain exemplars of richness and clarity. Mention should also be made of Fortune's (1942) linguistic monograph and his work in Highland New Guinea. According to J. B. Watson (1964: 1), 'Fortune's 1935 field work among the Kamano of the Eastern Highlands [Fortune 1947] should perhaps be considered the pioneer ethnography of the Highlands, although little of it has been published.'

[2] Macintyre's (1983a) *kula* ring bibliography contains 625 entries.

[3] In *The Elementary Structures of Kinship*, Lévi-Strauss maintains that 'It would then be false to say that one exchanges or gives gifts at the same time that one exchanges or gives women. For the woman herself is nothing other than one of these gifts, the supreme gift among those that can only be obtained in the form of reciprocal gifts. The first stage of our analysis has been intended to bring to light this basic characteristic of the gift, represented by women in primitive society, and to explain

More specifically, Williams analysed the structural consequences of one marriage form, patrilateral cross-cousin marriage, while Fortune, in a most remarkable, purely formal exercise, anticipated Lévi-Strauss's models of restricted, generalized, and discontinuous exchange.[4] At the time of his Malinowski Lecture, however, Forge (1972), arguing for the primacy of exchange principles over descent rules at all levels of New Guinea social structure, could cite only two comprehensive accounts of exchange systems in Melanesia: A. Strathern's (1971) and Young's (1971), on competitive exchange in the New Guinea Highlands and in the Massim, respectively. And the only structural models he mentioned were Bateson's (1958) 'diagonal and direct duality', general principles which may provide a broad characterization of big-man systems, or chiefdoms (Shore 1982), but which do not suffice for their analysis.

In the modern era, studies of exchange have proliferated as a result of research which was actually done in the years preceding Forge's lecture. Conceptual advances occurred on many fronts. One must cite in particular Barnes's (1962) critique of Africanist models and his recommendation that exchange and alliance be considered as significant aspects of Highland New Guinea social structure; Burridge's (1959) recognition of the inherent importance of siblingship and exchange over descent in Tangu; Rappaport's (1968) introduction of systems thinking to the study of ritual and environment interactions; R. Wagner's (1967) analysis of native symbolizations of social groups and alliance relations in Daribi; Schwartz's (1963) proposed models of areal networks; Harding's (1967) substantivist analysis of trade in the Vitiaz Strait; Sahlins's (1963) typological contrast between Melanesian big men and Polynesian chiefs; and the papers in the Glasse and Meggitt (1969) symposium on marriage and ceremonial exchange in the New Guinea Highlands. In Micronesian studies, mention must be made of Alkire's (1965) cultural ecological analysis of inter-island relations in the Western Carolines and Marshall's (1972) explicit

the reasons for this. It should not be surprising then to find women included among reciprocal prestations; this they are in the highest degree, but at the same time as other goods, material and spiritual' (Lévi-Strauss 1969: 65).

[4] Williams was keenly aware of his own originality in formulating an alliance theory of exogamy. In a footnote to the chapter 'Exchange Marriage and Exogamy' in *Papuans of the Trans-Fly*, he noted that 'A similar theory, to which, however, the writer owes nothing, is R. F. Fortune's in *The encyclopaedia of social sciences* ...' (1936: 171).

application of digraph theory to the analysis of transactions in land, marriage, and adoption on Namoluk in the Eastern Carolines.

Studies of exchange now cover a variety of subjects including kinship and marriage (Biersack 1982; Clay 1975; Kelly 1974; McDowell 1980; D. R. Smith 1983), ceremonial sequences (Meggitt 1974), communication networks (Hage and Harary 1981a), chieftainship (Brunton 1975; Persson 1983; Hage, Harary, and James 1986), social stratification (Friedman 1981), trade networks (Hage 1977; Irwin 1974), areal systems (Gewertz 1983), gender relations (A. B. Weiner 1976), ritual transactions (Herdt 1984), and pollution beliefs (Hage and Harary 1981b). For Melanesia, which dominates the literature on Oceania and therefore the ethnographic material in this book, there are reviews on special topics such as marriage exchange (M. Strathern 1984) and symposia on specific institutions such as the *kula* ring (Leach and Leach 1983) and the *hiri* (Dutton 1982). Comparative work has begun, most notably Rubel and Rosman's (1978) study of ceremonial feasting and marriage exchange in New Guinea. Many such studies, whether based on structuralism, dialectics, ecology, semiotics, or social network ideas, use mathematical concepts, formally or informally. Thus there are references to 'lattices' (Kelly 1974), 'structural isomorphisms' (Rubel and Rosman 1978), 'structural and numerical symmetry' (Tuzin 1976), 'equivalence relations' (Sahlins 1976), 'network distance' (Terrell 1986), 'network connectivity' and 'centrality' (Irwin 1978, 1983; Brookfield and Hart 1971), 'exchange matrices' (Gregory 1982), 'conditional probability models' (R. Wagner 1972), 'Hamiltonian cycles' (Schwimmer 1973), 'planar graphs' (Irwin 1974), 'bipartite' social organization (Serpenti 1965), and 'Klein groups' (Mosko 1985).

With the exchange paradigm finally and clearly in the ascendancy there is a definite need for an appropriate branch of applicable mathematics—one that is descriptively adequate, analytically rich, and accessible to anthropologists. We propose graph theory as a general model for the analysis of exchange structures for the following reasons. First, through its language and theorems, graph theory can accommodate the variety of forms actually found in Oceanic societies and provide accurate descriptions, sound classifications, and structurally informative analyses. For example, given the underlying graph of an exchange structure, one can say precisely what is meant by the word 'cycle', one can compare the cyclic properties of different structures, and one can make deductions about the cyclic properties of a particular structure. Secondly, graph theory contains techniques for calculating such quantitative features of exchange networks as 'centrality', 'betweenness', and 'rank'. Thirdly, through its associated matrix methods, graph theory provides mechanical procedures for the accurate and rapid analysis of large exchange systems. These methods also encourage complete descriptions of exchange systems which permit other investigators to rethink them, thereby overcoming a common deficiency of traditional ethnography properly lamented by Forge. Fourthly, graph theory is a natural source of models for simulating exchange processes where historical data are absent. Thus one can study the flow and distribution of goods in a network, or the alternative ways in which a given type of network could have evolved. Fifthly, graph theory offers techniques for the purely logical enumeration of structural forms, which means that all possible permutations of a structure can be determined in advance and studied as a group rather than in piecemeal fashion. Sixthly, graph theory offers succinct and precise notations for commonly occurring structures, thereby obviating unduly intricate and ambiguous verbal descriptions. Finally, graph theory, through its visual presentations, gives a clear, intuitive perception of the logical basis of exchange structures in place of makeshift adaptations of mathematical symbolism.

Primordial Structures

What is a graph? Intuitively, a graphical structure consists of a finite set of points joined by a set of lines. If the lines are directed (have arrows on them) the structure is called a directed graph D; if they are not, it is called an undirected graph G. The term graph is sometimes used generically to include either type, and also specifically, as an abbreviation for undirected graphs. The meaning should be clear from context. We can introduce graphs by distinguishing, first of all, the three primordial structures shown in Fig. 1.1. The first structure, a directed graph or digraph D, consists of two points, a transmitter (t) and a receiver (r), joined by a directed line or arc. This is the smallest (non-trivial) directed path (as just a single point constitutes the so-called trivial directed path). It is also the smallest (non-trivial)

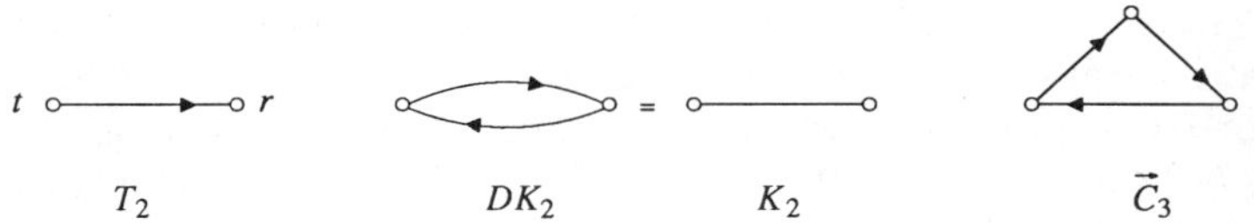

FIG. 1.1. The primordial exchange structures

asymmetric relation. For convenience it is designated as T_2. It is the building block of all directed graphs D. The second structure consists of a symmetric pair of arcs or, equivalently, for economy of depiction, of a single undirected line joining the two points, in which form it is a graph G. It is called K_2 and is the smallest path, the smallest symmetric relation, and the smallest non-trivial graph. It is the building block of all (undirected) graphs G. The third structure, $\vec{C}_3$, is the second smallest directed cycle; the smallest one is the symmetric pair of arcs DK_2, while $\vec{C}_3$ is asymmetric.

We call these graphs primordial because they combine to form larger graphs, and because they are minimum expressions of certain basic empirical structures. This is easily seen in the case of exchange relations. In Lévi-Strauss's (1969) classification of the elementary structures of kinship, K_2 is the simplest model of restricted exchange (the wife-giving relation is symmetrical), as in bilateral cross-cousin marriage, and $\vec{C}_3$ is the simplest model of generalized exchange (the wife-giving relation is asymmetrical), as in matrilateral cross-cousin marriage. In Sahlins's (1972) classification of economic exchange, T_2 is the model of generalized reciprocity, defined as a 'sustained one-way flow', and K_2 is the model of balanced reciprocity, defined as direct symmetrical exchange.

In *The Elementary Structures of Kinship*, Lévi-Strauss maintains that generalized and restricted exchange and also discontinuous exchange (patrilateral cross-cousin marriage, which is K_2 over two generations) are equally present to the human mind 'at least in unconscious form, and that it cannot evoke one of them without hinting of this structure in opposition to—but also in correlation with—the two others' (1969: 464). Godelier's (1982) work on the Baruya, a society in Highland New Guinea, contains some interesting support for this view and more generally for the ethnographic primordiality of all three of our graphs, K_2, T_2, and $\vec{C}_3$. Godelier first describes Baruya marriage as restricted exchange (K_2), under which he includes sister exchange and variant forms such as patrilateral cross-cousin marriage which are reducible to it. Like a number of other societies in New Guinea, Etoro for example (Kelly 1974), the Baruya consider that the life force or vital substance is contained in sperm. Sperm is exchanged when a senior male transmits it (orally) to a junior male to augment or ensure his growth (T_2). Godelier comments on the asymmetry of this exchange, noting its resemblance in this respect to generalized exchange:

Pour ceux qui opposeraient un peu trop facilement et, il faut le dire, mécaniquement, les sociétés à échange restreint des femmes et les sociétés à échange généralisé, comme si les premières ignoraient le principe sur lequel reposent les secondes, l'exemple des Baruya offre clairement le moyen de corriger ces vues. Car si les Baruya privilégient le principe de l'échange restreint des femmes pour instituer les rapports de parenté entre les lignages et les individus, ils appliquent une sorte de principe d'échange général de sperme entre tous les hommes qui n'appartiennent pas à la sphère de l'échange des femmes, pour fabriquer un homme et instituer la domination masculine. Les deux principes de l'échange existent donc dans leur pensée, mais sont mis en pratique dans des sphères distinctes de leur existence sociale (Godelier 1982: 95).

Godelier notes, however, that sperm exchange, which he also calls *échange général*, is not cyclical like *échange généralisé*.[5] But elsewhere in his book he describes a Baruya solution to a problem of debt which in fact shows that $\vec{C}_3$ is also present:

Chez les Baruya, en effet, il arrive souvent qu'un lignage A doive une femme à un lignage B, alors qu'au même moment ce lignage B doit une femme à un lignage C. Après accord entre ces trois groupes, le groupe A transfère directement une de ses femmes au lignage C, compensant d'un coup la dette de B envers C et de A envers B (Godelier 1982: 57).

We note one other interesting feature of Baruya exchange. Parallel to the exchange of sperm between men is the exchange of milk between women: just as young men drink from the penes of

[5] Like Godelier and practically everyone else, we define generalized exchange, here and elsewhere in the book, as cyclical. One reader, however, argues that generalized exchange can be cyclical but need not be: 'In a system of spouse-giving/spouse-taking inequality, we can have a pool of virgin princesses (or less likely princes) at one end of the chain and a pool of unmarried beggars (or beggaresses) at the other without any cycle being formed.' Such a possibility is suggested by our reference to sacred maids in Pukapuka in Chapter 8 and by Gough's (1955) description of Nayar. Lévi-Strauss (1963a) describes such a system as one which will either succumb to its contradictions or else be transformed temporarily or locally into a cyclical order.

senior men, so prepubescent girls may drink from the breasts of nursing mothers. Since, in Baruya conception, milk is really a product of sperm, this does not, according to Godelier, represent a *contre-modèle féminin* so much as a different but structurally related confirmation of male superiority. These two forms of exchange are related as sex duals—each digraph is obtained from the other by changing the sex of all the points as shown in Fig. 1.2.[6] In Chapter 6 we introduce sex duality in graphs as a method for generalizing certain types of kinship structure.

FIG. 1.2. Sex dual exchange structures in Baruya

Necessary Relations

Graphs can be represented by pictures. They are analysed by the application of theorems. By specifying properties of graphs that necessarily follow from given conditions, theorems enable one to draw conclusions about certain properties of a structure from knowledge about other properties. Thus the answer to many research questions depends not on the accumulation of more data but on the examination of the structural properties of graphs. We can give a brief example using the concept of a 'functional digraph' as a model of an informal exchange structure in New Guinea.

A function is conventionally defined as a relation whose ordered pairs (u, v) have the property that for each first element u, there corresponds a unique second element v. For example, the kinship relation 'has as mother' is defined as the set of all ordered pairs (u, v) such that u has v as mother, i.e. v is the mother of u. In a *functional digraph* D, there is exactly one directed line from each point. Thus D captures structurally the defining property of a function.

An interesting empirical realization of a functional digraph is a

[6] Another example of sex duality is *Bräutigamspreis* as reported by H. Thurnwald (1938) and later confirmed by J. Nash (1978) for the Nagovisi of south Bougainville.

1-choice structure which arises whenever each person in a group chooses exactly one other person according to some criterion. In *Exchange in the Social Structure of the Orokaiva*, Erik Schwimmer (1973) describes taro exchanges between pairs of households in a Papuan village. Taro gifts symbolize one household's desire for intimate relations with another household, and the quantity of the gifts reflects the level of intimacy desired. On the basis of transactions over a period of two months Schwimmer distinguishes first from second and third preferential partners. To illustrate, a fragment of the digraph of first choices of the households is shown in Fig. 1.3.

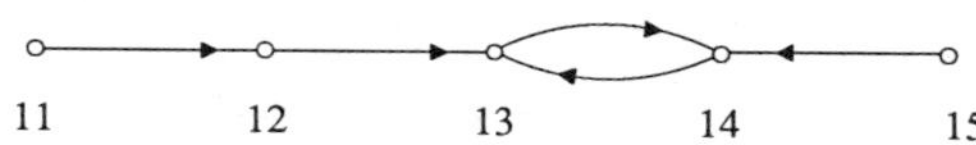

FIG. 1.3. A digraph of first preferential partners in taro exchanges between Orokaiva households

One can see that the digraph in Fig. 1.3 contains a 2-cycle, which is the structure DK_2 shown in Fig. 1.1, of intimacy, with all other choices directed towards it. One could perhaps regard the members of this cycle as the 'in-group'. In this particular case, it is known from Schwimmer's monograph that households 13 and 14 belong to leading families in the village, that 12 is an 'autocratic' type of leader with few intimates, and that 15 is someone anxious to stay on friendly terms with everyone. The question arises as to whether such a structure of intimacy is accidental or whether it is found generally. The answer is given by Theorem 2.4 on functional digraphs in Chapter 2. In order to describe it informally, we show all the eleven connected functional digraphs (those that hang together in one piece) with five points in Fig. 1.4. Although Fig. 1.3 is drawn differently, it is isomorphic to (structurally the same as) the second digraph in the second row of Fig. 1.4.

Informally stated, the criteria of Theorem 2.4 specify that each connected functional digraph either consists of a single cycle, like the last digraph of Fig. 1.4, or contains just one cycle with all paths directed towards it, as in all the remaining structures. Thus the

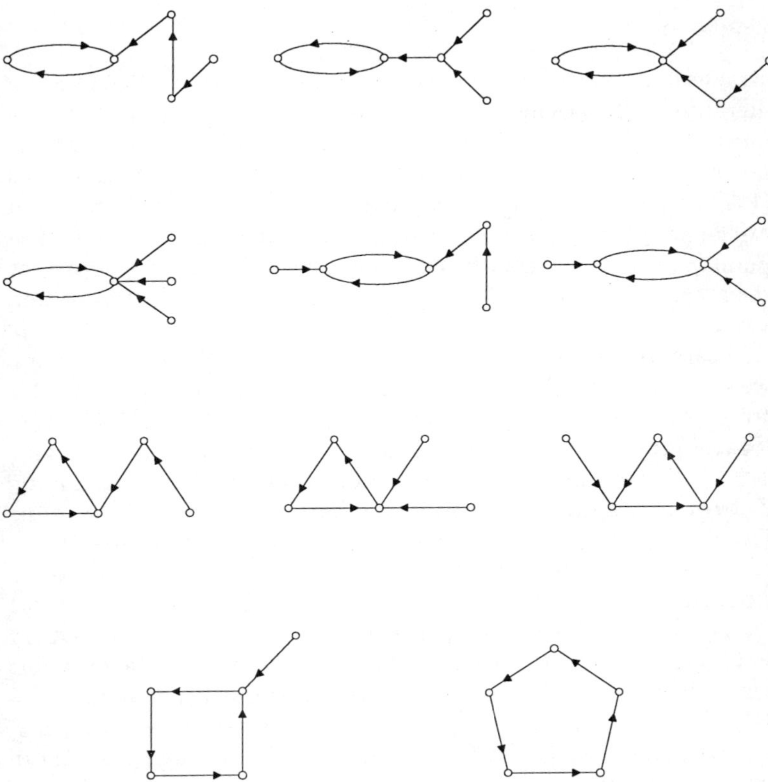

FIG. 1.4. The connected functional digraphs with five points

empirical possibilities of 1-choice exchange structures are limited by the nature of functional digraphs.

We will encounter functional digraphs again, in Chapter 5 on Markov chain models of the flow of valuables in ceremonial exchange networks, and in Chapter 6 on the enumeration of marriage exchange structures. The use of theorems is an essential part of any mathematical analysis. In Appendix 4, we explain the nature of a theorem and its proof and we give either the proof or the source of the proof for all theorems used.

Graphs and their Uses[7]

Graphs and some of their uses are already mentioned in the Oceanic literature. The following informal presentation will serve to illustrate different types of graphs and introduce the main topics taken up in the following chapters. Fig. 1.5 shows Malinowski's (1920) graph of the *kula* ring, the anthropological archetype of an exchange system. It is a map with a graph superimposed on it. The points in the graph are island communities and the lines represent the exchange of goods and two classes of shell valuables (*mwali* and *soulava*).

Clearly this is not a 'ring' in the sense of a single cycle, although we conjecture in Chapter 5 that it was so originally. Because this structure does 'ramify', certain communities are more favourably located than others. In an earlier application of graph theory (Hage 1977) it was shown that the dominant trading community, Tubetube, has the most central location. Like Mailu Island in the Coastal Papua network studied by the archaeologist Geoffrey Irwin (1974, 1978), Tubetube has an ecological inducement to trade extensively in the form of an insufficient carrying capacity to support its population. Like Mailu, Tubetube has specialized in pottery manufacture and, through the high development of a maritime technology, succeeded in exploiting its locational advantage in a large overseas network. In Chapter 4 we disentangle the concepts of centrality and connectivity as used in Irwin's (1983) analysis of both of these networks.

In an implicit graph theoretic analysis of the *kula* ring, Brunton (1975) has proposed that differential access to valuables—the *soulava* (necklaces) and *mwali* (armshells) that circulate in converse directions—is a determinant of chieftainship in certain communities. In order to simulate the flow and distribution of valuables in the *kula* ring, we model it as a Markov chain in Chapter 5 and use the results to give a generalization of Brunton's theory: we suggest that either of the extreme conditions of restricted or easy access to valuables could serve as an enabling condition for the emergence of political hierarchy.

The model for asymmetric exchange relations (no symmetric pairs

[7] In commemoration of the Norwegian graph theorist Oystein Ore, who wrote a very useful book with this title (Ore 1963). For an encyclopaedic account see Hage and Harary (1985).

FIG. 1.5. Malinowski's (1922) sketch map of *kula*. 'The dotted circles represent the *Kula* communities, the dotted squares represent the districts indirectly affected by the *Kula*.'

of arcs) is a directed graph or digraph D in which the lines have arrows, i.e. they are arcs. Roy Wagner (1967) uses such a model in his ethnography of the Daribi, a society in the Mount Karimui area of Papua New Guinea. According to Wagner, marriage exchange is organized by a rule of reciprocity between clans and constituent sibling groups called *zibi*. No two *zibi* should stand in the mutual relation of wife-taker and wife-giver, in order to avoid the contradiction of two groups pressing identical claims against each other. Through the transmission of maternal substance, 'blood', wife-givers own their sister's children and are thus entitled to continuing compensation payments from the husband's group. An asymmetrical relation between clans resulting from the superiority of wife-givers to wife-takers is avoided by the rule that every clan should 'back' or reciprocate a marriage by giving a wife in the opposite direction. This 'closes accounts' between them. Thus,

> if *zibi* A of Clan I gives a sister to *zibi* B of Clan II . . ., *zibi* B should reciprocate by giving a sister to Clan I, and she is given to *zibi* C [of Clan I], for *zibi* A are wife-givers. Thus, *zibi* B has discharged its obligation with regard to Clan I. *Zibi* A, however, has given a sister who has not been reciprocated, and *zibi* C has received a wife and must reciprocate by giving one to Clan II. It does this by giving its sister to *zibi* D . . . of the latter clan, for *zibi* B are its wife-givers, and thus discharges its obligation to Clan II. Now *zibi* D has received a wife from Clan I, and it discharges its obligation by giving a sister to *zibi* A . . ., thereby discharging the obligation of Clan II to *zibi* A, which began the cycle (1967: 155–6).

In short, the minimum model of Daribi marriage exchange is a directed 4-cycle consisting of two pairs of *zibi* from two different clans, as depicted in the digraph $\vec{C}_4$ of Fig. 1.6, based on a similar figure in Wagner (1967: 155).[8]

We note that graphical models refer only to patterns of relations, to the way in which pairs of points are joined (by lines or arcs), and not to their geometric arrangement. The digraph in Fig. 1.7 is thus isomorphic to that of Fig. 1.6. There are occasions when one drawing might be preferred to another. Fig. 1.6 for example emphasizes the exchange between a pair of clans while Fig. 1.7 emphasizes four groups on a single exchange cycle. J. F. Weiner (1982) uses a digraph like the first one in his observation that Daribi

[8] According to Wagner (personal communication), abundant statistical data on Daribi marriage show that the rules for *zibi* exchange are followed closely.

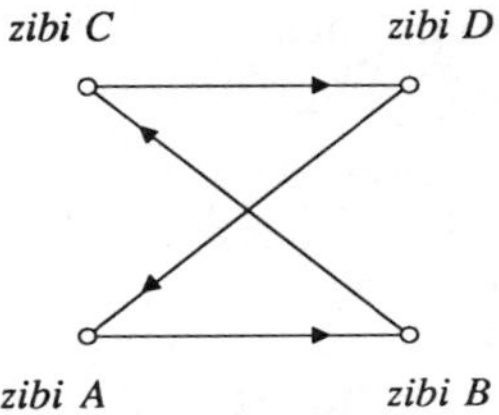

FIG. 1.6. A directed graph of Daribi marriage exchange (adapted from Wagner 1967)

marriage rules can be interpreted as 'dispersed affinal alliance' at the subclan (*zibi*) level, but as restricted exchange, i.e. the symmetric exchange of women between two groups, at the clan level. In his view, Daribi sister-exchange and bilateral and patrilateral cross-cousin marriage rules in New Guinea, when interpreted at this higher level, are all 'different expressions of a single underlying system of restricted exchange operating throughout the Highlands' (1982: 21). In Chapter 2 we show that the underlying graph of an exchange structure with dual divisions—whether it takes the form of a cycle, a tree, or a lattice—is a bipartite graph. These are all concepts we shall define later. We use the example of dual organization, but the model applies in general. We also clarify the cyclic structure of generalized, restricted, and discontinuous marriage exchange, with special reference to possible analogues in ceremonial exchange.

Exchange structures may vary in the strength, number, probabilities, or types of relations, in which case the appropriate model is a network. The best-known overseas exchange network in

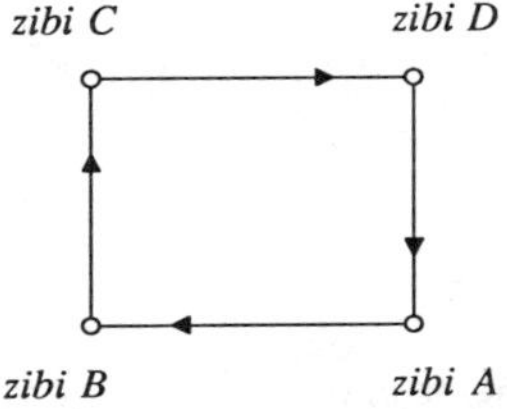

FIG. 1.7. An equivalent representation of the digraph in Fig. 1.6

Polynesia was centred on Tonga (see the map in Fig. 1.10). Jean Guiart describes it as an 'empire':

> A l'échelle de l'Océanie, l'ensemble tongien pourrait être décrit comme un empire insulaire, puisque le hiérarque, le Tui Tonga, recevait tribut des archipels ou îles suivantes: Niue à l'est, de Samoa au nord-est, Rotuma, Futuna et Uvea (Wallis) au nord-ouest, et que son influence s'étendait sur la partie orientale des Fiji, les îles Lau. Ce tribut n'était souvent versé qu'à la suite d'une expédition navale envoyée de Tonga à cet effet, suivant un processus similaire à celui de la monarchie assyrienne (Guiart 1963: 661).

P. V. Kirch (1984) depicts a segment of this ensemble by a diagram, which he calls the 'topologic structure of the Tongan maritime chiefdom', similar to the network shown in Fig. 1.8.

The dual sacred and secular chiefly head of the Tongan Empire, represented by the Tu'i Tonga (Lord of Tonga) and the Tu'i Kanokupolu (or *hau*, temporal lord), was located in Tongatabu. The Tongan Islands include the Ha'apai and Vava'u groups and the outliers include the more distant islands of Niuatoputapu, Niuafo'ou, and 'Uvea, later additions to the empire. Tongans gave Fijians whales' teeth, (Samoan) mats, and barkcloth, in exchange for canoes, red feathers, sandalwood, baskets, and pottery. And they

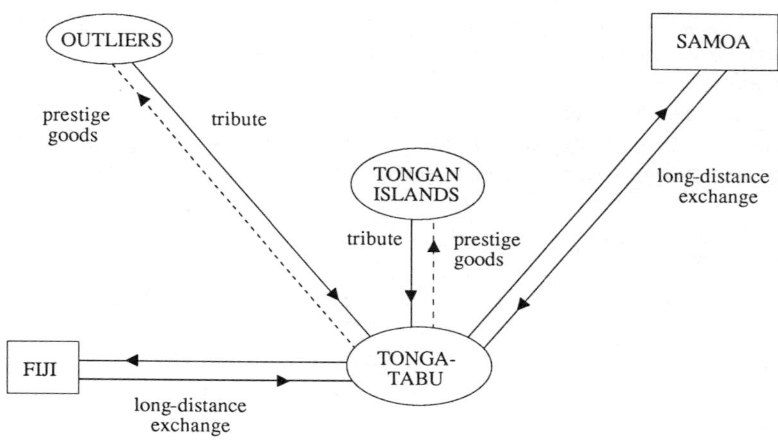

FIG. 1.8. The 'topologic structure' of the Tongan maritime chiefdom in Polynesia (after P. V. Kirch, 1984)

gave Samoans barkcloth, sleeping mats, and (Fijian) red feathers in exchange for fine mats. The tribute paid to the paramount chieftainship on Tongatabu consisted of agricultural produce (first-fruits) and also material items such as mats, barkcloth, pearl shells, and ironwood. Return gifts of foreign prestige goods, obtained from Fiji and Samoa, were given to chiefs on the tributary islands. One could add other lines to this network, in particular, lines representing kinship and marriage ties. Thus local chiefs on tributary islands were replaced by junior kinsmen of the paramount chiefs in Tongatabu, and Fijians and Samoans were 'spouse-givers' to Tongans (Kaeppler 1978) (see Chapter 8).

In considering the topological (graphical) structure of the Tongan Empire in Polynesia, one is irresistibly led to construct and to compare a similar model of the 'Yapese Empire' in Micronesia, an overseas network centring on the high island of Yap and the low islands of the Western and Central Carolines, as shown in Fig. 1.9 (see Map 3.1).

Yap, more precisely the Gagil District of Yap, was the landowning 'parent' of the low island 'children' of the Carolines, from whom it demanded gifts in the form of canoe tribute (*pitigil tamol*), land tribute or 'rent' (*sawei*), and religious tribute (*mepel*). The gifts from the low islanders consisted of woven fibre cloth, fine mats, coconut fibre rope, shell valuables, and various other items, reciprocated by 'optional' Yapese return gifts of food, canoe timber, ochre, and pots. The Gagil Yapese used the exotic low island goods to manipulate their alliance relations in Yap, and the precariously

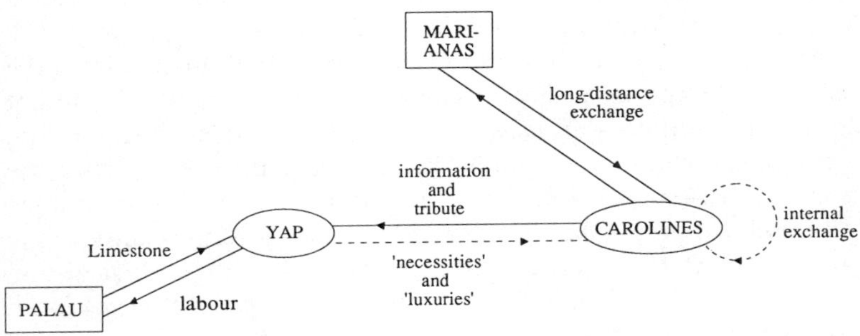

FIG. 1.9. The 'topologic structure' of the Yapese Empire in Micronesia

situated low islanders used the high island goods first of all to survive (Alkire 1965) and then to exchange and enhance status (Lingenfelter 1975). As Alkire (1980) has recently shown, the Yapese also benefited from the superior low island navigational knowledge which enabled them to sail to Palau. There they exchanged their labour for the argonite used to make the prestigious 'stone money' which played an essential role in Yapese ceremonial exchanges. The low islanders for their part sailed all over the Carolines from Truk and adjacent islands to Palau, and regularly sailed over 600 kilometres north to the Marianas for purposes of exchange.

Although they are broadly similar as 'prestige good systems' (Friedman 1981), the Yapese Empire was in certain specific respects quite the opposite of the Tongan Empire. The Tongans did most of the long-distance voyaging between Fiji and Samoa (because of their central position, according to Sharp 1964), while the Yapese sailed very little except to Palau. The Tongans controlled adjacent islands, at first through conquest and then through kinship connections, while the Yapese never exercised military control over the outer islands and their kinship links to them were purely metaphorical. The Yapese, unlike the Tongans, were culturally and linguistically distinct from the islanders in their domain. Most importantly, in the Tongan system, the tribute relation was instrumental in supporting an order of stratification, while in the Yapese system it had at least equal importance in promoting the survival of the tributaries. The real fascination of the Yapese system lies less in the relations between Yap and its tributaries than in the economic and political relations between the tributaries themselves. These relations depended on an intricate network of voyaging routes (sea-lanes) which connected all the islands of the Western and Central Carolines, the Marianas, Yap, and Palau, to each other.

In Chapter 3, we reconstruct this entire network mainly from the substantial but neglected reports of the 1908–10 Hamburg Südsee-Expedition and, using graph theoretic models of centrality, betweenness, and the neighbourhood of a point, we explicate the relations between advantageous network location and the economic and political dominance of particular low island communities. The analysis of this system, by virtue of its numerous ecological and political contrasts, helps to illuminate certain aspects of well-known Melanesian networks.

Fig. 1.10 represents a rather different kind of network, the 'Lapita voyaging network' reproduced from Roger Green (1979). Lapita is the name for a distinctive form of pottery found in archaeological sites in an area stretching from New Ireland and New Britain in Island Melanesia to Samoa and Tonga in Western Polynesia, with dates ranging from 2000–1900 BC in the west to 1600 to 500 BC in the east. Lapita has been variously interpreted as a cultural horizon or tradition, and in the east as ancestral to Polynesian culture. Green's hypothetical network connects islands (or parts of islands) in the west and east in which sites with Lapita pottery have been found, and also intervening islands (represented by ?) in which it has not yet been, but eventually should be found. With the exception of the New Hebrides–Fiji link, all pairs of islands lie within a 600-kilometre range, an assumed limit of regular two-way voyaging. The question marks on the points in this hypothetical network tell archaeologists where to look for new sites, and the distances on the lines help to account for cultural continuity and discontinuity, defined as a line of ceramic simplification which goes from west to

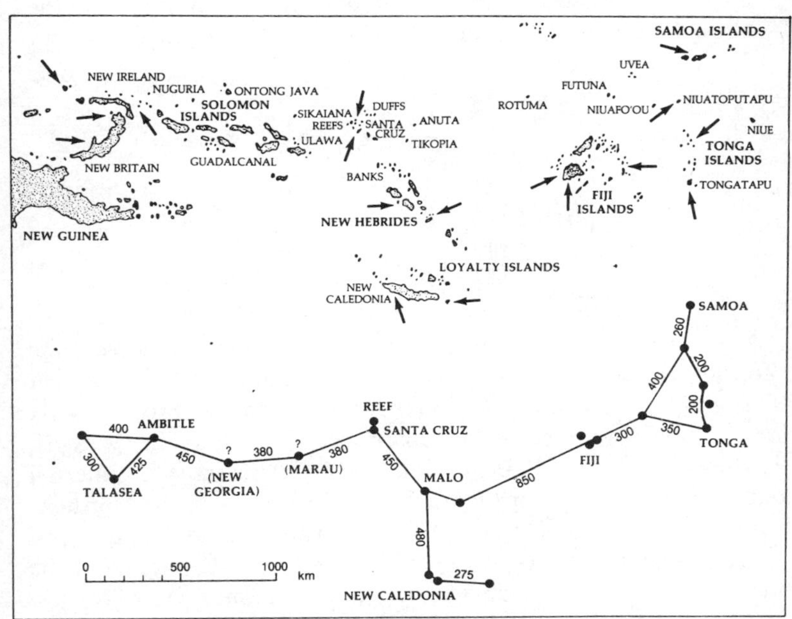

FIG. 1.10. The 'Lapita voyaging network' (from R. C. Green, 1979)

east along the whole network chain. One segment of this line, called eastern Lapita, starts in Fiji. The inference is that an eastern Lapita group, after reaching the islands of Fiji, Tonga, and Samoa, became relatively isolated from the west and was ancestral to Polynesian culture (with western and far western Lapita ancestral to other cultures throughout Island Melanesia). The linguistic evidence, which consists of a family 'tree' graph of the Austronesian language family, supports this archaeological inference (see Kirch 1984). Green's network model is a very simple but effective technique for constructing a connected graph on a set of points. At the end of Chapter 4, we discuss a number of graphical methods for constructing graphs of possible evolutionary sequences in the development of exchange networks.

It is often convenient to represent a graph or digraph as a square binary matrix, called an adjacency matrix, in which each point has a row and a column and in which the entries in the cells are either 1 or 0 to show the presence or absence of a line or arc joining a pair of points.

Thus the adjacency matrix A of the digraph in Fig. 1.3, the functional digraph of Orokaiva inter-household exchange, is:

$$A(D) = \begin{array}{c} \\ 11 \\ 12 \\ 13 \\ 14 \\ 15 \end{array} \begin{array}{c} \begin{array}{ccccc} 11 & 12 & 13 & 14 & 15 \end{array} \\ \begin{bmatrix} 0 & 1 & 0 & 0 & 0 \\ 0 & 0 & 1 & 0 & 0 \\ 0 & 0 & 0 & 1 & 0 \\ 0 & 0 & 1 & 0 & 0 \\ 0 & 0 & 0 & 1 & 0 \end{bmatrix} \end{array}$$

A number of early Oceanists used matrix representations, quite naturally it would seem, for the summary and display of ethnographic data. E. M. Loeb (1926), for example, gave a matrix presentation of inter-village marriages on Niue in Western Polynesia, and E. and P. Beaglehole (1938) did the same for paternal and maternal inter-lineage marriages on Pukapuka in the Northern Cooks. H. Powdermaker (1933) used the value matrix of a network to record the number of intermarriages between the eagle and the hawk clans of the matrilineal moieties in Lesu society in New Ireland. Since she considered only the symmetric relation of intermarriage between clans, and since there is a rule of moiety exogamy, she provided all the information needed by showing the upper right quadrant of a partitioned matrix like the following one, which is the standard form of the adjacency matrix of a bipartite structure.

		1	2	3	4	5	6	7	8
	1	0	0	0	0	2	0	1	3
eagle clans	2	0	0	0	0	1	4	0	0
	3	0	0	0	0	2	6	2	0
	4	0	0	0	0	1	1	1	3
	5	2	1	2	1	0	0	0	0
hawk clans	6	0	4	6	1	0	0	0	0
	7	1	0	2	1	0	0	0	0
	8	3	0	0	3	0	0	0	0

When moieties or other exchange groups are implicit or hidden, their existence can be revealed by such matrix partitioning. This is one of many elementary operations that can be used to find higher-order structural properties of exchange graphs. Some of these operations, including matrix addition, multiplication, and transposition, are already known to Oceanists through their presentation by Y. Lemaître (1970) in his article, 'Les Relations inter-insulaires traditionelles en Océanie: Tonga' (see Hage 1979a). In Chapters 4, 5, and 6 we use matrix methods to quantify and classify a variety of exchange structures.

It is sometimes asked how many exchange structures there are of a given type. We have already mentioned Lévi-Strauss's and Sahlins's theoretical classifications. An ethnographic example of structural enumeration is given in Leenhardt's (1980 [1930]) description of shell money payments between kinsmen in New Caledonia. His 'représentation graphique du mouvement des monnaies dans les échanges sociaux' is reproduced in Fig. 1.11. These structures are

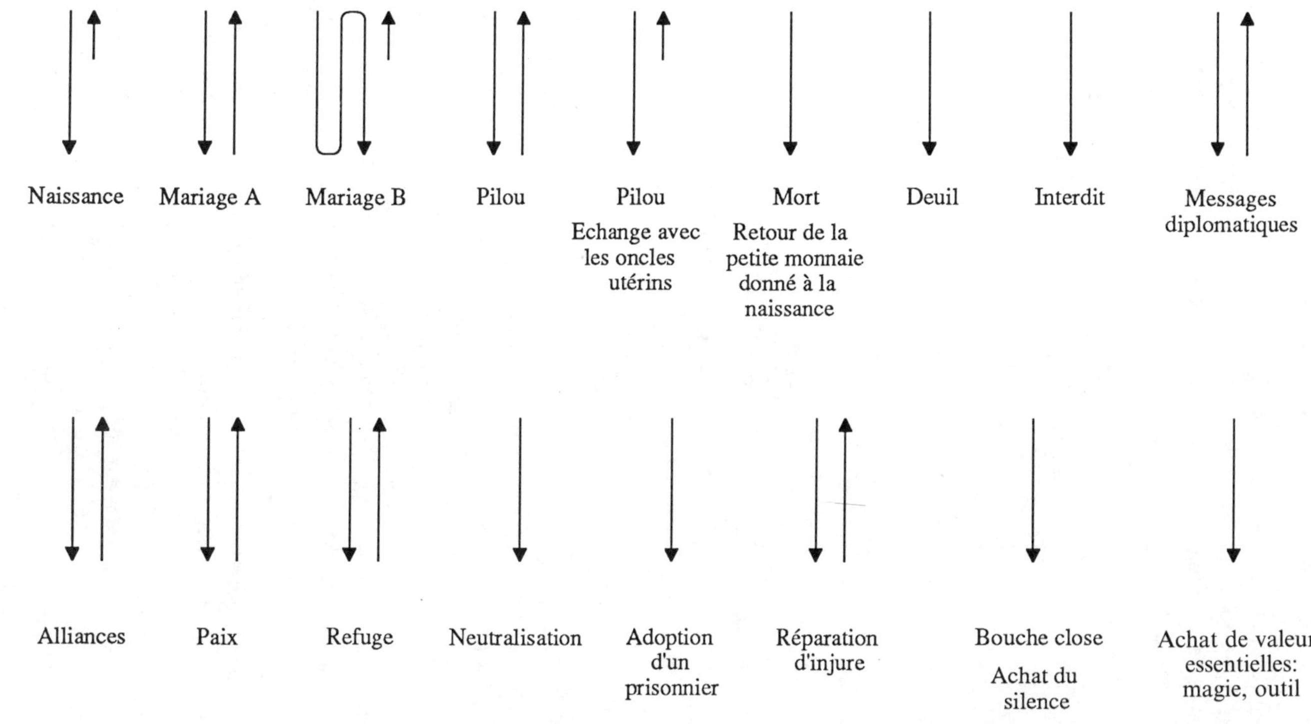

FIG. 1.11. Leenhardt's (1980 [1930]) directed networks of shell money exchanges in New Caledonia

implicit directed networks in which the length of an arc represents the length of the shell money string given in exchange.

Leenhardt's method is to lay out some of the different transactions in order to discover what the usage of this money corresponds to in native thought. At birth, for example, money is given to the maternal uncle who assures the nephew's breathing by blowing into his ear. The uncle signals his goodwill by receiving the money (along with other gifts) and giving a shorter length of money in return. At marriage (A) equal lengths of money accompany the equal exchange of cross-cousins between clans. The practice of *bouche close* refers to buying the silence of a witness to adultery, to preparations for war, etc. From this array of implicit directed networks, Leenhardt distinguishes three kinds of exchange relations: dependence, in which money is used as a symbolic and propitiatory object, as in payments at birth (the money also represents the infant's breath); agreement, as in reciprocal equal payments at marriage and the oath of alliance; and payment for highly valued goods such as tools, and magic, or favours, such as buying silence, or diminishing the gravity of an avoidance infraction.[9]

In Chapter 6 we take up, formally, the topic of structural enumeration, using a duality operation on graphs and a simple method from combinatorics. Enumeration has both ethnographic and theoretical value. Thus we show that a succession of studies of Tongan kinship, from Rivers (1910) and Hocart (1915) to Biersack (1982) and Bott (1981), are, in effect, individual choices made from an ideal repertoire of structural forms.

Exchange structures and associated symbolism can often be explicated using binary operations on graphs. Permutational structures, for example, which are often expressed in native models, can be represented by graphs which are based on 'the Cartesian product of two graphs'. A simple but pertinent example is provided by an Arapesh folk model of big men, *buanyins*, who are the focus of inter-moiety exchanges:

> Native capacity is roughly divided into three categories: 'those whose ears are open and whose throats are open', who are the most gifted, the men who understand the culture and are able to make their understanding articulate; 'those whose ears are open and whose throats are shut', useful quiet men

[9] See also Leenhardt (1937) on the classification of these shell money exchange forms.

who are wise but shy and inarticulate; and a group of the two least useful kinds of people, 'those whose ears are closed but whose throats are open' and 'those whose ears and throats are both shut'. A boy of the first class is specially trained by being assigned in early adolescence a *buanyin*, or exchange partner, from among the young males of a clan in which one of his elder male relatives has a *buanyin* (Mead 1963: 27–8).

The graph of this structure is shown in Fig. 1.12. It is the Cartesian product of the graph K_2 with itself. In Chapter 7 we use binary operations on graphs—the product, conjunction, and union—together with group models to analyse culturally defined anatomical and physiological concepts associated with different types of exchange structures. And we indicate the value of these operations for the representation and notation of social structures.

Anthropologists have on occasion used the names of relational concepts for idiosyncratically defined notions in describing formal properties of exchange structures. Sahlins (1985), for example, in an analysis of Fijian transactions, defines the relation between two opposing parties who exchange land things for sea things as 'transitively determined' by the relation of these two parties to a third party, a chief, who neither takes nor gives but mediates between them. Anthropologists sometimes use symbolic notation to define relations. Hocart (1952, 1970) used the symbols = and ∴ as part of a 'Euclidean' demonstration. Thus in *The Northern States of Fiji*, one encounters such propositions as 'Turtling celebrations = man slayer's consecration' or '∴ Chief's consecration = birth ceremony.' Such demonstrations enlightened Elkin (1937) but

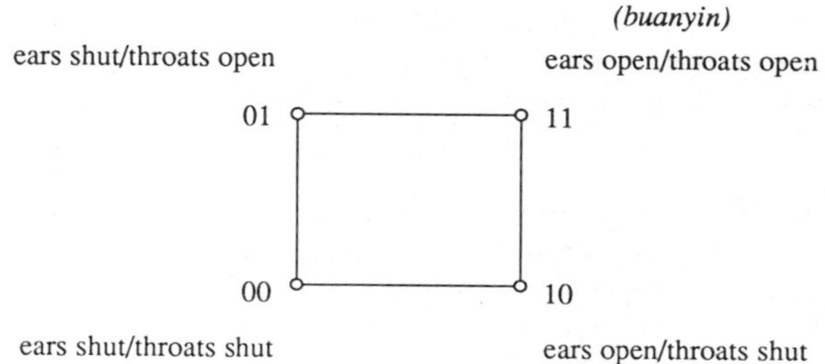

FIG. 1.12. A 'Cartesian product' model of the Arapesh theory of native capacity

confused Evans-Pritchard (1970). In his Editor's Introduction to *Kings and Councillors*, Rodney Needham sets out the abstract form of a Hocartesian analysis. According to Needham:

The typical form of a transitive analysis by Hocart is:

$$a = b$$
$$c = b$$
$$\therefore a = c \text{ (1970: xlvi).}$$

Unfortunately neither Sahlins nor Needham are describing a transitive relation as this term is used in logic.[10] In fact, the Sahlins–Needham error is just another version of the Kelly (1974) error[11] as discussed in Hage and Harary (1983a). It only says that two things have something in common by virtue of their relation to a third thing. As Lévi-Strauss (1984: 209) has remarked apropos of Kelly's 'transitive equivalence' analysis of Etoro siblingship: 'On peut évidemment définir de cette façon la relation entre des germains, mais aussi n'importe quelle autre et pas seulement dans l'ordre de la parenté; car elle exprime seulement le fait qu'envisagés sous un certain rapport, deux termes, deux individus, ou deux positions dans un réseau de relation quelconque se ressemblent.' It is clear, however, that relational properties do play a fundamental role in the analysis of exchange structures, as in Lévi-Strauss's distinction between restricted and generalized exchange. Since relation theory is coextensive with graph theory we conclude our presentation by defining, in Chapter 8, a repertoire of relational structures realized in a variety of exchange and communication structures.

Fig. 1.13 illustrates the graphical representation of logical relations and perhaps also the need for explicit definitions. The first two relations are transitive, while the third is intransitive. The second relation is symmetric, while the first and third are asymmetric. The first two relations are irreflexive and the third is reflexive. The fourth relation has none of these six properties. These are not the only relations with three elements having these properties.

As C. S. Peirce (1933) demonstrated, and as we will show in

[10] The following statement concerning the structure of cultural classification schemes is not an instance of a transitive relation either: 'In the Hawaiian, "chief" and "god" are transitively alike by opposition to men . . .' (Sahlins 1985: 147).

[11] Another case of independent discovery.

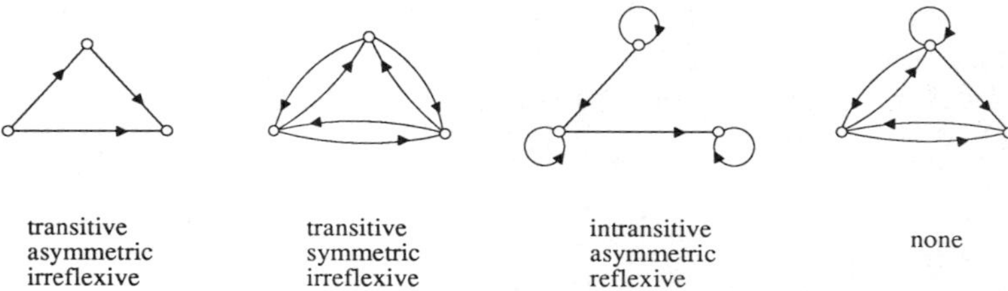

FIG. 1.13. Graphical models of logical relations, with their properties

Chapter 8, the properties of a relation can be determined by inspecting its matrix as well as its graph. A brief example may be given. In his definitive study of Samoan material culture Sir Peter Buck (Te Rangi Hiroa) describes the ceremonial division of flesh foods—the pig, turtle, bonito, and shark—among major social groups. This procedure dramatizes the opposition between the categories and grades of chiefs and, in some contexts, the identity within each class. Fig. 1.14 from Buck (1930) shows the division of the shark among high chiefs (*ali'i*), talking chiefs (*tulāfale*), and the village princess (*taupou*).

When there is a catch, members of the same class in different villages may ask their counterparts for their official shares, with talking chiefs asking each other for dorsal fins and so on. Thus each social category exchanges with itself. The reflexive property of this relation is shown by the loops at each point in Fig. 1.15, and equivalently by the 1s in every diagonal entry of the matrix to its

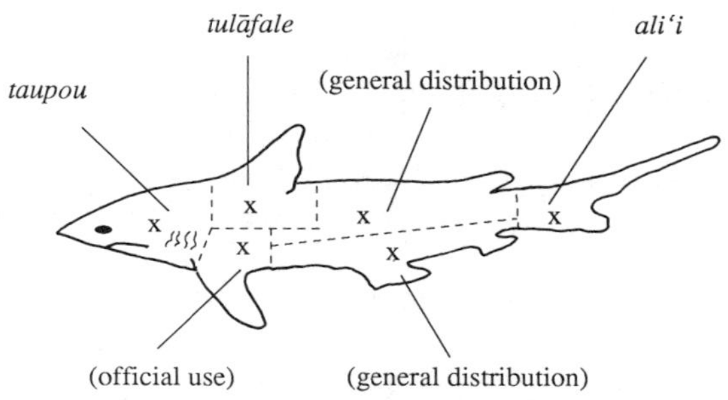

FIG. 1.14. Shark ceremonial divisions in Samoa (adapted from P. H. Buck, 1930)

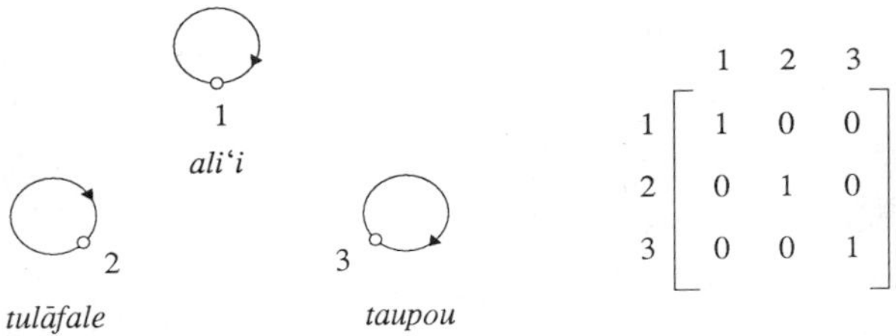

FIG. 1.15. A chiefly exchange structure in Samoa

right. A relation which has loops at every point and no arcs, or, equivalently, has 1 in every diagonal entry of its matrix and 0 elsewhere, is called, appropriately enough, the *identity relation*, having the identity matrix as defined in Chapter 4.

Set Theoretic Concepts

Since our definition of graph theoretic models is set theoretic, it will be convenient to conclude this chapter by summarizing a few basic concepts. We take 'set', 'element', and 'is an element of' as undefined terms with a natural intuitive meaning. The Venn diagrams in Fig. 1.16 should make these concepts clear.

The *universal set* U is the set of all elements under consideration. The *empty set* $\emptyset$ is the set which does not contain any elements. The set B is a *subset* of a set A, written $B \subset A$, if every element of B is in A. Two sets are called *equal*, written $A = B$, if each is a subset of the other. We say that B is a *proper subset* of A if B is a subset of A and B does not equal A. Clearly every set is a subset of the universal set.

There are several important operations on sets. The *union* of two sets A and B, written $A \cup B$, is the set consisting of all those elements which are in A or in B (or in both). The *intersection* of A and B, denoted by $A \cap B$, consists of those elements in both A and B. If the intersection of A and B is empty, the sets are said to be *disjoint*. The *difference* $A - B$ contains all elements of A which are not in B. The *symmetric difference*, written $A \oplus B$, is the set containing those elements in exactly one of the sets A and B. Their symmetric difference can be thought of as those elements either in A or in B but not in both:

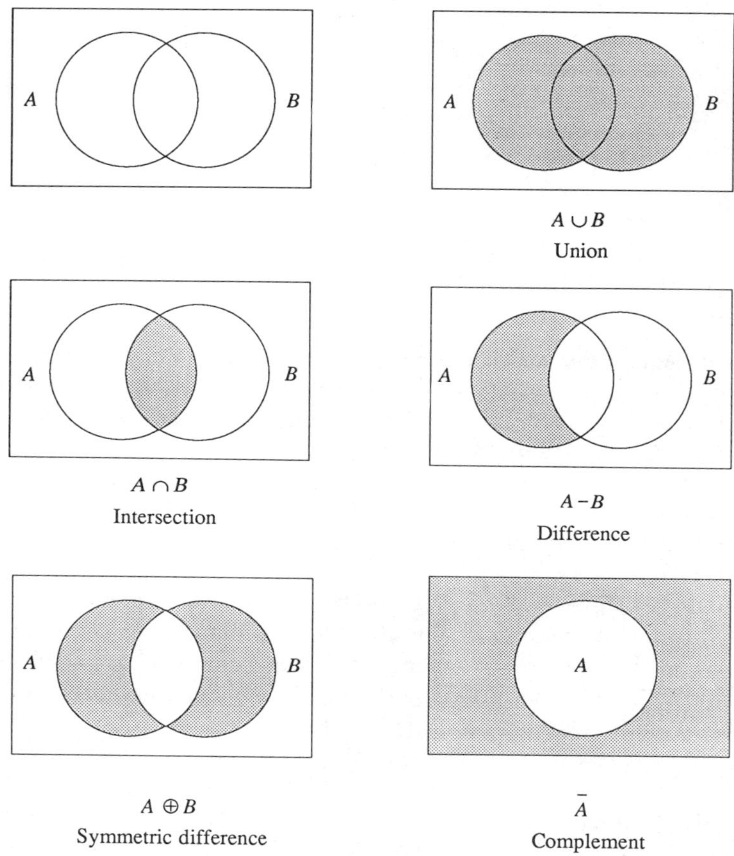

FIG. 1.16. Operations on sets

$$A \oplus B = (A \cup B) - (A \cap B).$$

It can also be thought of as the elements of A not in B together with those in B not in A:

$$A \oplus B = (A - B) \cup (B - A),$$

whence the name 'symmetric difference'. Finally, the *complement* of a set A, denoted by $\bar{A}$, consists of the elements of U not in A: $\bar{A} = U - A$.

Like the preceding graph models, many of these concepts are already known in Oceanic studies. Thus Rivers's (1914) reconstruction of the history of Melanesian society, which assumes successive waves of immigrants, is set theoretic: the 'sitting interment people' fuse with a set of original inhabitants to produce the 'dual people', who then combine in various ways with later arrivals, the 'kava people' and the 'betel people', who combine with or sometimes remain disjoined from each other. An ethnographic and less fanciful application is Lawrence's (1955) model of Garia 'security circles' as formalized by Brookfield and Hart (1971). Biersack (1984) has used set theoretic notation in describing an unusual form of social organization from New Guinea. According to Biersack, in Paiela society, exchange units consist not of discrete clans as is common in New Guinea, but of 'in-between people', i.e. those who are in a culturally recognized intersection of cognatic groups *A* and *B*.

Inevitably in the application of mathematics to social science, the latter will have the opportunity to repay part of its debt to the former, by suggesting natural and important new concepts. A modest example, the 'geodesic counting matrix', inspired by Chapter 3, is given in Appendix 3.

2

Paths, Cycles, and Partitions

> But over and above [marriage exchange] we have the striking institution of Sister Purchase. If you have no sister of your own then you must procure one. For if you want to marry you must have a girl to give away. That is to say, not only must you fulfil a contract of exchange already made, but you must be in a position to make exchanges. It is as if exchange were idealized, made an end in itself.
>
> F. E. Williams, *Papuans of the Trans-Fly*

Graphical models of exchange systems enable us to discover structural commonality beneath empirical diversity, and they provide for the coherent classification of structural forms. We begin with two analyses. First, we give a unitary definition of dual organization, a widely distributed and, it has been conjectured, archaic type of social structure in Melanesia. Rather than giving an ethnographic survey, we consider three radically different surface forms, all of which have the underlying structure of a bipartite graph. Then, using digraphs and networks, we clarify the application of models of restricted and generalized marriage exchange to the analysis of ceremonial exchange in New Guinea. We start with a set of mathematical definitions that are the foundation of all that follows.

Basic Definitions

Informally, a graph consists of a set of points, some pairs of which are joined by lines. Thus it is a form of abstract geometry. In the logic of relations a graph is defined as a symmetric irreflexive relation and in topology as a 1-dimensional simplicial complex. Algebraically, a graph can be characterized as a square symmetric matrix of zeros and ones, with only zeros on the main diagonal. Our formal definition is, as noted in Chapter 1, set theoretic.

A *graph* G consists of a finite non-empty set $V = V(G)$ of p *points* together with a prescribed set E of q unordered pairs of distinct points of V. We also write $G = (V, E)$. Each pair $e = \{u, v\}$ of points in E is a *line*[1] of G, and e is said to *join* u and v. We also write $e = uv$ and say that u and v are *adjacent points*; point u and line e are *incident* with each other, as are v and e. If two distinct lines are incident with a common point, then they are *adjacent lines*. A graph with p points and q lines is called a (p, q) *graph*. The $(1, 0)$ graph, consisting of just one point, is called *trivial*, mainly in order to exclude it by specifying that a graph be *non-trivial*.

It is customary to represent a graph by means of a diagram and to refer to the diagram as the graph. Thus in the graph G of Fig. 2.1, the points u and v are adjacent, but u and w are not; lines a and b are adjacent, but a and c are not. Although the lines b and c intersect in the diagram, their intersection is not a point of the graph. As we will see later, there are several equivalent ways in which to represent a graph, depending on one's purpose.

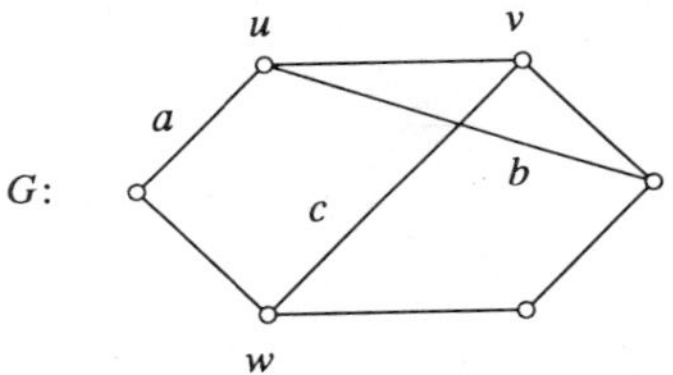

FIG. 2.1. A graph to illustrate adjacency

A *labelling* of a graph G is an assignment of labels 1 to p, or sometimes v_1 to v_p, to its points. The *degree* of a point v_i in a graph G, denoted by deg v_i or d_i, is the number of lines incident with v_i. In Fig. 2.2 there are two labelled graphs, G_1 and G_2, each of which has four points. Here G_1 is the complete graph and G_2 is obtained from it by removing the line $\{2, 4\}$. All points of G_1 have the same degree, 3; such a graph is *regular*. On the other hand, G_2 has two points of degree 2 and two of degree 3.

For any (p, q) graph G whose points have degrees $d_1, d_2, \ldots, d_p$, we have $\sum d_i = 2q$. Thus the sum of the degrees of the points of a graph is twice the number of lines. The simple proof of this assertion, known to Euler who discovered graph theory over 200 years ago, is that each line contributes 2 to this sum, 1 for each of its points.

[1] Common synonyms for point and line are vertex and edge or node and branch.

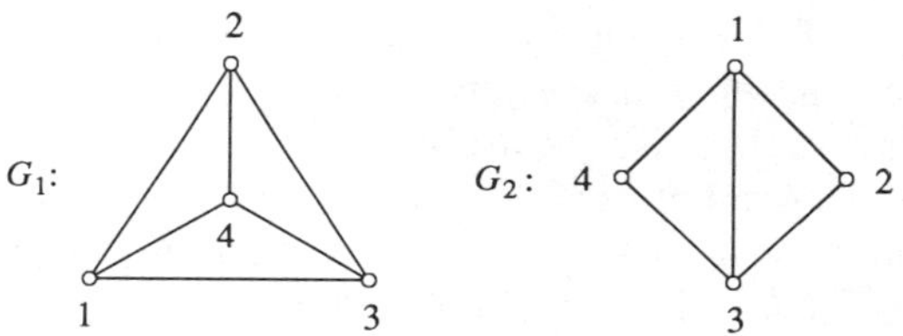

FIG. 2.2. Two labelled graphs

Theorem 2.1. The sum of the degrees of the points of any graph G is twice the number of lines of G.

Symbolically,

$$d_1 + d_2 + \ldots + d_p = 2q$$

or more concisely, using the usual summation notation,

$$\sum_{i=1}^{p} d_i = 2q.$$

Also, as the indices of summation $i = 1$ to p are considered to be understood, one often writes

$$\sum d_i = 2q.$$

This is illustrated in Fig. 2.3.

A *walk* of a graph G is an alternating sequence of points and lines $v_0, e_1, v_1, \ldots, v_{n-1}, e_n, v_n$, beginning and ending with points, in which each line e_i is incident with the two points v_{i-1} and v_i immediately preceding and following it. Both points and lines may occur more than once in a walk. The *length* of a walk is the number of occurrences of lines in it. This applies to each type of walk defined below. This walk *connects* v_0 and v_n, and may also be denoted by v_0, $v_1, v_2, \ldots, v_n$ (or more briefly simply by $0, 1, 2, \ldots, n$ when there is no confusion); it is sometimes called a v_0–v_n walk. It is *closed* if $v_0 = v_n$, and is *open* otherwise. It is a *trail* if all the lines are distinct (different) and a *path* if all the points (and thus necessarily all the lines) are distinct. If $n \geqslant 3$ and the walk has distinct lines and also distinct

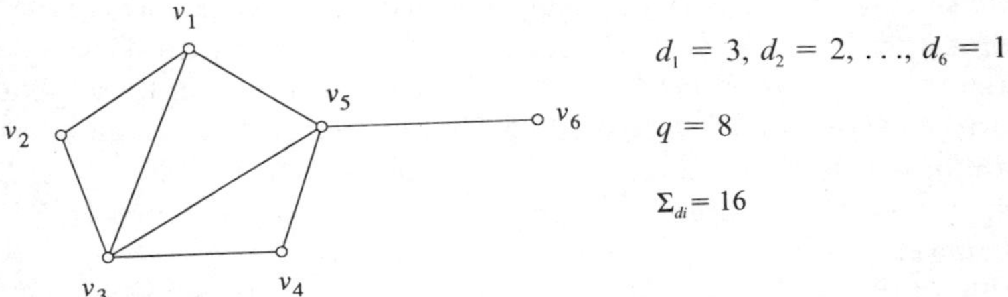

FIG. 2.3. A graph to illustrate the relation between degrees and the number of lines

points except for its end points $v_0 = v_n$, it is a *cycle*. By definition this cycle has length n; it is sometimes called an *n-cycle*.

In the labelled graph G of Fig. 2.4, 1, 2, 3, 2, 6 is a walk which is not a trail, and 1, 2, 3, 4, 5, 3 is a trail which is not a path; 1, 2, 3, 4, 5 is a path, and 2, 3, 5, 6, 2 is a cycle. Finally, 1, 2, 3, 4, 5, 6, 7 is called a *spanning path* as it is a path which contains all the points.

By the above definition, the length of a path is the number of lines in it. The *distance* between two points v_i and v_j, denoted by $d(v_i, v_j)$ or d_{ij}, is the length of any shortest path or *geodesic* g which joins them. Unfortunately, a limited number of letters must serve many needs; thus d_i is the degree of the ith point, whereas d_{ij} is the distance between it and the jth point. In the graph G of Fig. 2.4, $d_{13} = 2$ and $d_{45} = 1$, as points 4 and 5 are adjacent. The *length of a cycle* is the number of lines in it. In Fig. 2.4 there is one 3-cycle, 3, 4, 5, 3, and one 5-cycle, 2, 3, 4, 5, 6, 2, and also the 4-cycle mentioned above.

We denote by P_n the graph which is a path with n points (and hence has length $n-1$) and by C_n the cycle with n points (and length n).

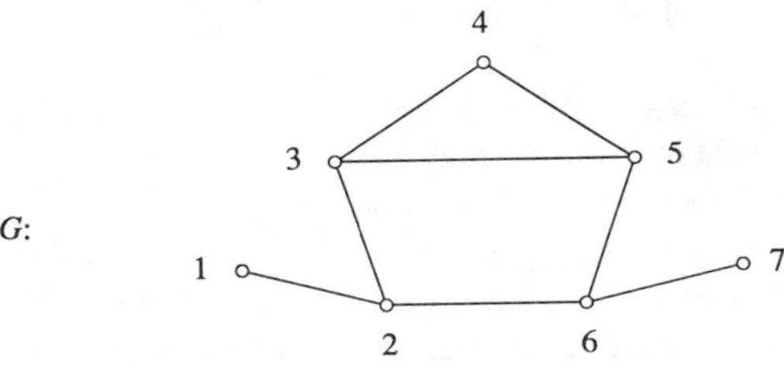

FIG. 2.4. A graph to illustrate kinds of walks

The concept of a maximal set is required in the context where that set is a graph. A set $S = \{x_1, x_2, \ldots, x_n\}$ is *maximal* with respect to some abstract property, denoted by P, if S satisfies P but no set that properly contains S (consists of all the elements of S and has additional elements) does. We will specialize this to the case where S is a graph (regarded as its set of points and lines) and property P is 'connected'.

A graph is *complete* if every pair of points are adjacent, and *connected* if every pair of points are joined by a path. A *subgraph* of G is a graph having all of its points and lines in G. A *spanning subgraph* of G contains all its points. A maximal connected subgraph of G is called a *connected component* or simply a *component* of G. In Fig. 2.5, the graphs to the right of the dashed line are connected, and those to the left are disconnected. The last graph is complete. The first graph consists of four (trivial) components, and the second has three components.

The established notation for the complete graph with p points is K_p, so that the last graph in Fig. 2.5 is K_4. It is easy to see that the number of lines in K_p is $p(p-1)/2$. For example, K_4 has $4 \cdot 3/2 = 6$ lines.

A *tree* is a connected graph with no cycles. Fig. 2.6 shows four of the eleven trees with seven points, the first being a path and the last a star.

A *cutpoint* of a connected graph is one whose removal (together with its incident lines) disconnects it, i.e. divides it into two or more subgraphs that have no lines joining them and hence are not connected to each other. A point v is an *endpoint* if $\deg v = 1$. All the graphs in Fig. 2.6 are trees and contain only cutpoints and endpoints. The *removal of a line e* from a graph G results in the spanning subgraph of G containing all the lines of G except e. A *bridge* of a connected graph G is a line whose removal disconnects it. In the graph of Fig. 2.4, the line $\{1, 2\}$ is a bridge; in a tree, every line is a bridge.

An *isomorphism* between two graphs $G_1 = (V_1, E_1)$ and $G_2 = (V_2, E_2)$ is a one-to-one correspondence between V_1 and V_2 that preserves adjacency. This is written $G_1 \cong G_2$ or sometimes $G_1 = G_2$. The two graphs in Fig. 2.7, although they appear to be different, can be proved isomorphic. Fig. 2.5 shows all of the graphs with four points. No two of these eleven graphs are isomorphic (collectively, they form a *non-isomorphic* set of graphs). Empirically interpreted, the

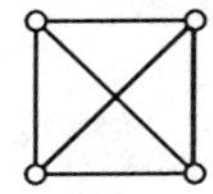

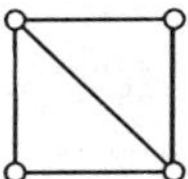

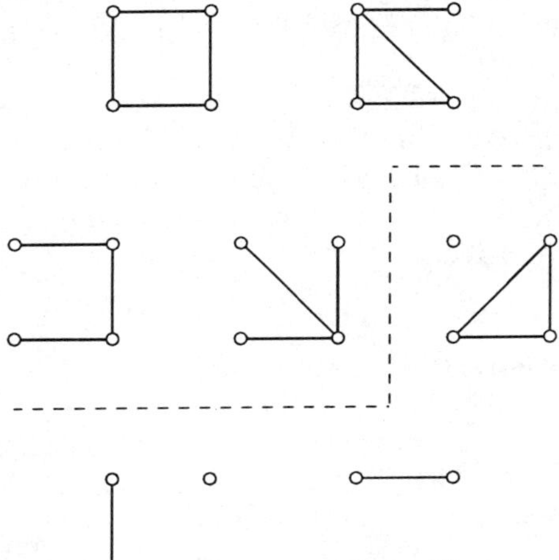

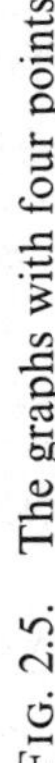

FIG. 2.5. The graphs with four points

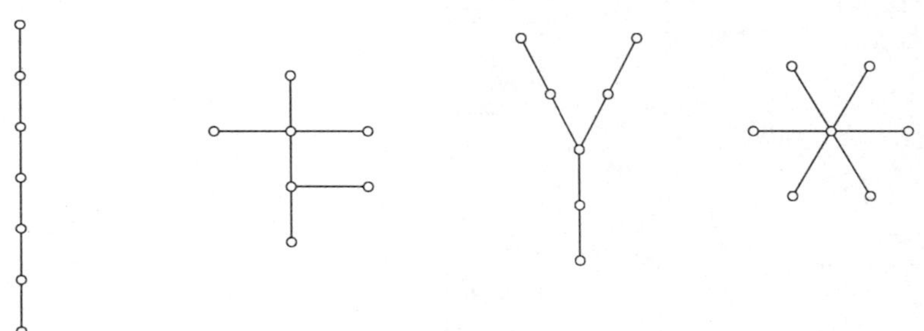

FIG. 2.6. Four trees with seven points

graph of any exchange system with four points will be isomorphic with one of these.

It is particularly easy to describe a disconnected graph all of whose $k > 1$ components are isomorphic to a connected graph G, for it is denoted by kG and is said to consist of k *copies* of G. For example, the *trivial graph* consisting of one point and no lines is K_1, and the first graph of Fig. 2.5 is $4K_1$.

Bipartite Graphs and Dual Organization

The division of a society into two groups or moieties, commonly called dual organization, can be accomplished in a variety of ways, in so many different ways in fact, especially when unlabelled and unconscious forms are considered, that one may ask whether this concept has a unitary definition or is not rather a family resemblance. We shall consider a series of forms of increasing complexity and decreasing explicitness but all of which have the

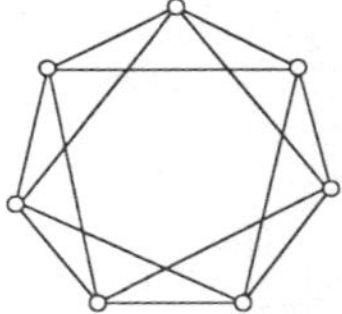

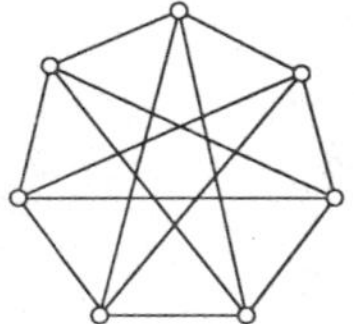

FIG. 2.7. Two isomorphic graphs

underlying structure of a bipartite graph. A bipartite graph is, as the name implies, a specific case of an n-partite graph, which is a general model for the study of societies partitioned into any number n of exchange groups. At the conclusion of this section we shall illustrate a tripartite social division using the model of a signed graph.

In his classic study, *The Melanesians*, R. H. Codrington (1891) emphasized the fundamental and ubiquitous division of societies into exogamous matrilineal classes:

> In the native view of mankind, almost everywhere in the islands . . ., nothing seems more fundamental than the division of the people into two or more classes, which are exogamous, and in which descent is counted through the mother. This seems to stand foremost as the native looks out upon his fellow men; the knowledge of it forms probably the first social conception which shapes itself in the mind of the young Melanesian of either sex, and it is not too much to say that this division is the foundation on which the fabric of native society is built up (1891: 21).

Codrington noted the presence of dual divisions in the Banks Islands and the New Hebrides (now part of Vanuatu) and he pointed out that such divisions need not be distinguished by name. Thus in Mota, each division is called a *veve*, a word that means division. Members of opposite *veve* refer to each other as *tavala ima*, those 'of the other side of the house'.

In *The History of Melanesian Society*, Rivers (1914) conjectured that dual organization was an archaic widespread form of social organization resulting from the fusion of a superior immigrant with a backward aboriginal population. As evidence he cited symbolic ascriptions of physical and mental differences to members of opposite moieties, for example light-skinned vs. negroid people, ignorant 'bush' vs. superior 'sea' people. Rivers considered matrilineal descent and exogamy to be fundamental to dual organization, and he explained these features by invoking presumed migrational and demographic causes. Although there is no linguistic evidence for Rivers's fusion hypothesis, Blust (1981) adduces linguistic data in support of a historical reconstruction of dual organization as a basic feature of early Austronesian societies in Melanesia, coastal New Guinea, Polynesia, and nuclear Micronesia. Blust does not cite matrilineal descent and exogamy as concomitant features, but he does suggest totemism, dualistic cosmological schemes, and residential separation.

In *The Elementary Structures of Kinship*, Lévi-Strauss defines dual organization as a 'system in which the members of the community, whether it be a tribe or a village, are divided into two parts which maintain complex relationships varying from open hostility to very close intimacy, and with which various forms of rivalry and cooperation are usually associated' (1969: 69). Moieties may be linked not only by the exchange of women, but also by the exchange of goods, services, and ceremonies, relations often characterized by a 'double attitude of rivalry and solidarity', as for example in competitive feasting. Regarded more as a principle of organization with variable applications than as a social institution, dual organization has no necessary correlates, such as exogamy or matriliny (although the latter is common). In Lévi-Strauss's view, it is basically a modality of reciprocity and thus referable in origin to fundamental mental structures, not to specific historical events or general evolutionary stages.

Regarding exchange as a structure of reciprocity, we note that it may take place not only between 'halves' but between subdivisions of each half, sometimes united in a single cycle. Thus Rivers's informant, John Pantutun, described moiety organization on Pentecost as based on six subdivisions:

> According to John Pantutun there are also subdivisions of the moieties of such a kind that a man of one subdivision is not free to marry any woman of the other moiety, but must take one of a given subdivision. Thus, he said that each moiety had three subdivisions which we may call A, B and C in one moiety, and D, E and F in the other. It was said that a man of A had to marry a woman of D, a man of B a woman of E, and a man of C a woman of F, while men of D, E and F divisions could not marry women of A, B and C respectively, but a man of the F division must marry a woman of A, and so on (Rivers 1914: 190).

If a man of E marries a woman of C and a man of D marries a woman of B, then marriage exchange between the moieties is organized as a (directed) 6-cycle.

Among the Etoro, the moieties are not subdivided into named intermarrying groups, but rather, the intermarrying groups produce an unnamed moiety division. According to R. C. Kelly (1974), the moieties are de facto groupings, in which membership is not determined by descent but is the outcome of a marriage rule—a complex system of restricted exchange organized as a lattice graph which consists entirely of cycles of variable, but even, length.

In a recent comparative study of the spatial correlates of dual organization, Rosman and Rubel (n.d.) describe both a conventional division into 'halves' and a 'chequerboard' pattern in which adjacent exchanging groups belong to opposite moieties.

Since dual organization has such a variety of forms and since many of these forms are implicit or hidden, it would be useful to have a general structural model. By reducing all forms of dual organization to an underlying bipartite graph, insight may be gained into its partitionable, cyclical, and relational properties.

A *bipartite graph* or *bigraph* G is a graph whose point set V can be partitioned into two subsets V_1 and V_2 such that every line of G joins (a point of) V_1 with (one in) V_2. The graph of Fig. 2.8a can be redrawn in the form of Fig. 2.8b to display the fact that it is a bigraph.

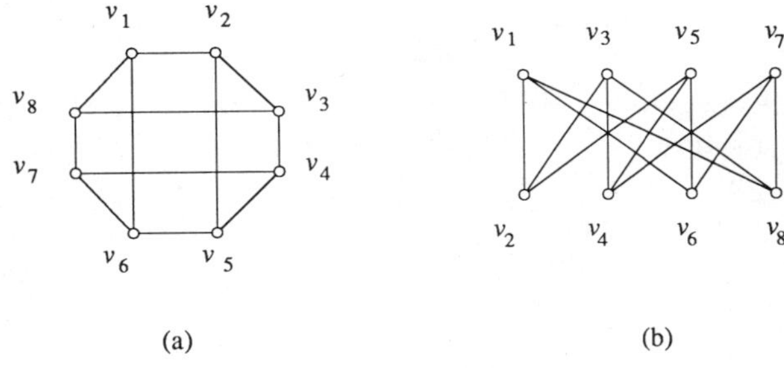

FIG. 2.8. A bigraph

If G is bipartite and contains every line joining each point of V_1 to every point of V_2, then G is a *complete bigraph*. If V_1 and V_2 have m and n points, we write $G = K_{m,n} = K(m,n)$. Note that a complete bigraph cannot be a complete graph unless $G = K_{1,1} = K_2$. A *star* is a complete bigraph $K_{1,n}$. Fig. 2.9 shows the complete bigraph $K_{3,3}$ and the star $K_{1,4}$.

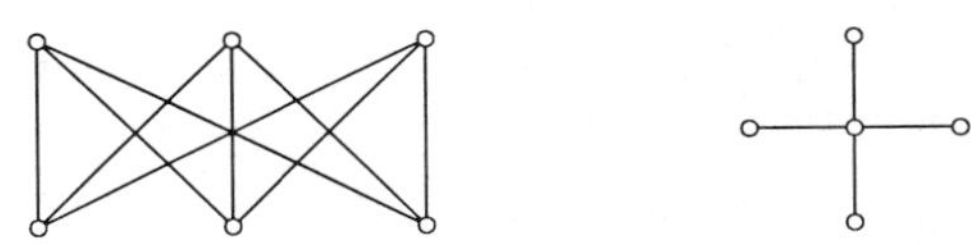

FIG. 2.9. The complete bigraph $K_{3,3}$ and the star $K_{1,4}$

We now apply this model to dual organization in three Melanesian societies: Arapesh, which has a conventional opposition between two named groups; Tanga, which has a deictic 'chequerboard' or, more accurately, a tree pattern; and Etoro, which has a lattice-graph. For Tanga we introduce an equivalent and spatially appropriate alternative characterization of a bigraph as a bicolourable graph. For Etoro we give a general theorem for bigraphs and we show the implication of the relational property of intransitivity in bigraphs of exchange structures.

Arapesh society, located in the Sepik River area of New Guinea (Mead 1938, 1940, 1947, 1961, 1963), is divided into two named patrilineal moieties represented by two eponymous birds, hawk (*kumun*) and cockatoo (*kwain*). Moiety symbolism expresses both the physical and mental opposition emphasized by Rivers and also the discrete/continuous opposition noted by Lévi-Strauss:[2]

iwhul people	*ginyau* people
taboo the cockatoo (*kwain*)	taboo the hawk (*kumun*)
eat their meat dry	eat their meat bloody
talk at night	talk at dawn
make lengthy speeches	make brief speeches
die sleeping with breath leaving through fingers and big toes	die standing up in a gasp

Arapesh moieties are not exogamous but instead regulate competitive feasting between rival big men, *buanyins*. *Buanyins* are members of different clans and opposite moieties. Their relation is by convention an aggressive one:

> It is the duty of *buanyins* to insult each other whenever they meet, to inquire sneeringly whether the other *buanyin* ever means to make anything of his life—has he no pigs, no yams, has he no luck in hunting, has he no trade-friends and no relatives, that he never gives feasts or organizes a ceremony? Was he born head first like a normal human being, or perhaps he came feet first from his mother's womb (Mead 1963: 28).

The graph of Arapesh dual organization is shown in Fig. 2.10. It is the smallest non-trivial complete graph, K_2, which is of course bipartite.

[2] Lévi-Strauss (1963b) interprets this opposition as a manifestation of a latent triadic structure in dual organization.

(*iwhul*) (*ginyau*)

$G = K_2$: 1 —— 2

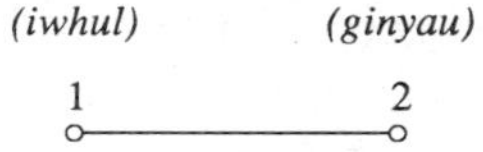

FIG. 2.10. The graph of Arapesh dual organization

The importance of moiety organization for intralocality feasts, and the flexible arrangements which can be made to reproduce this structure in spite of demographic and residential vicissitudes, are illustrated by the following account from Mead:

> Kobelen and Umanep, once said to have been parts of the same locality, split into two localities. This left Kobelen all members of *kumun* and Umanep all members of *kwain*. Each locality then redivided into *iwhul* and *ginyau*, for the purpose of local feasting reciprocities. However, the old idea of two birds associated with two halves again reasserted itself, or so it seems from the accounts, and each moiety in Kobelen took a hawk emblem, *iwhul* took the *wholowhepin* hawk and *ginyau* took the *genakaben* hawk (1947: 184).

Whatever the totems (it would be interesting to know the native distinction between the two types of hawks), the structure remains K_2.

We note that similar examples of moiety reproduction can be found elsewhere in Oceania. Sahlins describes an interesting case from Fiji:

> The village of Nuku ... has the usual dual organization of land and sea sections, although strictly speaking there has never been a single Land group in the community. Nuku was founded in the latter part of the nineteenth century exclusively by master fishers attached to the chiefs, Sea People par excellence who had migrated from the capital village of Navucinimasi and ulteriorly from the islands of Gau and Bau. Yet by the local conception, certain Nuku groups were Land People. If one suggests to Nuku villagers—as I often did—that all the local groups are Sea People, this is readily admitted. But it will also be explained that one body of the people was first to come to Nuku from the chiefs' village, that they *receive* the fish from the sea and are warriors (*bati*) for the later groups; that is, they are 'Land' in relation to the true Sea People who arrived afterward. This is an apposite example of 'stereotypic reproduction', in Godelier's (1972) phrase (Sahlins 1976: 41).

In his essay 'Do Dual Organizations Exist?' Lévi-Strauss (1963b) distinguishes two types of dualism: diametric, which can be thought of as the opposition between two sides or halves, and concentric, the opposition between a centre and a periphery. Geometrically, the first is modelled in a projective plane and the second in a cylinder. In some societies both forms exist, each with its own associated, often complementary, symbolism. Arapesh diametric dualism is expressed in the moiety structure and concentric dualism in the residential layout. According to Mead,

All level land and the center of the village are spoken of as *yapugenum* (literally, good . . .) while all precipitous land, especially the declivity characteristically surrounding a village, is called *yaweigenum*, or a bad place. On the steeply sloping 'bad place' are placed menstrual huts (*sho'wet*) . . .; on the slope and in such huts, in time of rain, infants are also born, so that the village may be protected from the blood of birth. These slopes are also used for latrines . . . They are also spoken of as the place of pigs. Here afterbirths are placed in trees and the special ceremonial meals are thrown away. Sections of these declivities may come to be associated with the supernatural, either with a *marsalai*, or with the wild taro which often grows there (1938: 206–7).

The village ground is kept bare of weeds and is periodically swept clean with brooms made of a dried *limbum* flower, *galo'it*. This is women's work (1938: 208).

The houses face either inward, towards the plaza, or are placed parallel with it . . . [The very large houses] are spoken of as yam houses, for here the yam crop is stored and here an important man will receive his guests. Cooking and sleeping are usually done elsewhere (1938: 204).

The symbolic oppositions in concentric dualism thus include:

centre	periphery
good	bad
male	female
public	domestic
yam	pigs
cultivated	wild
raw	cooked
men	spirits

If concentric dualism is defined as the opposition between a centre and a periphery, its graph is $K_2 = P_2$. If the centre is regarded as superior to the periphery then its graph is a directed arc and still bipartite. Lévi-Strauss evidently has the second model in mind when he says that 'any attempt to move from an asymmetric triad to a symmetric dyad presupposes concentric dualism, which is dyadic like the latter but asymmetric like the former' (1963b: 151). These operations are not explained, but we note with interest that these three models (assuming the triad is a directed 3-cycle as in generalized exchange) are the primordial structures defined in Chapter 1 and shown in Fig. 1.1.

The situation in Tanga in the Bismarck Archipelago is more complex than in Arapesh. Here the moieties are not named but deictically defined as an 'us/them' opposition. Each moiety consists of a group of matrilineal clans allied by the exchange of names and extension of sibling terms. Although they are 'enemies' of each other, pairs of clans in opposing moieties exchange women. The map in Fig. 2.11, adapted from F. L. S. Bell (1935), shows the spatial arrangement of the five clans on the island of Boeing in the Tanga group.[3]

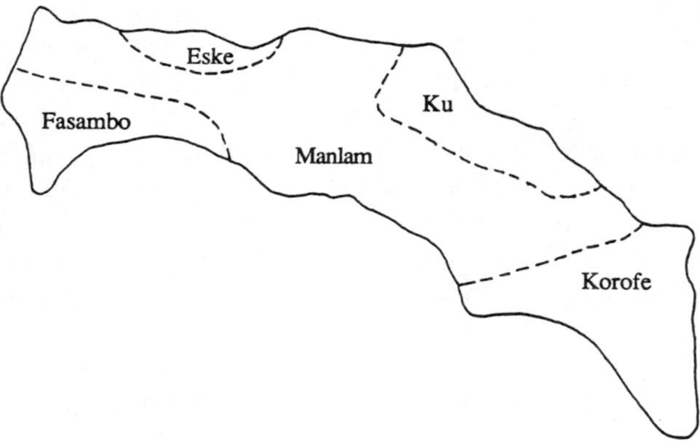

FIG. 2.11. A map of Boeing Island, Tanga group, showing the location of clans (from Bell 1935)

In Tanga contiguous clans marry and non-contiguous clans do not. The marriage exchange graph for Boeing is shown in Fig. 2.12, where the points represent clans and the lines marriage exchange. It is the star $K_{1,4}$, whose two point sets represent the two unnamed

[3] We have treated Filimat and Tasik as subsections of a main clan, Manlam, as Bell suggests.

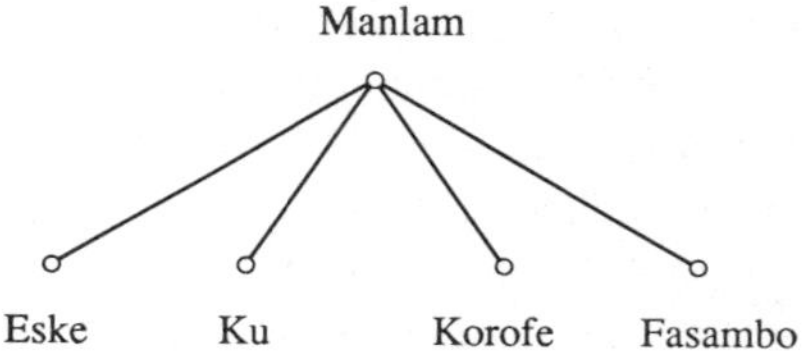

FIG. 2.12. The graph $K_{1,4}$ of Boeing dual organization

moieties. Every star is a tree and every tree is bipartite. Thus any exchange graph that has the structure of a tree, for example any of the graphs in Fig. 2.6, has a dual division.

The spatial arrangement of the clans on Boeing immediately suggests an alternative, formally equivalent definition of a bigraph as a bicolourable graph, and by implication of an n-partite graph as an n-colourable graph.

In colouring geographical maps, two countries with a common boundary must be given different colours in order to identify their territories unambiguously. When each country is replaced by one point, then two of these points are joined by a line whenever their countries have a common boundary. The resulting graph is called the *dual of the map* and suggests the next definition. A *colouring* of a graph is an assignment of colours to its points so that no two adjacent points have the same colour. A set S of points of a graph G is *independent* if no two of these points are adjacent. The set of all points with any one colour is independent and is called a *colour class*. An *n-colouring* of a graph G uses exactly n colours; it thereby partitions V into n colour classes. The *chromatic number* $\chi(G)$ is defined as the minimum n for which G has an n-colouring. A graph G is called *n-colourable* if $\chi(G) \leqslant n$ and is *n-chromatic* if $\chi(G) = n$.

Since G obviously has a p-colouring and a $\chi(G)$-colouring, it must also have an n-colouring whenever $\chi(G) < n < p$. The graph of Fig. 2.13 has $\chi = 2$; n-colourings for $n = 2, 3, 4$ are displayed, with positive integers designating the colours.[4] Clearly a non-trivial connected graph G is bipartite if and only if $\chi(G) = 2$.

The chromatic numbers for some of the familiar graphs are easily determined, such as those for a complete graph, $\chi(K_p) = p$, and for

[4] One of the most famous problems in all of mathematics was the conjecture, finally proven by Appel and Haken (1976), that every map which can be drawn in the plane is 4-colourable.

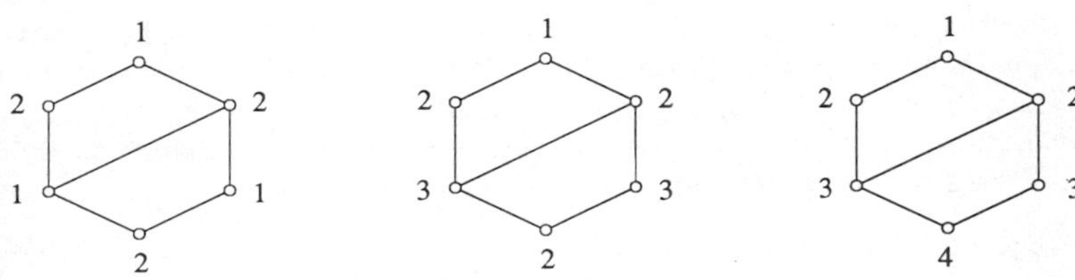

FIG. 2.13. Three colourings of a graph

any non-trivial tree T, $\chi(T) = 2$. The dual of the map of Boeing Island, which is a tree, requires only two colours, and the colour classes correspond to the two moieties. In general, for any bigraph G, $\chi(G) = 2$. Thus an exchange structure having the cyclic graph of Fig. 2.13 has a dual division.

J. A. Barnes has reminded us of a classic Australian example of the map-colouring, chequerboard model of marriage exchange. In their article on Kariera kinship, Romney and Epling analyse Radcliffe-Brown's (1913) map of local groups in the following way:

> An examination of Radcliffe-Brown's map (1913) . . . of the distribution of Kariera local groups reveals that the territory belonging to the local groups of 'our side' and 'the other side' was distributed in a checkerboard fashion. Thus, if one were to color territory of 'one kind' red and of 'the other kind' black on Radcliffe-Brown's map a rough approximation of a checkerboard would be obtained. One implication of this distribution is that each local group is adjacent to other local groups with whom marriage is allowed (Romney and Epling 1958: 61).

A more complex form of dual organization than in Arapesh or Tanga is found among the Etoro on the Papuan Plateau of New Guinea. In order to explicate their 'complex system of restricted exchange' Kelly (1974) uses a lattice-graph, which is based on two structural principles:

(1) Exchange is bilateral, i.e. symmetric.

(2) Any two lineages that exchange women with a common third lineage become brother lineages and cannot exchange women with each other.

Kelly's lattice graph is shown in Fig. 2.14a. The points represent lineages and the lines exchange relations. According to Kelly:

> All adjacent groups, on a horizontal or vertical axis, are major exchange groups and any pair of lineages which exchange women with the same third

descent group are brother lines. The four corner lineages each have two major exchange groups, those on the sides have three, and the interior lineages have four. These correspond to small, average, and larger lineages, respectively. The odd and even numbers represent the de facto moieties composed of groups of brother lineages (1974: 210).

By redrawing Fig. 2.14a as 2.14b, one sees immediately that the lattice graph, like the preceding dual organization graphs, is bipartite. Here the two point sets, or colour classes, consisting of odd and even numbers, correspond to two de facto (unnamed, unrecognized) moieties.

Kelly's second principle would prohibit lineage 8 from marrying with 2, 4, 12, 14 and 6, 10, 18. Such a large lineage's brother lineages are 'therefore nearly coextensive with the de facto moiety' (Kelly 1974: 210). Lineage 8 could, however, marry with 16 or 20 with which it has no exchange partners in common, but if it did, the graph

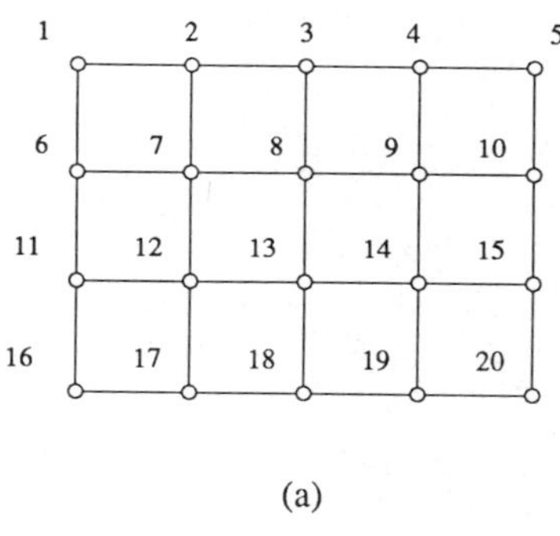

(a)

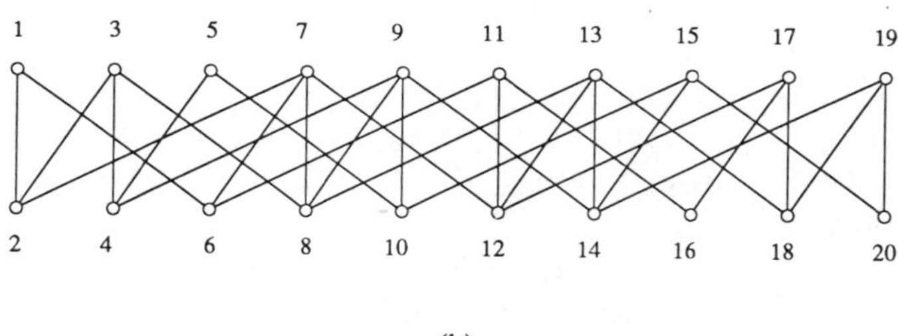

(b)

FIG. 2.14. The graph of Etoro dual organization

would no longer be bipartite, i.e. there would no longer be a moiety structure. In order to ensure a bipartite structure and to make every lineage's brother lineage completely coextensive with the de facto moiety, we hypothesize that the 'brother' relation in Etoro is transitive. Thus lineage 8 is a brother of 14 and 14 is a brother of 20, and 8 is therefore a brother of 20 even though 8 and 20 have no exchange partners in common. At least this can be deduced, since according to Kelly (1974: 97–8), 'Each lineage ... regards every other lineage either as a "brother" line or as a descent group with which it exchanges women.' Lineages 8 and 20 do not exchange women, and so by this statement they must be 'brothers'. (See also Kelly 1974: 117.)[5]

Since not all moiety arrangements take the form of a lattice graph or a tree, it would be useful to have a general criterion for a bipartite graph. The following theorem was given by König (1936); it appeared in the first book entirely devoted to graph theory.

Theorem 2.2. A graph G is bipartite if and only if it contains no cycles of odd length.

As we have shown, some forms of dual organization have a tree as their underlying graph. Theorem 2.2 has an immediate consequence for trees, which have no cycles at all.

Corollary 2.2a. All trees are bipartite.

Since there are no triangles in such a graph, the exchange relation must always be intransitive: if A gives to B and B gives to C then A cannot give to C. There may be any number of particular cultural ideologies which embody this abstract rule.

Actually Kelly's model of moiety formation is just the reverse of a well-known theoretical model which also happens to be unwittingly represented as a bipartite graph. In Kelly's model a moiety structure is generated by the cultural rules governing the relations between smaller exchanging groups (lineages). In Robin Fox's (1983 [1967]) model a moiety structure is preserved in spite of the internal segmentation, due to demographic increase, of two larger exchanging groups. According to Fox,

> We can ... imagine a moiety system having arisen in which the A's exchange women with the B's. If we are still at a rather primitive stage of hunting and

[5] The reasoning can be succinctly stated: if an element of a set $\{x, y\}$ is known not to be x, then it must be y.

gathering, then our local groups will still be small bands of males with their wives. The two moieties will then consist of a number of local bands A^1, A^2, A^3 ... A^n, and B^1, B^2, B^3 ... B^n. A local group then, may not just arrange exchanges with *one* group of the opposite moiety, but may have exchange arrangements with several. Thus A^1 may exchange with B^1, B^3, and B^5, while B^1 may exchange with A^1 and A^3 ... and so on (1983 [1967]: 182–3).

Fox represents this moiety structure by an implicit bipartite graph, drawn in the conventional zigzag way except that it is turned on its side as shown in Fig. 2.15. (In this figure we have replaced the capital letters and superscripts in the above quotation with lower case letters and subscripts.) Fox regards this graph as a useful simplification in his exposition of Australian marriage systems. (This particular type of drawing is also an excellent one to imagine in going through the proof of Theorem 2.2.)

Theorem 2.2 also applies to bipartite directed graphs. Thus the directed 6-cycle of marriage exchange described by Rivers's informant, John Pantutun, is bipartite.

We note that F. A. E. van Wouden (1968 [1935]), in his independent discovery of generalized exchange, perceived that dual organization requires even cycle length. Speaking of the special case in which a system of 'exclusive', i.e. matrilateral, cross-cousin marriage consists of a single (directed) cycle, which he regarded as its original form, van Wouden said:

Dual organization of the tribe is not required by the system, but can very well accompany it. The number of clans must total at least four, and must always be even: clan 1 takes wives from 2, 2 from 3, 3 from 4, and 4 finally

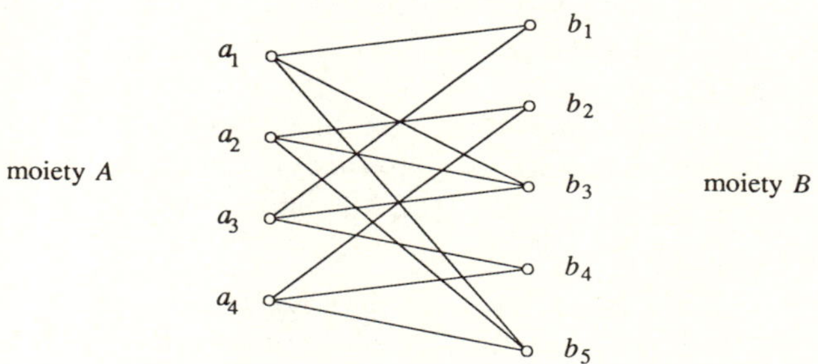

FIG. 2.15. Fox's (1983 [1967]) implicit bipartite graph model of moiety structure

takes from 1. The pair of clans 1 and 3, and that of 2 and 4, constitute exogamous groups, i.e. exogamous phratries. If the number of clans comes to more than four, e.g., to six, the division into two exogamous halves is not so directly apparent but is still very probable (1968 [1935]: 88–9).

Of course, the number of groups need not be even if there is more than a single cycle of exchange. (The number of points in a bipartite graph need not be even, e.g. $K_{2,3}$ has five points.)

It is commonly but not universally the case that a society is divided into two exchange groups or two sets of exchange groups. Thus the underlying graph may be 3- or 4- or in general n-partite rather than bipartite. Fig. 2.16 illustrates as a purely theoretical possibility a tripartite graph.

Colourable Signed Graphs and Competitive Exchange

Since we have defined a colouring of a graph G, it is a natural generalization to define a colouring of a signed graph S, following Cartwright and Harary (1968).[6] This is the model for exchange structures in which relations between groups are based on antithetical duality, that is, positive vs. negative as opposed to the presence vs. the absence of relationships. We shall give an example of an exchange structure whose signed graph is 3-colourable.

A *signed graph S* is obtained from a graph G when each line of G is designated either positive or negative. The following definition of

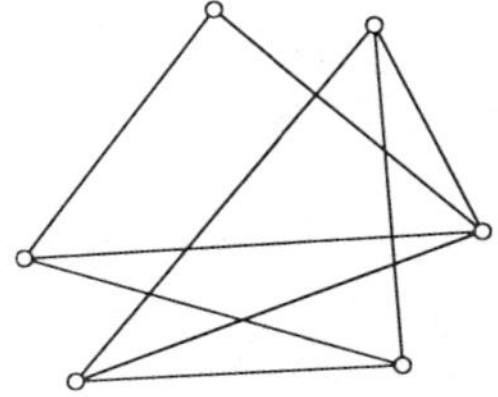

FIG. 2.16. A tripartite graph

[6] See Hage (1973) for an application of this model to alliance structure in Highland New Guinea.

colouring a signed graph S is different from the previous definition of colouring a graph G. An *n-colouring of a signed graph S* is an assignment of n colours to its points such that (1) every two points joined by a negative line are in different colour classes, and (2) every two points joined by a positive line are in the same colour class. We say that S *has a colouring* or is *colourable* if it has an n-colouring for some n. It follows immediately from these definitions that if a signed graph S has only negative lines the problem of colouring S is the same as that of colouring its underlying unsigned graph. If, however, S has some positive lines, it is not necessarily colourable.

Let S^+ be the spanning subgraph obtained by removing all negative lines from S. The *positive components* of S are just the components of S^+. It follows from this definition that two distinct points of S are in the same positive component if and only if they are joined by a path consisting entirely of positive lines (called an *all-positive path*). Clearly, the positive components of S partition $V(S)$ into subsets such that each positive line joins two points in the same subset, and S has exactly one such partitioning.

The following theorem (Cartwright and Harary 1968) presents a criterion for a signed graph to have a colouring.

Theorem 2.3. A signed graph S has a colouring if and only if S has no cycle with exactly one negative line.

This theorem is illustrated in Fig. 2.17, which shows a colourable signed graph S (in which negative lines are represented by dashes). Its three positive components are evident in S^+, and its point set $V(S)$ can be partitioned into the colour classes $\{v_1, v_2\}$, $\{v_3, v_4, v_5\}$, $\{v_6\}$. Clearly, S has no negative lines joining two points of the same positive component, nor does it have a cycle with exactly one negative line.

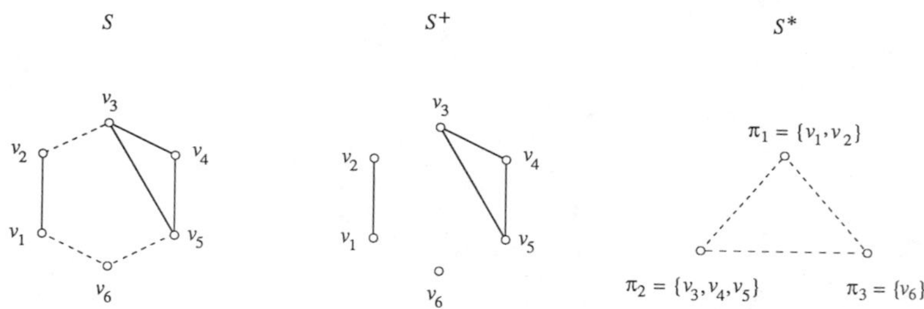

FIG. 2.17. A (uniquely) 3-colourable signed graph

The *condensation of S by its positive components*, denoted by S^*, is the signed graph whose points are the subsets $\pi_1, \pi_2, \ldots, \pi_n$ determined by (contracting or shrinking to a point) the positive components of S and whose lines are determined as follows: in S^* the points π_i and π_j are adjacent if and only if there is at least one negative line in S joining a point of π_i and one of π_j. The construction of S^* from S is illustrated in Fig. 2.17.

A signed graph S has a *unique colouring* if there is only one partition of $V(S)$ into $\chi(S)$ colour sets. When $\chi(S) = n$, it is called a *unique n-colouring*.

The signed graph S in Fig. 2.17 is uniquely 3-colourable. Further, the signed graph S_1 in Fig. 2.18 is uniquely 2-colourable, as its point set $V(S)$ can only be partitioned into two colour classes as follows: $\{v_1, v_2\}$, $\{v_3, v_4, v_5, v_6\}$. The signed graph S_2 in Fig. 2.18 is not colourable.

On Goodenough Island in the D'Entrecasteaux Group off the eastern coast of New Guinea, competitive exchange between agnatic descent groups, represented by rival big men, is based on a structure of traditional enmity and friendship. Enemies exchange food, which, however, they cannot consume but must pass on to their friends. Thus every group must have both friends and enemies.

We model this competitive exchange system by a signed graph S_1 in which the points represent exchanging units and the lines exchange relations: negative lines for food enemies and positive lines for food friends. The ideal structure, as described by Young (1971), consists of four groups A, B, C, D as illustrated in Fig. 2.19a. Continuing, Fig. 2.19b shows a signed graph S_2 with six points

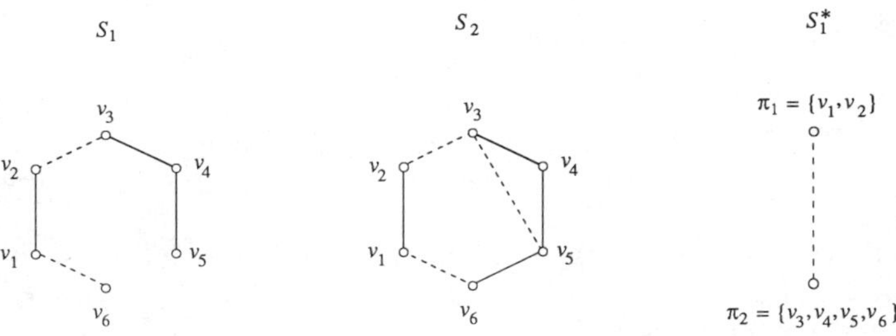

FIG. 2.18. A (uniquely) 2-colourable signed graph and a non-colourable signed graph

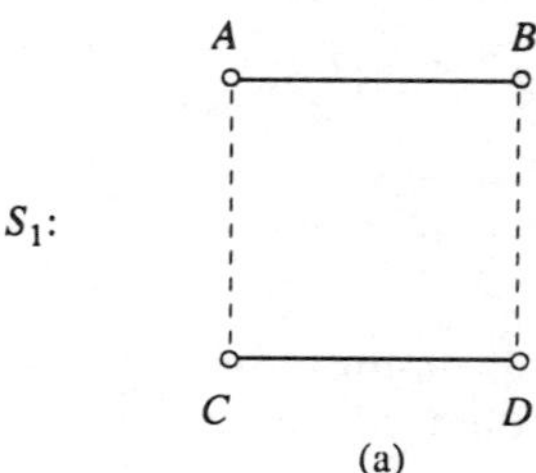

(a)

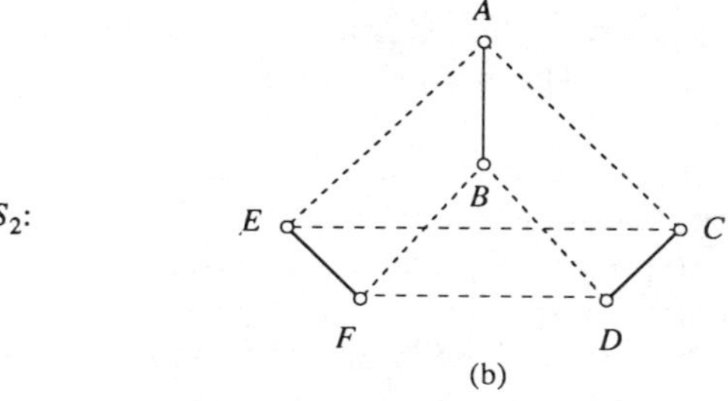

(b)

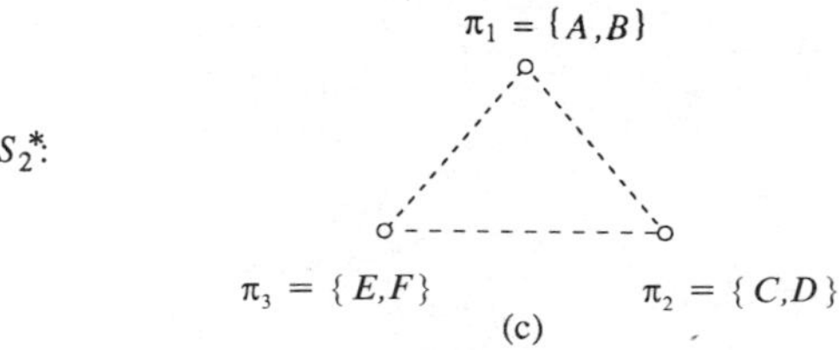

(c)

FIG. 2.19. A signed graph illustrating competitive exchange

which contains three signed subgraphs isomorphic to this ideal structure. Fig. 2.19c shows the condensation S_2^* of Fig. 2.19b by its positive components.

Fig. 2.20a shows a signed graph based on Young's description of the village of Kalauna. The points represent clans or major clan segments (which contain sets of smaller exchanging groups), and the positive and negative lines represent friend/enemy relations. It departs from the ideal arrangement, although each clan or clan segment does have at least one food enemy and one food friend, which guarantees that the system works.

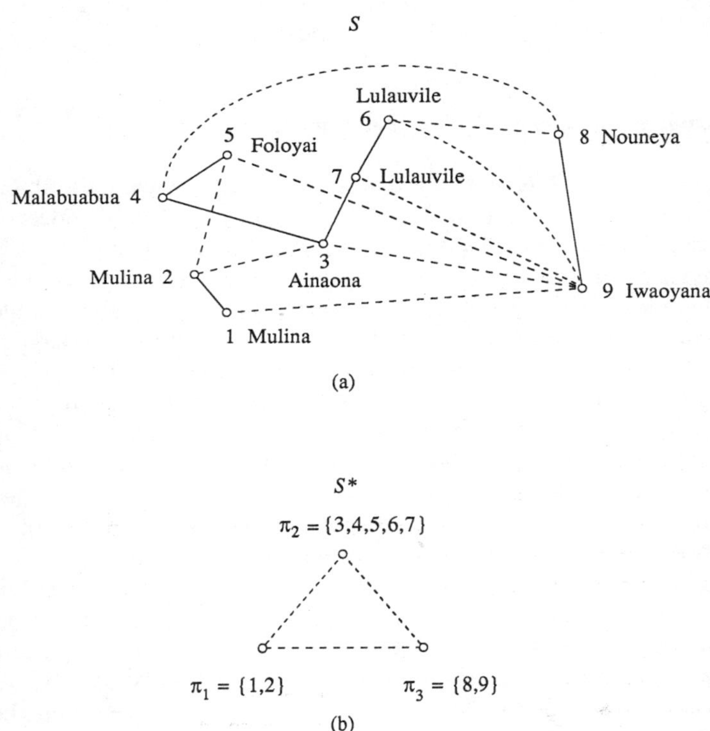

FIG. 2.20. Competitive food exchange on Goodenough Island

The signed graph S in Fig. 2.20a is uniquely 3-colourable, as shown by its condensation in Fig. 2.20b. We note in general that this colouring characterizes clan alignments, and in particular that the all-positive path joining the points 3, 4, 5, 6, 7 predicts support in clan confrontations. According to Young there are two major antagonists in the village, Iwaoyana (9) and Lulauvile (6 and 7).

Since Nouneya is Iwaoyana's [friend], Malabuabua sides with Lulauvile in opposition to its enemy Nouneya. Foloyai and Ainaona, being Malabuabua's [friends], are also drawn in to oppose Iwaoyana and Nouneya. Mulina, with traditional enemies in both camps, and no [friendship] obligations beyond itself, tends to remain 'uncommitted', which effectively means helping both sides (Young 1971: 74).

Young concludes that clan alignments are determinate and that there is a 'general balance of power' among the clans.

Ceremonial Analogues of Marriage Exchange

In a major contribution to comparative studies in Melanesia, P. G. Rubel and A. Rosman (1978) adapt Lévi-Strauss's (1969) models of marriage exchange to the analysis of ceremonial exchange in New Guinea. From a study of thirteen societies in the Lowlands and Highlands they propose an evolutionary taxonomy of exchange systems in which an *Urform* based on restricted exchange undergoes successive transformations in the direction of generalized exchange. They hypothesize that with expansion from the Lowlands into the Highlands of New Guinea, there was a (non-unilinear) progression from exchange based on dual organization to exchange based on long chains of delayed reciprocity as found in the Mae Enga *te* cycle. There is, however, a double confusion in their model, first in their interpretation of the *te* as structurally equivalent to patrilateral cross-cousin marriage, and second in their classification of patrilateral cross-cousin marriage as a form of generalized exchange rather than discontinuous exchange as in Lévi-Strauss (1969). There have been other attempts to apply Lévi-Strauss's models to ceremonial exchange in Melanesia, for example Feil (1980) on the Tombema Enga *te* and Damon (1980) on the *kula* ring, and there have long been divergent interpretations of these models, for example Homans and Schneider (1955), Maybury-Lewis (1965), Josselin de Jong (1952), and Ekeh (1974). It would be helpful, therefore, to clarify the models of marriage exchange, determine what, if any, application they have to ceremonial exchange, and then, if necessary, provide structural models which correctly describe the various forms of exchange actually found in Melanesia. To do this we will require a number of concepts from the theory of directed graphs.

Digraphs

A *directed graph* or *digraph* D consists of a finite set V of points and a collection of ordered pairs of distinct points. Any such pair (u, v) is called an *arc* or *directed line* and will usually be denoted by uv. The arc uv goes from u to v and is *incident* with u and with v. We also say that u is *adjacent to* v and v is *adjacent from* u. When arc $x = uv$, we say $u = fx$, the first point of x, and $v = sx$, its second point.

In a graph G, each point has a degree, but in a digraph D it has both an outdegree and an indegree. The *outdegree* $od(v)$ of a point v is the number of points adjacent *from* it, and the *indegree* $id(v)$ is the number of points adjacent *to* it. Fig. 2.21 shows all the 16 non-isomorphic digraphs with three points. In the last digraph of the third row, there is one point with $od = 1$ and $id = 2$, one with $od = 2$ and $id = 2$, and one with $od = 2$ and $id = 1$.

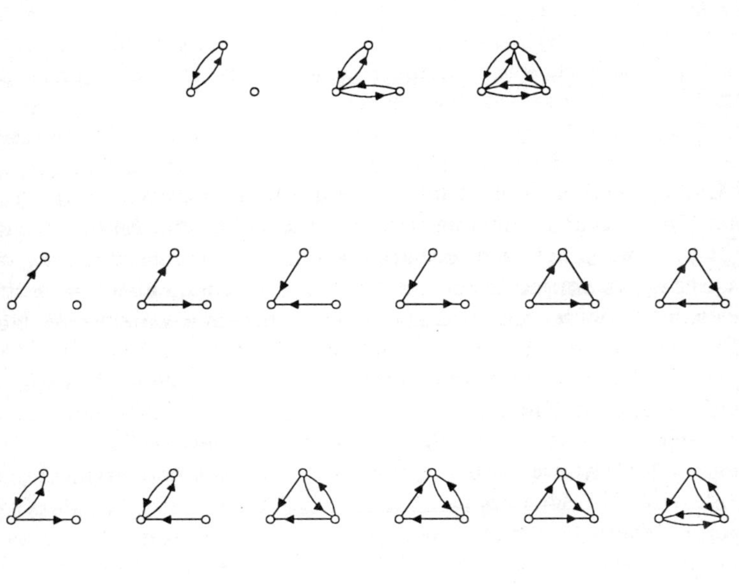

FIG. 2.21. The digraphs with three points

Every point of a digraph can be classified according to the combination of its indegree and outdegree. An *isolate* is a point whose outdegree and indegree are both 0. A *transmitter* is a point whose outdegree is positive and whose indegree is 0. A *receiver* is a point whose outdegree is 0 and whose indegree is positive. A *carrier* is a point whose outdegree and indegree are both 1. Any other point is called *ordinary*. In Fig. 2.22, point 1 is a transmitter, 2 is a carrier, 3 is a receiver, 4 is ordinary, and 5 is an isolate.

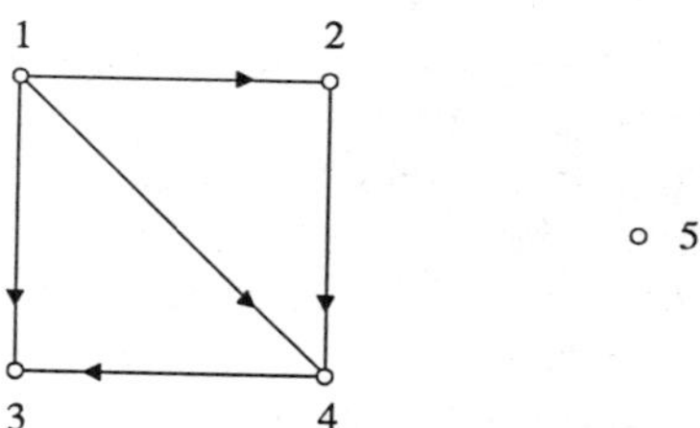

FIG. 2.22. A digraph to illustrate the classification of points

Corresponding to the definition of a walk in a graph, a (*directed*) *walk*[7] in a digraph is an alternating sequence of points and arcs, v_0, x_1, v_1, . . ., x_n, v_n, in which each arc x_i is $v_{i-1}\,v_i$. Note that points as well as arcs are permitted to repeat. For brevity we may again write the sequence of points v_0, v_1, . . ., v_n to indicate the same walk. The *length* of such a walk is *n*, the number of occurrences of arcs in it. Thus the length of each type of walk described below is well defined. A *closed walk* has the same first and last points, an *open walk* does not, and a *spanning walk* contains all the points. A (directed) *path* is a walk in which all points are distinct; a (directed) *cycle* is a non-trivial closed walk with all points distinct except the first and last. The directed cycle of length *n* is denoted by $\vec{C}_n$. If there is a path from *u* to *v*, then *v* is said to be *reachable from u*, and the *distance d*(*u*,*v*) from *u* to *v* is the length of any shortest such path. There may be more than one such path. In Fig. 2.23, the sequence of points 1, 2, 3, 4, 2, 3, 5 is a directed walk, the sequence 1, 2, 3, 5 is a path, and 2, 3, 4, 2 is a cycle. There is no spanning walk in this digraph.

[7] A directed walk was called a 'sequence' in Harary, Norman, and Cartwright (1965), but the former is the currently accepted term.

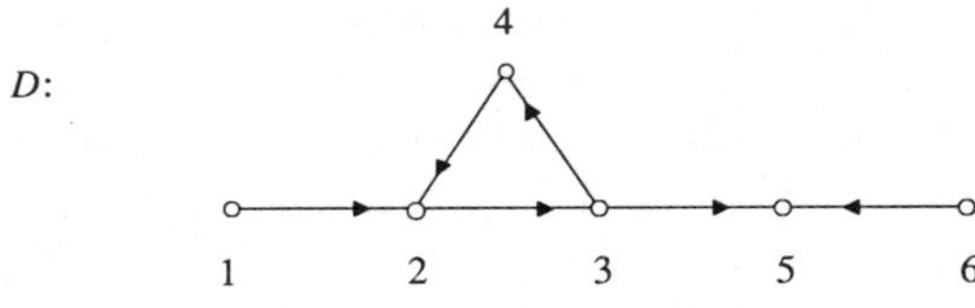

FIG. 2.23. A digraph to illustrate walks

Each walk in a digraph is directed from the first point v_0 to the last v_n. We also need a concept which does not have the property of direction and is analogous to a walk in a graph. A *semiwalk* is again an alternating sequence $v_0, x_1, v_1, \ldots, x_n, v_n$ of points and arcs, but each arc x_i may be either $v_{i-1}v_i$ or v_iv_{i-1}. Then a *semipath* is a semiwalk in which all points are distinct, and hence a path is a semipath with consistent direction. A semipath that is not a path is called a *strict semipath*. In Fig. 2.23, points 1 and 6 are *joined* by a semiwalk 1, 2, 3, 4, 2, 3, 5, 6, and by a semipath 1, 2, 3, 5, 6.

A *semicycle* is obtained from a semipath on adding an arc joining the end points of the semipath. Every cycle is a semicycle since every path is a semipath, but not conversely. The digraph D of Fig. 2.24 contains three semicycles: $Z_1 = 2, 3, 1, 2$; $Z_2 = 1, 3, 4, 1$; and $Z_3 = 1, 2, 3, 4, 1$. Clearly, Z_1 is not a cycle whereas Z_2 and Z_3 are; Z_3 is a *spanning cycle*, because it contains all the points of the digraph.

Since every digraph is a relation, digraphs may be described in terms of the properties of relations, as fully explicated in Chapter 8. For present purposes, a digraph is *symmetric* if for every arc uv there is an arc vu. It is *asymmetric* if every arc uv precludes an arc vu. A

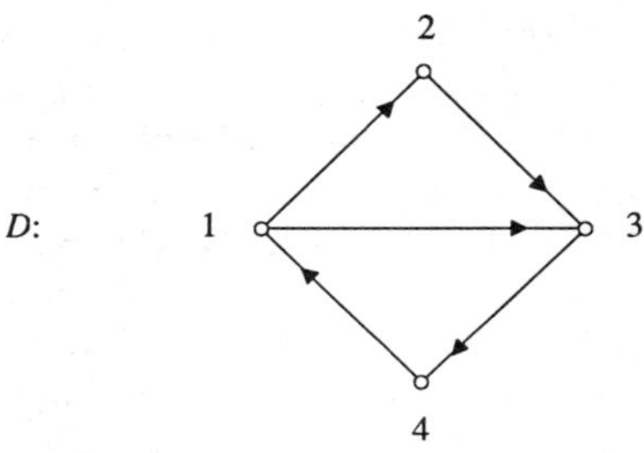

FIG. 2.24. A digraph to illustrate cycles and semicycles

digraph is *complete* if every pair of points u and v are joined by at least one arc. The first row of Fig. 2.21 shows all the (non-vacuously) symmetric digraphs and the second row all the (non-vacuously) asymmetric digraphs with three points. The digraphs in the third row are neither symmetric nor asymmetric. The last digraph in the first row, the last two in the second row, and the last four in the third row are complete. An asymmetric, complete digraph is called a *tournament*, exemplified by the last two digraphs in the second row of Fig. 2.21.

In Fig. 1.1 the second primordial graph is K_2, which has two points and one line, and is also expressible as the digraph DK_2 which has two points and a *symmetric pair of arcs*. Similarly, for each undirected graph G, we have the *corresponding digraph* DG obtained from G on replacing each line of G by a symmetric pair of arcs. For example the path graph P_n thus leads to the *symmetric path* DP_n and the cycle C_n goes to the *symmetric cycle* DC_n.

A graph G is either connected or not, but a digraph D may be connected in three different ways. A digraph is *strongly connected*, or more briefly *strong*, if every two points are mutually reachable; it is *unilaterally connected*, or *unilateral*, if for any two points, at least one is reachable from the other; and it is *weakly connected* or *weak*, if every two points are joined by a semipath. Sometimes a digraph is called *connected* when it is weakly connected. Clearly, every strong digraph is unilateral and every unilateral digraph is weak, but the converse statements are not true. A digraph is *disconnected* if it is not even weak. We note that the trivial digraph consisting of exactly one point is (vacuously) strong because it does not contain two distinct points.

There are four different 'connectedness categories' C_i of a digraph D which indicate its level of connectedness. Fig. 2.25 shows digraphs that are disconnected (C_0), weak but not unilateral (C_1), unilateral but not strong (C_2), and strong (C_3). In contrast to the connected components of a graph, there are three different kinds of digraph components: strong, unilateral, and weak subgraphs that are maximal with respect to these properties. By analogy to graphs, a disconnected digraph having $k > 1$ weak components all isomorphic to D is written kD and is said to consist of k copies of D.

We pause just briefly to give the structural criteria (from Harary 1959a and Harary, Norman, and Cartwright 1965) for the functional digraphs discussed in Chapter 1. We require one further

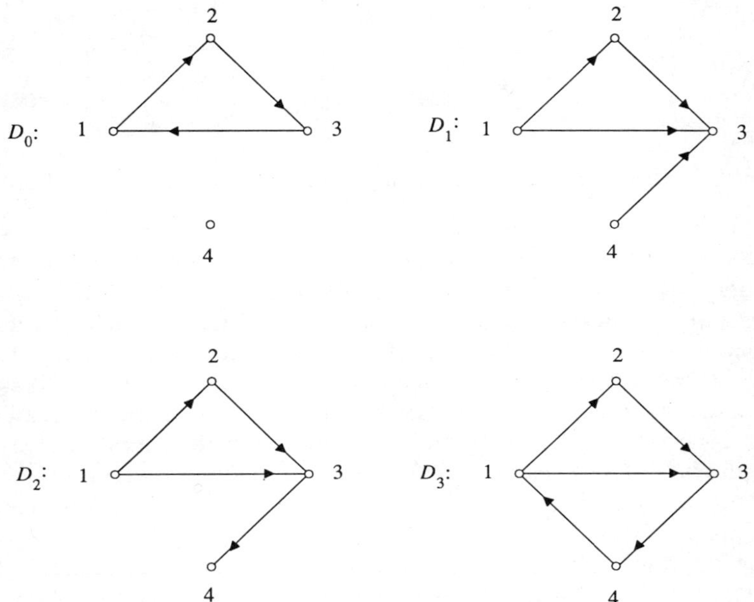

FIG. 2.25. Digraphs to illustrate connectedness

definition. A *tree toward a point* v is a weak digraph with no semicycles in which all paths are directed toward v, for example the third digraph in the second row of Fig. 2.21.

Theorem 2.4. The following statements are equivalent for a weak digraph D:

(1) D is functional.
(2) There is an arc x on a cycle of D such that $D - x$ is a tree toward fx.
(3) D has exactly one cycle Z, and after deleting its arcs each weak component of the resulting digraph consists of a tree toward a point of Z.

Networks

Finally, in order to accommodate exchange structures whose relations differ in strength, number, or type, we use the model of a

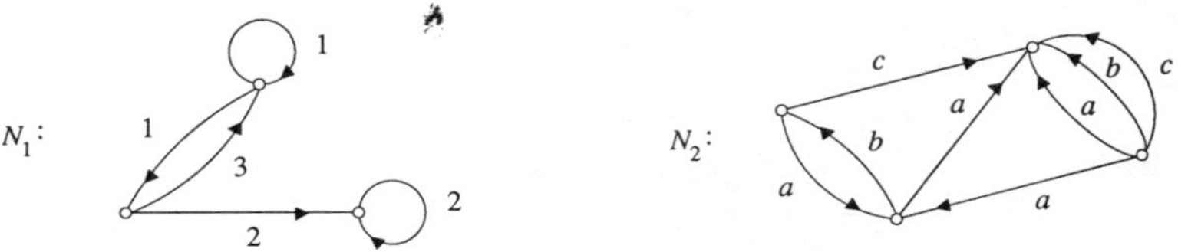

FIG. 2.26. A numerical and a non-numerical network

network. A *network* N consists of a relation (V, E) together with a 'value' assigned to each line or arc (if directed, including loops). If this value is always 1 we then have a relation. Thus relations form a special class of networks. The set S of values can be numerical or non-numerical as illustrated in Fig. 2.26 for two networks whose value sets are $S = \{1, 2, 3\}$ and $S = \{a, b, c\}$. In this chapter we shall be concerned with non-numerical networks whose value sets S represent different relations. The number of kinds of relations in the network is called the *type* of the network. Thus N_2 in Fig. 2.26 is a network of type 3.

Using these concepts we can define in graph theoretic terms the basic models of restricted, generalized, and discontinuous exchange. In a lucid discussion of Lévi-Strauss's theory, J. A. Barnes (1971) points out that the contrast between restricted and generalized exchange is based on three interrelated criteria: a genealogically specified rule of marriage; the number of groups in the exchange system; and symmetrical vs. asymmetrical exchange.

Lévi-Strauss defines restricted exchange as follows:

> The term 'restricted exchange' includes any system which effectively or functionally divides the group into a certain number of pairs of exchange-units so that, for any one pair X–Y there is a reciprocal exchange relationship. In other words, where an X man marries a Y woman, a Y man must always be able to marry an X woman. The simplest form of restricted exchange is found in the division of the group into patrilineal or matrilineal exogamous moieties. If we suppose that a dichotomy based upon one of the two modes of descent is superimposed upon a dichotomy based upon the other the result will be a four section system instead of a two-moiety system. If the same process were repeated, the group would comprise eight sections instead of four (1969: 146).

Restricted exchange is first of all represented by a symmetric digraph. The second part of the definition describes Australian marriage class systems where the number of intermarrying pairs is 2,

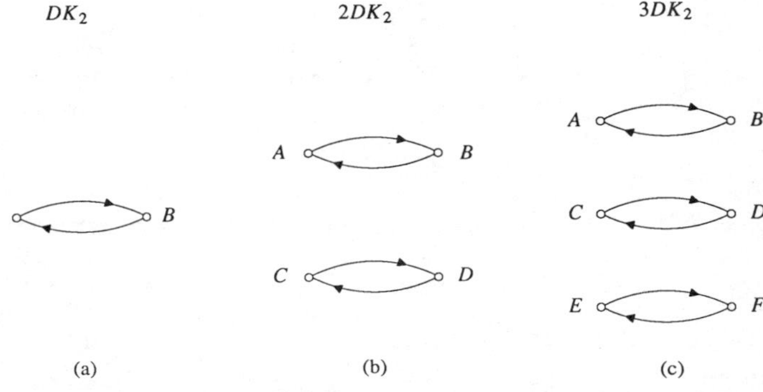

FIG. 2.27. Three digraphs of restricted exchange

4, or 8. The first part of the definition, however, only specifies a division of society into a 'certain number of intermarrying pairs'. Fig. 2.27 shows three graphical models of restricted exchange: a moiety division, a division into two exchanging pairs, and a division into three exchanging pairs. The graphical definition of restricted exchange is thus k copies of DK_2, a symmetric pair of arcs written $k\,DK_2$.

Generalized exchange is more complex:

> Generalized exchange establishes a system of operations conducted 'on credit'. *A* surrenders a daughter or a sister to *B*, who surrenders one to *C*, who, in turn, will surrender one to *A*. This is its simplest formula. Consequently, generalized exchange always contains an element of trust (more especially when the cycle requires more inter-mediaries, and when secondary cycles are added to the principal cycle). There must be the confidence that the cycle will close again, and that after a period of time a woman will eventually be received in compensation for the woman initially surrendered (Lévi-Strauss 1969: 265).

On the basis of this characterization generalized exchange is first of all an asymmetric digraph, as no two groups are both wife-givers and wife-takers to each other. In the simplest case the model of generalized exchange is a directed 3-cycle, as noted in Chapter 1.

Livingstone (1969) has proposed two digraphic models of generalized exchange. The first is a *strong tournament*, a complete, asymmetric, strongly connected digraph as illustrated in Fig. 2.28. This tournament appears in *The Elementary Structures of Kinship* as

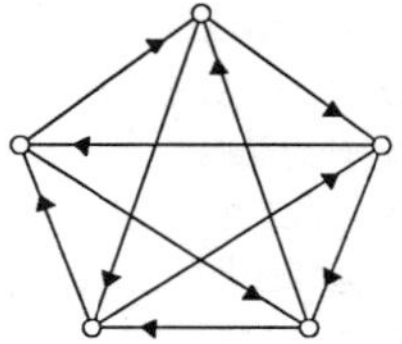

FIG. 2.28. A strong tournament model of generalized exchange

a model of Kachin marriage exchange.[8] Like all strong tournaments, it contains a spanning cycle as well as shorter cycles of varying length. Theorem 2.5 (from Harary and Moser 1966 and Harary *et al.* 1965) characterizes the cyclic structure of all such tournaments.

Theorem 2.5. If a tournament T is strong, then it contains a cycle of each length $k = 3, 4, \ldots, p$.

Livingstone considers that a strong tournament is too restrictive to serve as a universal model of generalized exchange, for it requires every group to be a wife-giver or wife-taker to every other group. He suggests as a less stringent model, an isograph as defined in Harary *et al.* (1965). An *isograph* is a digraph in which for every point v, $id(v) = od(v)$. The equality of indegree and outdegree would reflect the equality of the sexes in any group. Fig. 2.29 shows three isographs. The first two could serve as models of generalized exchange, but the third could not. The second isograph is in fact used implicitly in Kelly's (1968) model of generalized exchange in Dobu, as shown in Chapter 6 (Fig. 6.19). Further, since all symmetric digraphs are isographs, not all isographs can model generalized exchange.

In order to give a universal definition of generalized exchange we need to define a Hamiltonian digraph. A digraph D (always with p points) is called *Hamiltonian* if it contains a spanning cycle $\vec{C}_p$. Interpreting Lévi-Strauss's phrase 'principal cycle' in this way, we define generalized exchange as an asymmetric Hamiltonian digraph.

[8] Actually, Fig. 46 in *The Elementary Structures of Kinship* shows the husband-giving relation. Its converse and isomorphic digraph is the wife-giving relation. Lévi-Strauss (1969: 253) emphasizes that this particular model is 'only a particular case of a more general formula which seems to function at all levels of the social reality. We have already cited evidence establishing that the division into *mayu ni* [wife-givers] and *dama ni* [wife-takers] is the rule, not only for the five "tribes" as most writers call them, but for the smaller units, the subgroups and families, into which they are divided.'

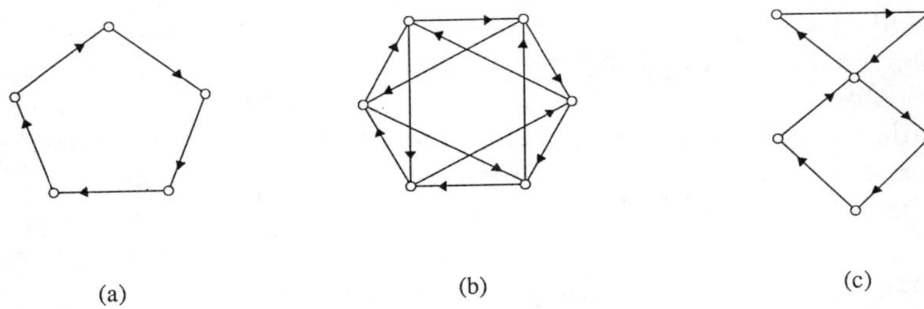

FIG. 2.29. Isographs: (a) and (b) are models of generalized exchange but (c) is not

A digraph consisting of a single cycle is Hamiltonian, but so are all strong tournaments. Unfortunately there is no general criterion for such a digraph,[9] but fortunately for anthropological analysis, most marriage exchange systems are small enough to determine whether or not they contain a spanning cycle.

Lévi-Strauss has emphasized that generalized exchange is cyclical and expandable. We note that T. C. Hodson (1925) in an early independent discovery of this system emphasized just these properties. Hodson pointed out that matrilateral cross-cousin marriage 'requires as a minimum basis three exogamous divisions', and he observed: 'An interesting feature about the single cousin marriage system is that it is capable of almost indefinite expansion according to this formula:— (A + B) (B + C) (C + D) . . . [(R − 2n) + (r − n)] [(R − n) + r] (R + a)' (Hodson 1925: 172).

Hodson's idea is clear even though his algebraic notation is puzzling. It obviously does not entail the numerical differences and products of integer-valued variables A, B, C, . . ., R (why R?). Apparently also R = r and A = a, for what else could a and r mean? Rather it seems that (A + B) (B + C) . . . (R + A) is intended to indicate symbolically that there is a wife-giving arc from group A to group B, and another from B to C, and so on, and finally such an arc from R to A, forming in all a directed cycle.

For the sake of completeness we define explicitly three types of cross-cousin marriage. In all three cases ego is a male. In matrilateral cross-cousin marriage ego marries his mother's brother's daughter,

[9] In fact this is a celebrated extremely difficult problem, needed in operations research to handle the practical 'Travelling Salesman's Problem'.

as illustrated in Fig. 6.14 in Chapter 6. In the patrilateral case ego marries his father's sister's daughter, as illustrated in Fig. 6.15. Finally, in the bilateral case, ego's mother's brother and ego's father's sister are married and ego marries their daughter.[10] Restricted and generalized exchange are directly expressed in bilateral and matrilateral cross-cousin marriage respectively. Discontinous exchange is expressed in patrilateral cross-cousin marriage. This is defined as a '*closed structure*, within which a cycle of exchange opens and closes in the following manner: a woman is ceded in the ascending generation, a woman is acquired in the descending generation, and the system returns to a point of inertia' (Lévi-Strauss 1969: 444). Closed structures therefore do not perpetuate themselves, and do not form a global or overall structure:

> Instead of constituting an overall system, as bilateral marriage and marriage with the matrilateral cross-cousin each do in their respective spheres, marriage with the father's sister's daughter is incapable of attaining a form other than that of a multitude of small closed systems, juxtaposed one to the other, without ever being able to realize an overall structure (1969: 445).

Patrilateral cross-cousin marriage is a short-term arrangement between pairs of groups. Its graphical model, based on Fig. 6.15, is a set of connected 2-cycles, more precisely a symmetric path DP_n as shown in Fig. 2.30.[11]

There has been considerable confusion concerning the presumed structural equivalence of cycle length in matrilateral and patrilateral

FIG. 2.30. A digraph of patrilateral cross-cousin marriage (discontinuous exchange)

[10] Sir Basil Thomson (1908) in his book on Fiji designated bilateral cross-cousins, whose marriage is culturally defined as right and proper, as 'concubitants': 'The word "Concubitant" is adopted because, besides being a fair translation of the Fijian word *vei-ndavolani* (*vei* = reciprocal affix, *ndavo* = to lie down), it expresses the Fijian idea that persons so related ought to cohabit' (1908: 182). Together with Fison, Thomson worked out a model showing that concubitancy results when the entire tribe is divided into two exogamous intermarrying classes.

[11] Diagrams similar to Figs. 6.14 and 6.15 are given in Lévi-Strauss (1969). In Chapter 6 we give a more complex graphical model of patrilateral cross-cousin marriage based on F. E. Williams's concept of a 'multiplicity of reciprocating pairs'.

cross-cousin marriage. These forms are clearly different in the graphical models, but less evidently so in conventional kinship diagrams like the ones from Fortune (1933) reproduced in Chapter 6 (Figs. 6.14–17). By a *short cycle* in a digraph is simply meant a symmetric pair of arcs, while a *long cycle* $\vec{C}_n$ is a directed cycle with $n \geqslant 3$. Lévi-Strauss interprets the kinship diagrams as showing a contrast between the long cycle in matrilateral and the short cycle in patrilateral cross-cousin marriage. A graphical model such as the one in Fig. 2.30 shows this clearly. Matrilateral and patrilateral cross-cousin marriage are distinguished precisely on the basis of cycle length: at least 3 in the former vs. exactly 2 in the latter. Notice in Fig. 2.30 that the walk $A, B, C, D, E, D, C, B, A$ is not a cycle but a closed walk in the form of a symmetric path DP_n. In a cycle, as opposed to a symmetric path having $n \geqslant 3$ points, only the first and last points are repeated.

Some authors deny that these forms of marriage exchange differ in terms of cycle length. Referring to a kinship diagram of patrilateral cross-cousin marriage (like the one in Fig. 6.15), Maybury-Lewis states:

It will be seen from [the] figure that a descent group A which gives women to another descent group B in one generation will receive women from B in the next. Lévi-Strauss referred to this as discontinuous exchange and contrasted it with the generalized exchange effected by matrilateral cross-cousin marriage. He suggested that discontinuous exchange effected a series of short marriage cycles, whereas generalized exchange brought about a single unifying cycle in the society which practiced it (Lévi-Strauss 1949: 562). Homans and Schneider pointed out, however, that patrilateral cross-cousin marriage systems require a cycle of descent groups linked by marriage in the same way as with matrilateral cross-cousin marriage (1955: 13). The difference between the two lies not in the length of the cycle but in the fact that a matrilateral prescription ensures a unidirectional flow of marriages from group to group in successive generations, whereas the patrilateral one results in a change of direction of this flow with each generation (Maybury-Lewis 1965: 215–16).

Homans and Schneider say that:

father's sister's daughter marriage, just like mother's brother's daughter marriage, requires at least three lineages; any one lineage is linked by marriage to two others in the ring, and the ring can be lengthened indefinitely. On all these counts it meets Lévi-Strauss's requirements for generalized exchange. The only difference is that the men of B lineage . . .

give women alternately to the As and to the Cs instead of always getting them from one and giving them to the other (Homans and Schneider 1955: 13).

By the digraph concept 'ring', Homans and Schneider mean simply a symmetric cycle DC_n, which can be obtained from the symmetric path DP_n in Fig. 2.30 by joining the first and last points with a symmetric pair of arcs. Now there are two long, in fact spanning, cycles: *A*, *B*, *C*, *D*, *E*, *A*, and *A*, *E*, *D*, *C*, *B*, *A*. But even under this interpretation, which Lévi-Strauss (1969: xxxiv–xxxv) regards as mistaken and as showing a wilful disregard of the philosophy of the societies which either practise it or condemn it, patrilateral would still be distinguished from matrilateral cross-cousin marriage on the basis of cycle length. For in addition to the spanning cycles joining all groups, there would remain the short cycles joining all pairs of groups.[12] The supposed equivalence of these two marriage forms is thus based on conceptual confusion as well as optical illusion.

We can now take up Rubel and Rosman's taxonomy of ceremonial exchange in New Guinea. They distinguish four basic types, which we shall summarize and put into a standard graphical form. The first type, the simplest structurally and first evolutionarily, is dual organization, as exemplified by Keraki society west of the Fly River in southern New Guinea (Williams 1936). Here exogamous (patrilineal) moieties regulate both ritual and marriage exchange. Men of opposite moieties who exchange sisters are *tambera*, 'exchange partners' to each other. Each, as mother's brother to the other's child, performs a series of ritual services for his sister's son associated with the exchange of feasts and gifts. The importance of inter-moiety exchange is clearly underlined in marriage by such practices as sister purchase, in which a man may buy a classificatory sister from the same moiety of another group to exchange for a wife from the opposite moiety of his own group, and in initiation rituals in which a youth must be sodomized (to promote his growth—Hage 1981) by a senior male of the opposite moiety. The exchange digraph of Keraki dual organization is the directed 2-cycle DK_2 as shown in Fig. 2.31a. (We note that other examples are regarded as closer to or further from the prototype according to whether marriage and ceremonial exchange are identical as in Keraki or separate as in Arapesh.)

[12] See also Needham (1958) on this point.

D: 1 2

(a)

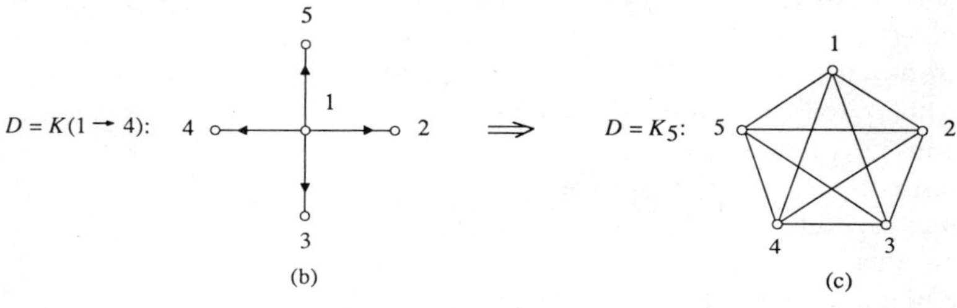

(b) (c)

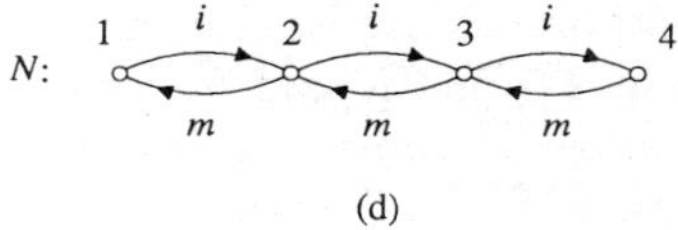

(d)

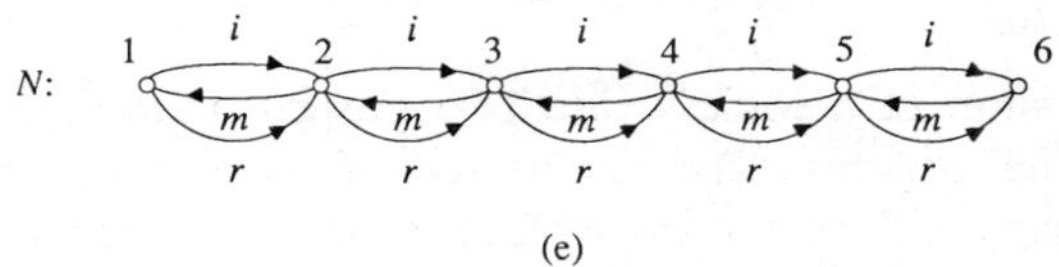

(e)

FIG. 2.31. Graphs of ceremonial exchange structures in New Guinea

Rubel and Rosman call the second type of ceremonial exchange structure a 'star formation'. It is exemplified by Maring society in the Central Highlands (Rappaport 1968). Periodically, one clan acts as host of a feast to all surrounding clans. Since each clan has a turn as host, the exchange digraph is in fact an outstar $K(1 \rightarrow n)$ on any given occasion. As shown in Fig. 2.31b, a complete, symmetric digraph is obtained for the entire system, since DK_{n+1} is the union of n subgraphs, an outstar from each of the $n+1$ points. For clarity in Fig. 2.31c we show K_p instead of DK_p. Rubel and Rosman regard this structure as a transformation of the prototype, because, while exchange relations are dyadic, there are more than two groups.

The third type has no generic label, but is regarded as a transformation 'in the direction of generalized exchange' (Rubel and Rosman 1978: 284), because exchange can be organized as a 'chain' sequence. It is exemplified by *moka*, a system of reciprocal and competitive exchange in Mount Hagen in the Western Highlands of New Guinea (A. Strathern 1971). The principal organizers and transactors are rival big men in different patrilineal clans, who are usually related as affines and close cognates. An exchange sequence consists of two steps: an initiatory gift from *A* to *B* followed by a larger main gift from *B* to *A*. The increment in the main gift, called *moka*, results in the superiority of *B* to *A*. One sequence may follow another with a reversal of directions: an initiatory gift from *B* to *A* and a main gift from *A* to *B*. If the main gift in the second sequence matches that in the first, *A* is equal to *B*; if it is greater, *A* is superior to *B* until the next sequence, when the directions are again reversed. Strathern aptly describes the system as one of 'alternating disequilibrium'. There are two types of *moka*: shell *moka*, which consists of gifts of shells and pigs, and pig *moka*, gifts of cooked pork or pigs. *Moka* exchange may unite more than two groups, when it is organized into 'chains' such that *A* gives an initiatory gift to *B*, who gives one to *C*, who gives one to *D*, and at the end of the chain, *D* gives a main gift to *C*, who gives one to *B*, who gives one to *A*. The directions are reversed in the next sequence. The exchange digraph of *moka* is the network *N* of type 2 shown in Fig. 2.31d, where *i* and *m* stand for two different relations: the exchange of the initiatory gift and the main gift. The walk 1, 2, 3, 4, 3, 2, 1 tells the sequential order of the exchanges.

The fourth type of exchange structure again has no generic label, but is regarded as 'an example of generalized exchange structurally identical to patrilateral cross-cousin marriage' (Rubel and Rosman 1978: 290). This is because it is organized into permanent chains rather than situationally defined ones as in *moka*. *Te* exchange as practised by the Mae Enga (Meggitt 1974) in the Western Highlands of New Guinea is more elaborate than *moka* with respect to the number of steps in each sequence, the types of goods exchanged, and the greater length of the chains. In the *te* the initiatory gift consists of pigs, pork, salt, shells, plumes, and, in modern times, money. The main gift coming from the opposite end of the chain consists of pigs. This is followed by a return gift in the opposite direction consisting of cooked pork. It takes about four years to complete a sequence uniting a large number of patrilineal clans. In each sequence the directions of the three gifts alternate. The exchange graph of the *te* is the network *N* of type 3 adapted from Meggitt (1974) and shown in Fig. 2.31e. Happily, the letter *r* preserves the alphabetical order established by *i* and *m* which reflects the temporal order of the gifts. (We note that *te* chains may have peripheral branches feeding into the main flow, and that the main flow itself may branch and converge, but this would not change anything essential about its graphical structure.)

A comparison of the previously defined marriage exchange digraphs with the ceremonial exchange digraphs makes it clear that the two sets of structures are, with the single exception of restricted exchange and dual organization, not equivalent. The 'star formation' does not consist of *n* copies of DK_2. In the case of Rubel and Rosman's third and fourth types, marriage exchange is based on a single relation, the exchange of women, and is therefore a digraph, while *moka* and *te* are based on multiple relations and are therefore networks. The distinction between a digraph and a network is clear in Lévi-Strauss's comment on Leach's interpretation of marriage exchange: '. . . Leach seems to attribute to me the absurd idea that, in Kachin society, women are exchanged for goods. I have never said any such thing. It is clear that, as in all other social systems, women are exchanged for women' (1969: 238).[13]

But even ignoring this distinction, the *te* and *moka* structures are unlike generalized exchange (matrilateral cross-cousin marriage), because their cycles are always of length 2. They are also unlike the digraphs of discontinuous exchange (patrilateral cross-cousin marriage), because the exchanges in *te* and *moka* occur in prescribed sequences of arcs in their networks, which do define a global system of exchange.

The conclusion we draw from this analysis is that ceremonial exchange structures in New Guinea should be studied directly rather than purely analogically. Instead of concentrating on family resemblances between marriage and ceremonial exchange, it would be better to study these structures for the properties they actually

[13] Quite idiosyncratically Leach (1983) has defined all exchange forms as either 'symmetrical' (an *i* for an *i*) or 'asymmetrical' (an *i* for a *j*), thereby ruling out generalized exchange in Lévi-Strauss's sense. Fortune (1933), in his independent discovery of generalized exchange, emphasized along with Lévi-Strauss that women are exchanged for women (see Chapter 6 below).

have. Given the richness of exchange forms in New Guinea, one could expect important theoretical contributions to general anthropology.

Exchange Quartets[14]

The digraphs in Fig. 2.31 do not of course constitute a taxonomy of New Guinea ceremonial exchange structures but represent only a few major types. There are many directions in which further research might proceed. One direction, suggested by one of the criteria of marriage exchange forms, would be to study the interactions between structure and number. While the structural necessity of four clans in Tikopia is a matter of interpretation (Hooper 1981; Firth 1981), many systems in New Guinea are necessarily composed of four groups, or a minimum of four groups. The underlying structural models may take the form of signed graphs, undirected graphs, or networks, as we shall very briefly indicate. On Goodenough Island, the smallest connected exchange structure in which every group has both an exchange friend and an exchange enemy is a signed undirected 4-cycle, as shown in Fig. 2.19a. In Arapesh society, interlocality exchanges between a pair of rival big men, *gabunyans*, depend on preceding intralocality exchanges between a pair of local big men, *buanyins*, of opposite moieties, producing an undirected 4-cycle. Such feasting systems have a property often noted of New Guinea marriage systems—that of widening the net of exchange to include many groups. In Bánaro, a society located on a tributary of the Sepik River (Thurnwald 1916; Rubel and Rosman 1978), marriage and ceremonial exchanges are integrated into a 4-point network of type 2. There is a division into two groups (gentes) A and B, each split into two parts x ('left') and y ('right'). Marriage is between different groups of the same side, $AxBx$ and $AyBy$. Men belonging to the two sides of each gens are united as *mundu* and perform sacred rituals for each other in initiation, marriage, and death, giving the relations $AxAy$ and $BxBy$. The network of this system is shown in Fig. 2.32, in which the double and dotted lines respectively represent the exchange of women in marriage and the exchange of ceremonial services in rites of passage.

[14] For an extensive analysis of four-element structures in Bush Mekeo culture see Mosko (1985).

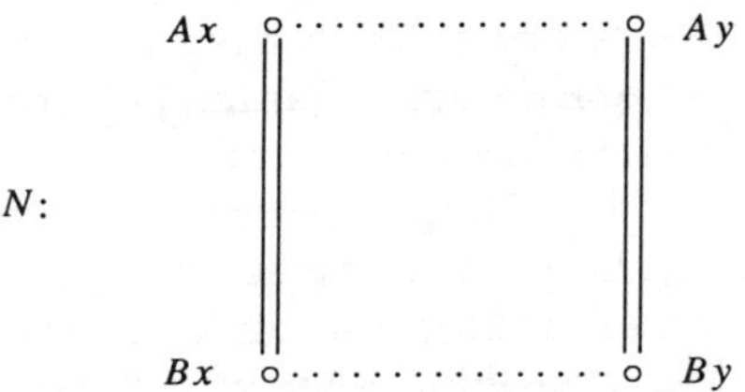

FIG. 2.32. A network model of Bánaro marriage and ceremonial exchange relations

A further example of an exchange quartet is given in Chapter 7, where Tangu siblingship and marriage as described by Burridge (1958) are analysed in terms of two binary operations on complete graphs.

3
Centres, Neighbourhoods, and Roots

> Von allen Zentralkarolinern waren die Polowat-Leute die besten und verwegensten Seefahrer. Ihr Wikingergeist erfuhr immer wieder neuen Antrieb durch das starke Bedürfnis nach fremden Waren (Kleidung, Schmuck u.a.) vor allem aber durch ihren ausgeprägten Familiensinn. Dieser liess sie gemeinsame Besuchsfahrten zu den Nachbarinseln . . ., wo sie Verwandte wussten, veranstalten. Man nahm dazu die Frauen und auch die Töchter mit, um sie der Verwandtschaft vorzustellen.
>
> Hans Damm and Ernst Sarfert, *Inseln um Truk*, Ergebnisse der Südsee-Expedition 1908–10

Recent work on Oceanic exchange systems suggests that network position may outweigh environmental and demographic variables in determining the economic status and political organization of island communities. Irwin (1978), in an archaeological reconstruction of the Mailu system off the south coast of New Guinea, concludes that increasing network centrality, as defined by distance in a graph, was the major factor in the rise of Mailu Island as a pottery manufacturing centre and middleman in regional trade. Brunton (1975) in a reinterpretation of *kula* ring ethnography argues that limited access to the flow of valuables, as inferred from the degrees of points in a graph, rather than high economic productivity or population density, led to the rise of chieftainship in the northern Trobriand Islands. Hage, Harary, and James (1986), using a Markovian simulation of network flow, propose that either of the extreme conditions of restricted or general access to valuables could lead to political hierarchy in a *kula* community, depending on associated environmental conditions and local definitions of exchange practices. This is a promising area of research, of interest to economic and political anthropologists, archaeologists, and network specialists alike. It can be advanced in two ways: empirically, by expanding the range of network structures studied, in particular by including the hierarchical systems found outside of Melanesia, and theoretically, by applying suitable models for the analysis of locational advantage.

We have a double objective. The first is to reconstruct the network of voyaging and exchange which traditionally connected most of the islands in the Carolines and Marianas in Western Micronesia. Previous work has concentrated on political aspects of exchange, the 'Yapese Empire' (Lessa 1950), on local subsystems of exchange (Alkire 1965), and on the navigational principles and voyaging techniques which enabled long-distance travel (Gladwin 1958, 1970; Goodenough 1953; Riesenberg 1972). Ethnographies of particular islands (Burrows and Spiro 1957; Lessa 1966) and ethnohistories (Hezel 1983; Lessa 1962, 1975) afford glimpses of the system in its original state. The essential general ethnographies are the contributions of Damm and Sarfert (1935) and Krämer (1935, 1937) to the reports of the 1908–10 Hamburg Südsee-Expedition (Thilenius, ed.). These authoritative but relatively unexploited sources survey, at least briefly, most of the islands in the Carolines with special attention to navigation and trade and provide the descriptive basis of our presentation.

The second objective is to elucidate the structural basis of trading success and political stratification among low islands in the Carolines. Unlike the linguistically and culturally more heterogeneous Melanesian networks, in the Massim (Malinowski 1922), the Huon Gulf (Hogbin 1947), and the Vitiaz Strait (Harding 1967), trading relations in the Carolines system were not confined to geographically adjacent or near neighbours but ranged far and wide, while political relations transcended informal horizontal linkages to form nested hierarchies. The economic and political centres of the Carolines did not completely coincide, and they appear to have had somewhat different structural origins. To account for the pre-eminence of communities in trading activity we use the model of distance (median) centrality (Ore 1962; Hakimi 1964) as a measure of the ease of access to resources which each community has compared to all others. To account for political stratification we use the model of betweenness centrality as an indication of the potential for control of communication. The betweenness model was first formulated by Freeman (1977, 1979) and first applied in anthropology by Hage and Harary (1981a) in an analysis of informal power and mediated communication, in a New Guinea big-man system. Because distance centrality and betweenness models deal only with global properties of a graph, we supplement them by introducing the concept of a point and its neighbourhood. This third model departs from the usual social network concern with the degree of a point, which is concerned with the number, not the particular characteristics of points adjacent with a given point. This may be a crucial consideration in the study of island networks. Using these models, we show that network location is a better predictor of social stratification on low islands than either economic productivity (Mason 1968) or population size (Orans 1966) and we conjecture that the exploitation of favourable location in relations between islands leads to the elaboration of internal political hierarchy.

At the conclusion of this analysis we show the implications of network analyses for ecological studies of coral island settlement patterns in Micronesia, using the model of a rooted graph. Whether an island can be settled in the first place depends not only on its size and productive potential but on the type of exchange network it is in.

Voyaging, Exchange, and Stratification in the Caroline Islands

The Caroline Islands stretch in an east–west band of about 3,200 kilometres just above the equator in Micronesia, roughly from 3° to 10° north latitude and from 131° to 163° east longitude (Map 3.1). This analysis is concerned with the Western and Central Carolines from Palau and Yap to Truk and Lukunor and with their relation to the Marianas some 600 kilometres to the north. These islands were connected by a network of regular and frequent voyaging and thus separable from the 'westernmost Carolines' (Lessa 1983), and from Kosrae, the Greater Ponape (now known as Pohnpei) area, and the Polynesian outliers of Nukuoro and Kapingamarangi. There are two types of islands in this group: high islands, which include Babelthuap in the Palau group, Yap, and Truk (actually a cluster of high islands in a large lagoon), and all of the Marianas including Guam, Rota, Tinian, and Saipan; and low coral islands, either raised formations or atolls, which include all the rest. High islands are large, environmentally rich, and, with the partial exception of Truk, socially stratified. Low islands are small, sometimes consisting of less than one square mile of land, with relatively infertile coralline soil supporting a limited range of crops—typically coconut, bread-fruit, and taro—and small populations usually numbered in the hundreds. Chieftainship is present but variably elaborated. The languages of the low islands together with Truk are part of a single

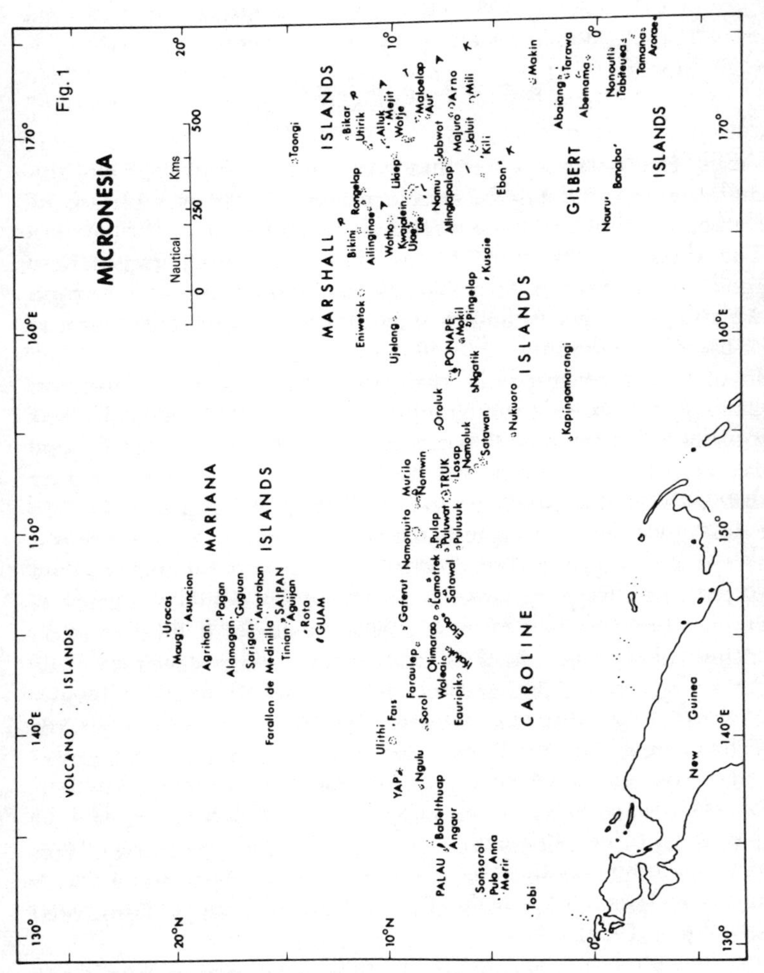

Map 3.1. Micronesia (from Alkire 1980)

dialect chain (described in Chapter 8), while those of the high islands, Yap, Babelthuap, and the Marianas, belong to three different groups. The low islands of the Carolines are vulnerable to frequent and devastating tropical storms, an average of 25 typhoons each year (Manchester 1951), that destroy life, food supplies, and garden land. As William Alkire (1965) makes clear, survival depends on common membership in an overseas exchange network which can offset natural disaster by redistributing resources and occasionally even populations.

Although the economic complementarity between high and low islands dominated much of traditional exchange, almost all voyaging was done by low islanders.[1] Navigation was in the hands of specialists, *pelu*, who underwent a lengthy and intensive period of formal instruction during which they learned sailing directions for every island in the Carolines and Marianas and even beyond as far as the Marshalls and Philippines, where accidental voyages sometimes ended up. Whereas Trobriand Islanders were only vaguely aware of *kula* ring communities beyond their neighbours' neighbours, Carolinians had a rather detailed knowledge of their entire network.

Navigation was based on dead reckoning, which uses estimated distance, time, and direction travelled to determine present position. The rising and setting positions of stars were used to determine direction, and heavy reliance was placed on perceptual cues such as reef formation, swell direction and pattern, and the flight of birds, to determine location. Voyaging was routine but hazardous, with canoes often lost even on short trips between islands. Practical knowledge was richly supplemented by divination techniques and weather magic (Lessa 1959). To minimize risk, voyaging was conservative, based whenever possible on island-hopping. Any long trip was broken up into a succession of shorter ones between adjacent pairs of intervening islands. Although north-east trade winds predominate, there is a long period of light and variable winds, which enabled Carolinians to reach all islands with greater ease (Finney 1985).

In cross-section, Carolinians imagined their islands as a set of columnar structures resting in sockets on the ocean floor, like masts perhaps, rising just above the surface as shown in the Puluwatese

[1] Similarly, in trade between low island Tuamotuans and high island Tahitians in eastern Polynesia, the former did most of the voyaging (Oliver 1974, vol. 1, chapter 8).

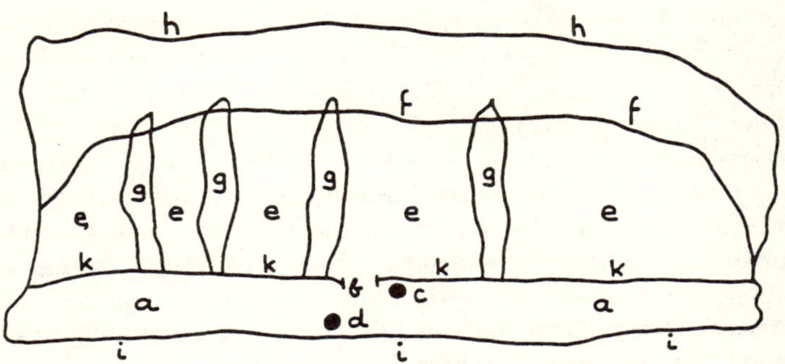

FIG. 3.1. The universe of the Puluwat Islanders (after a native drawing, Damm and Sarfert 1935).
Legend: h = sky; e = ocean; f = ocean surface; g = islands; k = ocean floor; a, i = interior and underside of earth; c = home of Ligobub, creator and ruler of the underworld; d = home of Saulal, god and creator of fish; b = Saulal's door for observing islands and men.

drawing in Fig. 3.1 from Damm and Sarfert (1935).[2] In the horizontal plane they conceived of these islands as nodes in a network of sea-ways:

> When a Puluwatan speaks of the ocean the words he uses refer not to an amorphous expanse of water but rather to the assemblage of seaways which lie between the various islands. Together these seaways constitute the ocean he knows and understands. Seen in this way Puluwat ceases to be a solitary spot of dry land; it takes its place in a familiar constellation of islands linked together by pathways on the ocean (Gladwin 1970: 34).

As long as the navigator stayed on these sea-lanes (or sea-ways) and did not stray into the open sea, the *metau*, he was assured of reaching his destination. Sea-lanes were individually named, so that a *pelu* could recount a trip (and mystify the uninitiated) simply by listing them in sequence, never mentioning any islands by name. Graphically, a path is thus specified by its sequence of lines without stating its points.

Exchange in the Carolines was based on travel over this network, which unfortunately has never been fully described. Alkire (1965) in his monograph on Lamotrek Atoll lists 21 sea-lanes in the Western

[2] Tuamotuans may have had a similar conception (Danielsson 1956).

Carolines. Sohn and Tawerilmang (1976) in their Woleai dictionary list 14 for the same area. Damm and Sarfert (1935) give 40 from Satawal, Krämer (1937) gives 13 from Faraulep, and Müller-Wismar (1918) in his contribution to the Hamburg Südsee-Expedition gives another 24 from Yap. Gladwin's unpublished field notes from Puluwat contain over 40 sea-lanes covering the entire Truk area of the Central Carolines. These sources taken together yield what appears to be a complete and, judging by their numerous correspondences, a reliable picture of the overseas network. The combined list, including connecting links to the Marianas, is given in Appendix 1.

Regarding each island as a point and each sea-lane as a line, the graph *G* of this network is shown in Fig. 3.2. The figure is roughly geographical with the points numbered in the following way: 1–15 are all the islands in the Yapese Empire, the tribute system headed by the high island of Yap and encompassing the fourteen low islands to its east; 1–26 are all the inhabited islands of the Carolines; 27 is a formerly inhabited island, never resettled after its population perished in a storm around 1800; 28–31 are small uninhabitable but productive islands, the first three of which are strategically located as stop-overs in travel between the Carolines and the Marianas; finally 32–5 are islands in the Marianas.

The conservative nature of voyaging is illustrated by sailing directions in Burrows and Spiro's (1957) monograph on Ifaluk (island 8). Thus the first course from Truk to Ifaluk is:

1. Truk to Satawal; (under the guiding star) *Tubwulimailap*.
2. Satawal to Ifaluk; *Tubwulimailap*.

But the second course is 'recommended because the gaps are shorter'. It proceeds strictly along a sequence of the sea-lanes shown in Fig. 3.2:

1. Truk to Puluwat; *Tubwul'ulu*.
2. Puluwat to Satawal; *Tubwulielieli*.
3. Satawal to Lamotrek; *Tubwul'ulu*.
4. Lamotrek to Elato; *Tubwul'ulu*.
5. Elato to Ifaluk; *Tubwulielieli*.

Only in sailing north to the Marianas are there no intermediate stops, and then the uninhabited islands of Gaferut (28), West Fayu (29), and Pikelot (30) are used as jumping-off points for Guam (32) and Saipan (35).

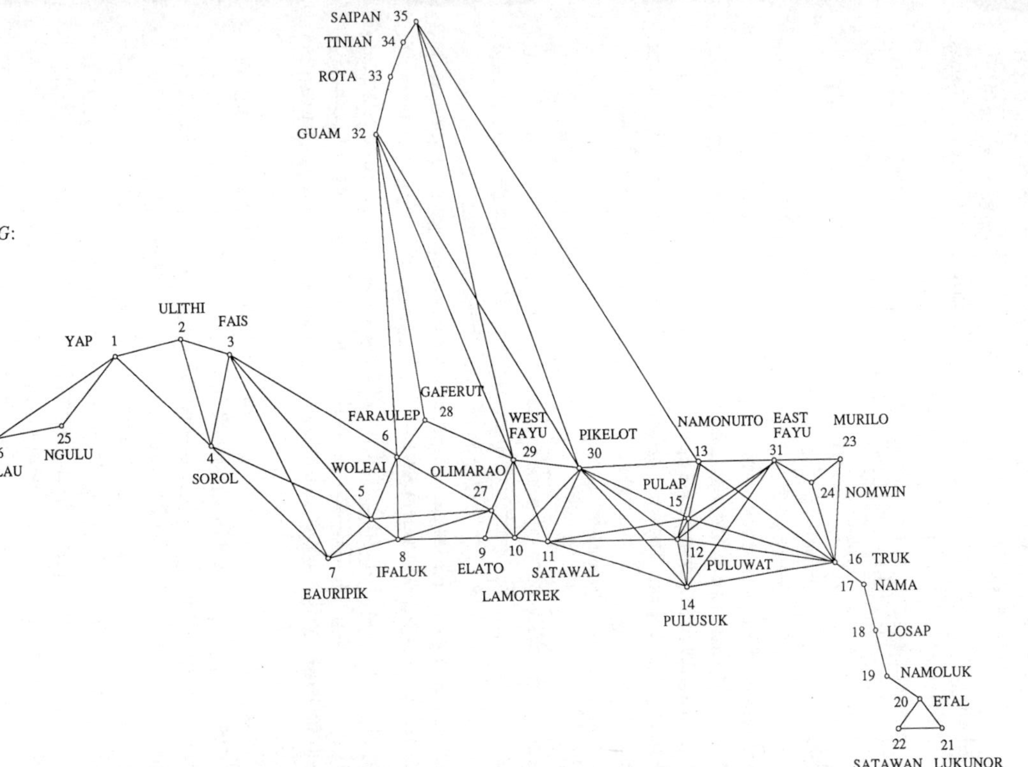

FIG. 3.2. The graph G of named sea-lanes in the Carolines and Marianas

Given the urgency of inter-island communication, together with fluctuating relations of amity and enmity and significant complementarities in the distribution of goods and resources, one would expect alternative as well as conservative courses, geography permitting. This is illustrated by sailing directions for Yap to Ifaluk trips (Burrows and Spiro 1957: 338). The first course is:

1. Yap to Ulithi; (under the guiding star) *Tagal'ulu.*
2. Ulithi to Fais; *Tagalielieli.*
3. Fais to Woleai; *Tagatumura.*
4. Woleai to Ifaluk; *Tagalielieli.*

The second course is:

1. Yap to Sorol; *Tagalielieli.*
2. Sorol to Woleai; *Tagalielieli.*
3. Woleai to Ifaluk; *Tagalielieli.*

The third course is:

1. Yap to Sorol; *Tagalielieli.*
2. Sorol to Eauripik; *Tagatumura.*
3. Eauripik to Woleai; *Tagalimel.*
4. Woleai to Ifaluk; *Tagalielieli.*

This does not exhaust the number of possible itineraries. One could, for example, sail from Yap to Sorol to Eauripik to Ifaluk (course 4) or from Yap to Ulithi to Sorol to Woleai to Ifaluk (course 5). The flexibility of course-structure in this network can be succinctly characterized by the application of a theorem on the class of graphs called blocks.

A *block B* is a non-trivial connected graph *G* containing no cutpoints. Fig. 2.9 gives a block, $K_{3,3}$, and a non-block, $K_{1,4}$, which has a cutpoint. In *G* all the islands lying between and including Yap, a cutpoint in the west, Truk another cutpoint in the east, and Saipan in the north are in the same block—the point set {1–16, 23, 24, 27–35}. There are 'disjoint' courses between every pair of islands in this block. Two *u*–*v* paths in a graph *G* are *point-disjoint* (or simply *disjoint*) if they have no points in common, other than *u* and *v*. They are *line-disjoint* if they have no lines in common. The fourth and fifth courses between Yap and Ifaluk are line-disjoint, while the first and fourth courses are point-disjoint (and therefore line-disjoint).

The following theorem gives a structural characterization of a block in terms of points lying on the same cycle.

Theorem 3.1. A graph G with $p \geqslant 3$ is a block if and only if every two points of G lie on a common cycle.

Corollary 3.1a. A graph G with $p \geqslant 3$ is a block if and only if for every two distinct points u and v of G, there exist two disjoint u–v paths.

By Theorem 3.1, for the majority of islands in the Carolines—all those in the largest block described above—every pair of islands lie on (at least) one common cycle and are joined by (at least) two courses which have no intermediate islands or sea-lanes in common. This assures great flexibility in communication. This observation is a special application of Menger's Theorem (Harary 1969: 47) which characterizes graphs in which every pair of points are joined by at least n point-disjoint (or line-disjoint) paths. Only in sailing through and beyond Yap and to the south of Truk are courses not disjoint. In spite of this flexibility, however, certain islands are more centrally located than others and thus have a structural advantage in the ease or in the control of communication.

Exchange and Central Location

Exchange in the Carolines was based on the complementary distribution of natural resources, especially, although not exclusively, in the relation between high and low islands, and on economic specialization, culturally determined as well as environmentally constrained. Thus the high island of Yap provided the surrounding low islands with yams, timber, ochre, turmeric, pots, and basalt in exchange for mats, woven fabrics, and coconut fibre rope (sennit), while the low islands exchanged tobacco, mats, textiles, canoes, ornaments, and numerous other items, including those obtained from high islands, with each other. Trade was frequent, far-ranging, and strongly motivated by the pursuit of 'luxury goods'—tobacco, shell ornaments, and especially turmeric from Truk and Yap, and iron tools and tobacco which became available in the Marianas with the arrival of the Spaniards in the sixteenth century. The peace of the trade was promoted by kinship bonds, real between low islands, metaphorical between Yap and its satellites, and by superordinate political structures. All island communities were active in trade, but not to the same degree nor apparently with the same success.

Alkire (1965) describes three forms of overseas exchange for Lamotrek, which probably obtained in general in traditional times. Barter usually takes place between distant islands, for example between Truk and Lamotrek, where common kin may not be present. Lamotrekans might give sennit and woven fabric (*tur*) for Trukese women's shell belts (*tügakh*) and turmeric. Such exchanges are immediate, with no bargaining or haggling, based instead on fixed rates. Contractual exchange is between kinsmen on neighbouring islands. Lamotrekans might commission relatives on Satawal, where timber is more abundant, to build a large sailing canoe, paying with *tur*, sennit, or other goods. Gifts are given in a variety of contexts. They are exchanged between kinship groups (matrilineal lineages and clans here as elsewhere in the Carolines[3]), given as a preliminary to contractual exchange, in recognition of a favour, and as an expression of political subordination, as when the neighbouring islands of Elato and Satawal pay semi-annual tribute to Lamotrek. Alkire also mentions formal friendship in individual exchange.

The exchange between high and low islands, with the latter as main beneficiary, is usually emphasized in Carolinian ethnography, e.g. Lessa (1950), Fischer (1957), but it is clear from Damm and Sarfert (1935) and Krämer (1935, 1937) that exchange between the physiographically similar low islands was much more elaborate than one might expect. There were environmental predispositions to economic specialization, such as good timber for sailing canoes found on Ulithi (2), Woleai (5), Satawal (11), and Puluwat (12), better soil for tobacco on Fais (3) and Lamotrek (10), and good supplies of red spondylus shells on Eauripik (7) and in the Lower Mortlocks (19–22). But as in the case of Melanesian trade networks in the Massim (Persson 1983), the Vitiaz Strait (Harding 1967), the Santa Cruz Islands (Davenport 1962), and the Lau Islands, Fiji (Sahlins 1962), the division of labour was not simply a reflex of the geographical distribution of resources used in manufacture but was in large part culturally determined in such a way as to maximize the exchange potential of individual island communities. Ulithi (2), for example, finished belts obtained from Ngulu (25) for use in exchange with Woleai (5), and it made canoes but imported paddles from Sorol (4) 'because it is said that Sorol [had] the time and labor to

[3] There may or may not be double descent on Yap (Schneider 1962, 1984; Labby 1976).

make them' (Lessa 1950: 47). The extent of economic specialization is indicated in Fig. 3.3, a subgraph of Fig. 3.2, consisting of the inhabited islands of the Carolines and Marianas with a partial list of items produced on each.

Fig. 3.3 shows some of the goods available only on the high islands and, for Yap and Truk, their large inventories of manufactured goods. It also shows that while each low island did not have a type of product which uniquely differentiated it from all others, in most cases each one produced or excelled in the production of at least one item which differentiated it from all adjacent islands, thereby providing a basis for direct exchange. The complexity, volume, and importance of exchange were enormously increased by indirect transfers. Thus Puluwat (12) traded sailing canoes for paddling canoes with its neighbour Pulap (15), but it also traded conus armshells obtained from Woleai (5) for red spondylus shells in Lukunor (21). Pulap in turn got some of its sailing canoes from Ulithi (2) in exchange for mats obtained in Truk (16).

Certain island communities had greater range and functioned more importantly in the role of middlemen in the trade than others. Foremost among these communities was Puluwat. In accounting for Puluwat's renown, Damm and Sarfert (1935) emphasize seafaring ability and a pronounced sense of kinship ties. Puluwat was known everywhere for its navigational expertise. According to Thomas Gladwin (1970: 201–2), Puluwatese navigators 'historically had a reputation as the best in the entire Caroline area. Their skills were said to link the more parochial knowledge of navigators on the islands to the east with those to the west . . .' The average range of kinship connections in other communities was undoubtedly much greater in the Carolines than in most Melanesian networks. In the Huon Gulf, for example, voyaging range was severely restricted by kinship range:

> The natives are thoroughly at home on the sea and handle their craft with such skill that journeys from places as far apart as Siboma in the extreme south and the Tami Islands of the far east would present no insuperable problems of navigation. Kinsfolk are confined to villages fairly close at hand, however, and each community restricts its voyages to its own section of the Gulf (Hogbin 1947: 248–9).

A hospitable reception on distant islands ensured by common kinship was a vital social component in overseas voyaging (Gladwin

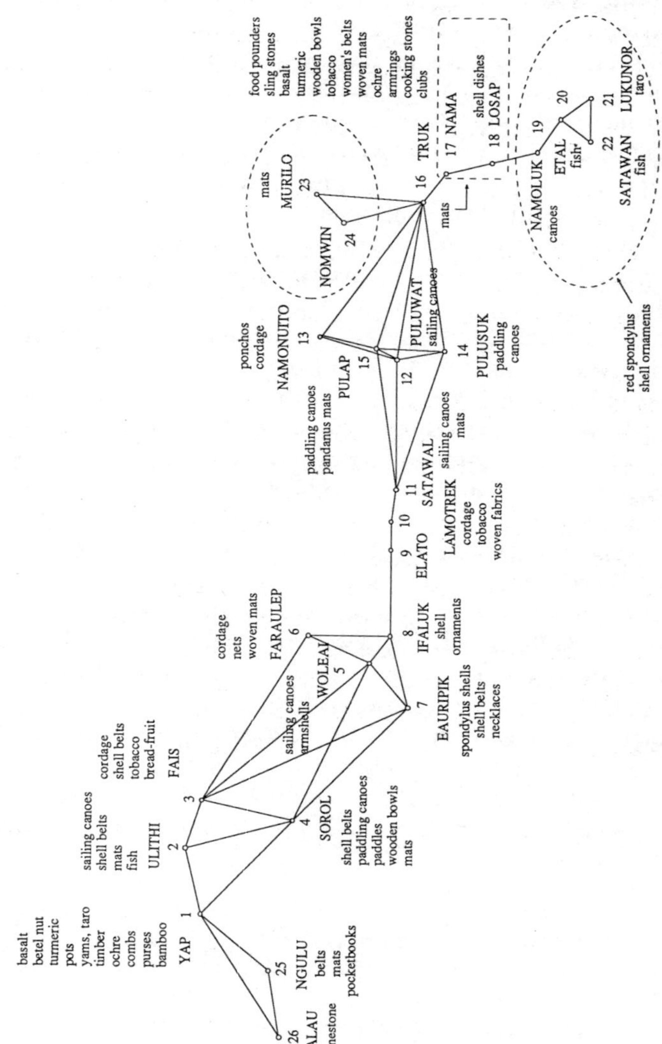

FIG. 3.3. A graph of inhabited islands with some of the goods produced at each

1958). Murdock and Goodenough's (1947: 331–2) observation on Truk is probably true in general: 'Besides the regulation of marriage, sibs possess only a single important function, namely, the channeling of hospitality. A native visiting another island or atoll for trade or any other peaceful purpose resorts to a member of his own sib, from whom he can expect shelter, food and (formerly) sexual hospitality.' The extent of Puluwat's kinship connections was great, perhaps above average, as suggested by a chart in Damm and Sarfert (1935) giving the representation of Puluwatese clans on a sample of other islands in the Carolines. Puluwatese clans were found all the way from Lukunor to Ulithi and, as the scattergram in Fig. 3.4 shows, there was no precipitate decline in clan representation on other islands with increasing graphical (sea-lane) distance from Puluwat.

But the structural basis of Puluwat's success as a trader and middleman and the probable cause of its *Wikingergeist* and *Familiensinn* was its location in the overseas network: its closeness to all islands and its adjacency to certain islands. According to Damm and Sarfert:

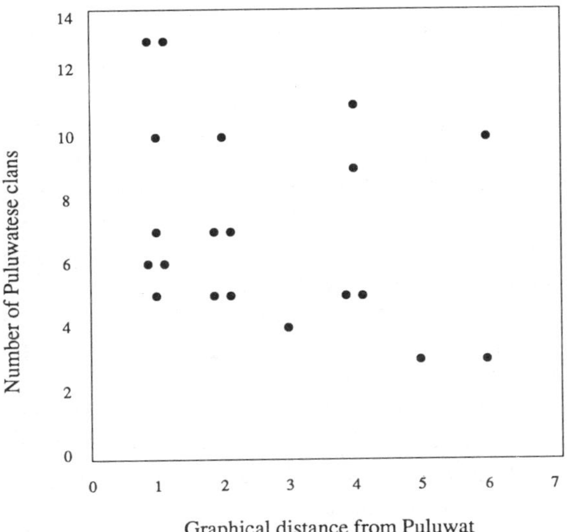

FIG. 3.4. The distribution of Puluwatese clans on a sample of other Caroline Islands (based on Damm and Sarfert 1935)

The Puluwatese had a certain importance as 'middlemen'. People on the neighbouring islands gave up trips to Truk, for example, and obtained their goods through the Puluwatese, who were closer and who made regular trading voyages there to fetch valued cosmetics (turmeric and oil). The Puluwatese also sailed regularly to Lukunor in order to trade for the much desired red spondylus shell ornaments (*faibobo*). And they travelled no less often to Satawal to barter for mats. These trips were made until recent times . . . The strongest magnet for the Central Carolinians was the island of Guam. It was here that they first contacted Europeans and became acquainted with iron. For many of the Central Carolinians the Guam trips became a tradition. Freycinet (1833) learned of these trips without, however, suspecting their significance. According to Finsch (1900) the main participants were the people of Woleai, Faraulep, Lamotrek, and Puluwat. They gathered at West Fayu and sailed together, taking about eight days to cover the approximately 300 sea-mile stretch to Guam . . . Noteworthy is a remark by Finsch, which aptly characterizes the entrepreneurial spirit of the Puluwat and Pulusuk people: 'The most industrious and at the same time the best sailors are the inhabitants of *Suk* (Pulusuk or Puluwat), who distribute their goods (in earlier times especially the knives and axes bartered for in Guam) eastwards to Truk and westwards to Woleai' (Damm and Sarfert 1935: 82; our translation).

Closeness, measured in number of sea-lanes, obviously conferred an advantage in trade, affecting both the existence and frequency of travel between communities. Most voyages to the Marianas seem to have originated from islands adjacent to the departure points of Gaferut, West Fayu, and Pikelot (Lewis 1972). It is not specified which of Puluwat's neighbours gave up travel to Truk, but elsewhere (Damm and Sarfert 1935: 56) it is said that Puluwatese sailed there twice as often as Satawalese, who were not adjacent but two steps removed. Lamotrek got shell ornaments from Lukunor, but mainly through Puluwat which was closer, and Ulithi got iron tools from the Marianas through Woleai. Lamotrekans sailed to Truk for turmeric, but still got some of their supply through Puluwat, just as people from the Lower Mortlocks sailed to Truk but also got Trukese goods through Losap, a recognized middleman (Krämer 1935). The relative ease of access which each community had to goods distributed over all other islands is indicated by the distance definition of centrality.

Recall that the distance between two points in a graph v_i and v_j, denoted by d_{ij}, is the length of any shortest path (geodesic) that joins them. In a connected graph G, every distance d_{ij} between two distinct points is a positive integer. For each point u, we define the *distance sum* of u, written $s(u)$, as the sum of the distances between u and all other points. Thus

$$s(v_i) = \sum_{j=1}^{p} d_{ij}.$$

In an early paper (Harary 1959b), $s(u)$ was called the status of u. Now the *distance centre* of G (often called its *median* in the mathematical literature) is defined as the set of all points u of G such that $s(u)$ is minimum. In the graph in Fig. 3.5 the distance sum of each point is shown in parentheses. Point 3 is most central, and points 1 and 4 are most distant or peripheral.

Considering sea-lanes as lines in a path, the distance between Lamotrek and Lukunor is 8, while between Puluwat and Lukunor it is 6. Table 3.1 shows the distance centrality, as measured by distance sum, of all the points in the Caroline Islands subgraph of the graph in Fig. 3.2 (points 1–31). The uninhabited islands, points 27–31, are included in this subgraph because they were either voyaging stopovers in the Carolines or departure points for the Marianas or both. These islands were also important sources of marine products (Alkire 1965, 1978; Fischer 1957; McCoy 1974). There were disputes over the ownership of islands 29 and 30, as discussed below.

According to Table 3.1 Puluwat, along with Pulap (the mythological source of seamanship lore),[4] is the third most central of

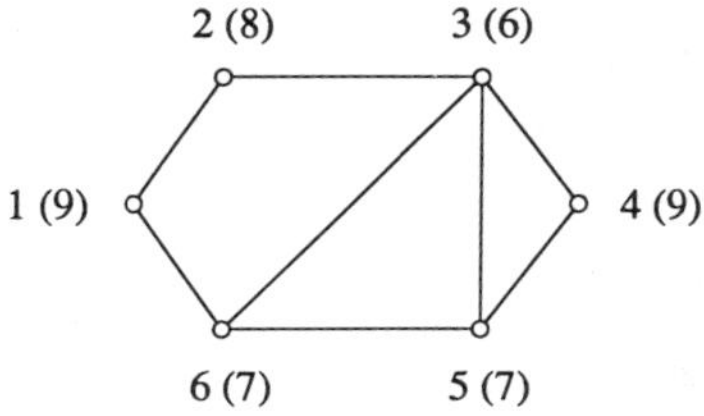

FIG. 3.5. A graph illustrating distance centrality

[4] Carolinian navigators held that seamanship lore originated at Pulap. According to Komatsu (1985), Pulap was a sacred island with a shrine to Warrieng, the god of navigation. In the myth 'How the lore of seamanship came to Ifaluk' (Burrows and Spiro 1957: 89–90), this lore spread out from Pulap (Bwennap) to adjacent, and then to successively more peripheral islands.

TABLE 3.1. *Distance centrality of Western and Central Caroline Islands*

Island	d_{ij}	Rank (all islands)	Rank (inhabited islands)	
1. Yap	161	19	17	Inhabited Carolines
2. Ulithi	160	18	16	
3. Fais	138	15	13	
4. Sorol	137	14	12	
5. Woleai	118	10=	8	
6. Faraulep	122	11	9	
7. Eauripik	139	16=	14=	
8. Ifaluk	123	12	10	
9. Elato	116	9=	7	
10. Lamotrek	100	3=	2	
11. Satawal	99	2	1	
12. Puluwat	103	4=	3=	
13. Namonuito	105	6	5	
14. Pulusuk	104	5	4	
15. Pulap	103	4=	3=	
16. Truk	112	8	6	
17. Nama	131	13	11	
18. Losap	152	17	15	
19. Namoluk	175	20	18	
20. Etal	200	22	20	
21. Lukunor	228	23=	21=	
22. Satawan	228	23=	21=	
23. Murilo	139	16=	14=	
24. Nomwin	139	16=	14=	
25. Ngulu	189	21=	19=	
26. Palau	189	21=	19=	
27. Olimarao	106	7		Uninhabited Carolines
28. Gaferut	116	9=		
29. West Fayu	100	3=		
30. Pikelot	98	1		
31. East Fayu	118	10=		

all inhabited islands in the overseas network. Puluwat and Pulusuk, which is fourth most central among inhabited islands, are described as the most industrious traders and best sailors in the Carolines. Sorol and Fais, by contrast, which rank twelfth and thirteenth, are the two islands where navigation and voyaging ceased, at an early date long before the deterioration of the system. These two islands became dependent on Ulithi for overseas transport. Ifaluk, which ranks a modest tenth, is described as an active trading community but not in the same league as Puluwat (Krämer 1937).

We note that in this economy centrality is positively correlated with seafaring. The same is true in the *kula* ring (Hage 1977) and in Mailu (Irwin 1974) in Melanesia, and in the Fiji–Tonga–Samoa network in Polynesia (Sharp 1964). But the association is not inevitable: in some other parts of the world, for example the Hebrides and Western Norway, it is the outermost islands that have the highest reputation for intrepid seafaring (J. A. Barnes, personal communication).

Puluwat was actually one of a cluster of highly central communities over which it had structural or physical advantages or both. Puluwat's structural advantage lay in its neighbourhood. The *neighbourhood* of a point u is the set $N(u)$ consisting of all points v which are adjacent with u. The *closed neighbourhood* includes u; it is $N[u] = N(u) \cup \{u\}$. In the sea-lane graph of Fig. 3.2, Eauripik's neighbourhood consists of Ifaluk, Woleai, Fais, and Sorol. It overlaps with Faraulep's neighbourhood but not with Lamotrek's.

By its adjacency to both Pikelot (30), a departure point for the Marianas, and Truk (16), Puluwat had ready access to the two supremely valued goods, turmeric and iron tools. Turmeric was used as a cosmetic in the Carolines on all ceremonial occasions. It figured prominently in mediated transactions, reaching all islands from Truk (and to a lesser degree from Yap). Iron tools, axes and knives, from the Marianas similarly diffused throughout the entire system. Puluwat's proximity to the sources of both goods gave it an advantage over Satawal and Lamotrek, which rank first and second most central among inhabited islands of the network, and which are also adjacent to Pikelot (and West Fayu) but not to Truk.

Puluwat's advantage over the other highly central communities of Pulap, Namonuito, and Pulusuk, which also have ready access to both Truk and the Marianas, was probably physiographic. Gladwin (1970) emphasizes Puluwat's excellent harbours and its abundance of timber for building large ocean-going canoes. Thus it may have had an advantage over Pulusuk, a raised coral island without a lagoon, over Pulap and Pulusuk, exporters of smaller paddling canoes, and over Namonuito, an exporter of ponchos and cordage.

Stratification and Central Location

In Melanesia favourable location in an overseas exchange network is an enabling condition for internal stratification. In the *kula* ring, chieftainship is based either on the aristocratic monopolization of scarce valuables or the entrepreneurial manipulation of abundant valuables flowing through particular communities (Hage, Harary, and James 1986). In the Vitiaz Strait, successful traders become prominent big men in centrally located island societies (Harding 1967). Although explanations based on economic productivity (Mason 1968) and population size (Orans 1966) have been offered to account for stratification within low islands, it turns out that in the Carolines, network location is a better predictor. Here environment sets limiting conditions for the exploitation of locational advantage.

Our conjecture is that the control and the ease of communication provided the structural bases for hierarchical relations between islands, which in turn contributed to the differential elaboration of chieftainship on particular islands.

As a network the Carolines are best known for the Yapese Empire, the system described in Chapter 1, in which the high island of Yap (Gagil District) demanded, at irregular intervals, gifts from all the low islands from Ulithi (2) to Pulap (15) inclusively (Lessa 1950, 1966; Lingenfelter 1975; Labby 1976; Alkire 1977). In a replication of the kinship idiom of its own social structure in which high-caste 'parents' owned the land used by low-caste 'children', Yapese 'parents' owned the land of their outer island 'children'. Specifically, the *sawei* or 'rent' relation, mentioned in Chapter 1, defined individual patrilineal lineages on Yap as owners of individual outer islands or districts of outer islands. The *sawei* exchange was in effect a symmetrical one, in which the low islanders obtained goods uniquely available on a high island and the Yapese obtained exotic low island goods for use in manipulating their own internal alliance system.[5] There was no question of Yapese military

[5] Bellwood (1979) has asked why Yap rather than Truk was at the head of such an empire given the close cultural and historical ties between Truk and the low islands.

control of the outer islands: Yap's main and credible sanction, given the location of the Carolines, was its possession of powerful magic, especially weather magic with which it could send typhoons. Lessa (1950) states that Ulithi had a *sawei* relationship with the other low islands, standing to them in the relation of 'parent to child', but this had no implication of caste: whereas low island men were prohibited from marrying upper-caste Yapese women, Ulithians intermarried freely with people of other low islands.[6] Neither did this relation have any implication of political control. Lessa (1966) regards Ulithi's *sawei* gifts as compensation for administrative services on Yap's behalf, resulting from Ulithi's proximity to Yap and intermediate position in communication between Yap and the low islands.

Less well known than the Yapese Empire are the political subsystems within the low islands. Thus Pulap, Pulusuk, and Namonuito paid tribute (food gifts) to Puluwat. And Puluwat and some of its subordinates, in turn, paid tribute to Lamotrek, which was originally superior to these islands and perhaps even to islands further east, and to all the low islands west to Ulithi (Lessa 1962). Kotzebue (1821) in fact refers to Lamotrek as 'king of the islands'. Hierarchy in this subsystem may well have been based on military force and locational advantage.

The prerogatives of Lamotrek chiefs appear to have been highly developed. According to Damm and Sarfert, who cite the early and, it is generally agreed, accurate descriptions of Carolinian culture by the Jesuit priests Cantova (1728) and Clain (1700), Lamotrek chiefs sat on a throne, they were approached in a bent-over posture, their hands and feet were kissed, and they wore beards as a special sign of prestige. They called meetings and directed communal work such as clearing of agricultural land, boat-building, and fish drives. They had the right to mete out punishments, and they may have had talking chiefs as in Polynesian societies. From Alkire (1965) it appears that chiefs were distinguished as a class, by endogamy and by the differential application of sanctions for offences. In his careful assessment of early descriptions of Carolinian culture, Lessa (1962) suggests that there may have been much greater stratification on low

Part of the answer undoubtedly lies in the greater political fragmentation and frequency of warfare that prevailed on Truk in traditional times. (See Alkire 1977.)

[6] But a Yapese man could marry any Ulithi woman, including a 'child' from a *sawei* group (Lessa 1950).

islands than reported for modern times, and he notes in particular the position of Lamotrek in the inter-island network:

> Apparently, certain islands held wider political power over other islands than we know of from contemporary situations. As far back as 1696 Lamotrek, even though an atoll, had a superordinate position over a large number of islands, including Ulithi, Fais, Sorol, Eauripik, Woleai, Satawal, Ifaluk, Faraulep, Etal and Elato (Lessa 1962: 351).

Most significant is Clain's report that Lamotrek rulers could come to power in war. There are indeed numerous references to warfare in the Carolines, for example, Puluwat's invasions of Pulusuk, Namonuito, and Nomwin in historical times. And there are legends of conquest, such as Ifaluk's colonization of nearby islands (Burrows and Spiro 1957).

Given the absolute necessity of inter-island exchange and the restriction of travel to the network of sea-lanes, any island that could control communication between other islands would be in a powerful position, able to demand tribute or other payments or gifts. Locally, Yap, as a cutpoint in the graph, was in such a position in the communication between Ulithi (2) and Ngulu (25), and in fact whenever people from these two islands visited each other they had to obtain the permission of, and render payment in the form of fine fibre loincloths to, the paramount chiefs of Yap (Lessa 1950). As another example, Lamotrek at one time claimed ownership of the uninhabited islands of West Fayu (29) and Pikelot (30), way stations to Guam, and whenever Puluwat sailors wished to sail to Guam via these islands they had to obtain Lamotrek people's permission (Damm and Sarfert 1935). An example from the modern era is mentioned by W. Alkire (personal communication): 'In sailing from Satawal to Elato or Woleai [assuming the trip was not made via West Fayu] one was expected to stop at Lamotrek, but on occasion it could be bypassed as long as some later explanation-cum-apology was offered—perhaps a reflection of the diminishing power of Lamotrek chiefs?' Although no island could uniquely control the communication between all pairs of islands in this system (this could only happen if its graph were a star), some islands did have more potential for control than others in the sense that they lay on a higher proportion of geodesics between all other pairs of islands. This is the basis for the betweenness definition of centrality.

Formally, *betweenness* refers to the frequency of occurrence of each point on the geodesics between all pairs of points in a graph, and is an index of the potential for control of communication. To illustrate, there are three geodesics between Ifaluk and Satawal in the graph of Fig. 3.2: (8, 27, 29, 11), (8, 27, 10, 11), and (8, 9, 10, 11). Lamotrek (10) and Olimarao (27) lie on two of them, while the other islands, Elato (9) and West Fayu (29), lie on only one. With respect to this point pair, Lamotrek and Olimarao are most between and thus have the greatest potential for the control of communication. For each pair of points in such a graph we enumerate all the shortest paths having two or more lines. Then the points that lie on the highest proportion of these paths are relatively more between than the others.

In general, if the graph of an exchange structure is a star $K_{1,n}$ then there is a single community, represented by the single cutpoint of the star, which can control exchange between all pairs of communities. If it is a complete graph K_p, then no community is between any other pair of communities. If the graph consists of a single cycle C_p then all communities have the same amount of betweenness. The graphs of most trade networks fall between these extremes, like the one in Fig. 3.2 for example. In order to compute the betweenness value of each point in a graph and the betweenness of the entire graph, the following definitions from Freeman (1977, 1979) are necessary.

Let g_{ij} be the number of i–j geodesics and let $g_{ij}(v_k)$ be the number of these geodesics containing v_k. Then the *betweenness value of* v_k *with respect to the point pair* v_i, v_j is the ratio

$$b_{ij}(v_k) = \frac{g_{ij}(v_k)}{g_{ij}},$$

which is of course the probability that a randomly selected i–j geodesic contains point k. In terms of these probabilities the *partial betweenness value* of v_k (independent of any one pair of points) is defined by the formula

$$C_B(v_k) = \sum_{j=1}^{p} \sum_{i=1}^{j-1} b_{ij}(v_k).$$

Whenever v_k lies on every i–j geodesic, the point pair i, j contributes 1 to the sum $C_B(v_k)$. The particular case that there is a unique path (and hence geodesic) joining each pair of distinct points occurs only when G is a tree. When there are alternative geodesics, $C_B(v_k)$ grows in proportion to the frequency of occurrence of v_k among these alternatives.

Among all p-point graphs, the maximum possible betweenness value of a point, $C_B(v_k)$, is $(p-1)(p-2)/2 = (p^2-3p+2)/2$ and is attained by the central point of a star. The *relative betweenness* $C'_B(v_k)$ is the ratio of $C_B(v_k)$ to this maximum expression, so that

$$C'_B(v_k) = \frac{2C_B(v_k)}{p^2-3p+2}.$$

The unique point (if any) with the highest value of $C'_B(v_k)$ is called the *most between point* of the graph. The *betweenness centrality* of a connected graph is 'the average difference between the relative centrality of the most central point, $C'_B(v^*)$, and that of all other points', giving:

$$C_B(G) = \frac{\sum_{i=1}^{p} [C'_B(v^*) - C'_B(v_i)]}{p-1}.$$

Note that $C_B(v_k)$ measures the betweenness of a single point, whereas $C_B(G)$ measures that of the entire graph.

To illustrate these measures, Fig. 3.6 shows a graph G together with its betweenness centrality $C_B(G)$ and the partial betweenness value $C_B(v_k)$ and the relative betweenness centrality $C'_B(v_k)$ of each of its points. As a further illustration of betweenness centrality, when G is a star, $C_B(G) = 1$, and when G is complete, $C_B(G) = 0$.

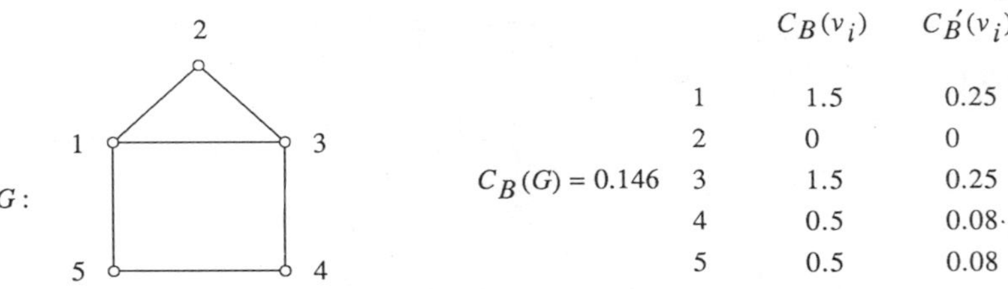

	$C_B(v_i)$	$C'_B(v_i)$
1	1.5	0.25
2	0	0
3	1.5	0.25
4	0.5	0.08·
5	0.5	0.08

FIG. 3.6. The measures of betweenness in a graph

Table 3.2 shows the betweenness measures for all the points in the Caroline Islands subgraph of the graph in Fig. 3.2. In this table $C_B(v_k)$ gives the exact partial betweenness value of each island, while $C'_B(v_k)$ gives the relative betweenness of each. The latter is not affected by the size of the network and can be used to compare graphs of different size. The centrality of the entire graph, $C_B = C_B(G)$, is 0.265, indicating only modest centralization, in contrast, say, to a star-shaped and therefore maximally centralized system like the one in the Vitiaz Strait (Harding 1967).

The following observations can be made on Table 3.2.

(1) *Network position and external stratification.* Lamotrek is the eighth most between of all islands in the network, and fourth most between of all inhabited low islands in the Carolines. Among all islands to which it was politically superordinate, islands 2–15, only Woleai is more between, ranking fifth among all islands and second among all inhabited low islands. When one considers Lamotrek's neighbourhood, one can see its true potential for the control of communication. By claiming ownership of the two uninhabited islands of West Fayu (29) and Pikelot (30) to which it is adjacent, Lamotrek placed itself in a position to control all travel between the Eastern and Western Carolines.

Formally, the lines incident with Lamotrek, West Fayu, and Pikelot form a *cut set*, a collection of lines of a connected graph whose removal results in a disconnected graph. A question arises, however, as to the extent and permanence of Lamotrek's control over those two islands.

A consideration of overlapping neighbourhoods in a graph suggests a historical speculation on political relations among low islands of the Carolines. It can be seen from Fig. 3.2 that West Fayu and Pikelot are also in Satawal's neighbourhood, and Pikelot is in Puluwat's neighbourhood. And in fact, according to Damm and Sarfert (1935: 56), although Pikelot once belonged to Lamotrek, it was subsequently claimed by Puluwat. In a modern context, it appears from Low's film, *The Navigators*, that the Satawalese now claim ownership of West Fayu and Pikelot. In earlier times this part of the network may have been politically somewhat unstable. Instability here would accord with Carolinian tradition of shifts in centres of power. According to Senfft (1904) the chiefs of Satawal (fifth most between of all inhabited low islands) once ruled together

TABLE 3.2. *Betweenness, size, and population of Western and Central Caroline Islands*

Island	Betweenness			Area (sq.mi.)[a]		Population[b]		
	$C_B(v_k)$	$C'_B(v_k)$	Rank	Land	Lagoon	1890–1909	1935 (native)	1973
1. Yap	56.000	0.1287	12=	38.670	10±		3,030	4,584
2. Ulithi	3.500	0.0080	23	1.799	209.560	797	433	710
3. Fais	23.404	0.0538	17	1.083	none	300	334	205*
4. Sorol	81.763	0.1880	10	0.361	2.740	156	8	12*
5. Woleai	112.410	0.2584	5	1.749	11.354	661	570	608
6. Faraulep	30.474	0.0701	15	0.163	0.902	155	291	122
7. Eauripik	4.517	0.0104	22	0.091	2.286	70	110	127
8. Ifaluk	18.075	0.0416	19	0.369	0.939	208	388	314
9. Elato	9.367	0.0215	21	0.203	2.888		71	32
10. Lamotrek	103.607	0.2382	8	0.379	12.166	300	192	233
11. Satawal	86.786	0.1995	9	0.505	none	224	287	359*
12. Puluwat	49.224	0.1132	13=	1.313	0.600		335	435
13. Namonuito	23.929	0.0550	16	1.710	723.900	294	303	640
14. Pulusuk	48.690	0.1119	14	1.083	none	177	194	320*

15. Pulap	49.224	0.1132	13=	0.383	12.093	500	257	470
16. Truk	165.200	0.3798	1	38.560	822±		10,344	20,105
17. Nama	125.000	0.2874	3	0.289	none		406	1,127*
18. Losap	104.000	0.2391	7	0.396	10.577	430	570	664
19. Namoluk	81.000	0.1862	11	0.322	2.972		294	263
20. Etal	56.000	0.1287	12=	0.731	6.252		255	266
21. Lukunor	0.000	0.0000	24=	1.090	21.246		888	913
22. Satawan	0.000	0.0000	24=	1.757	147.524		1,057	1,881
23. Murilo	0.000	0.0000	24=	0.497	135.082		339	383
24. Nomwin	0.000	0.0000	24=	0.716	121.573		106	472
25. Ngulu	0.000	0.0000	24=	0.165	147.707		53	8
26. Palau	0.000	0.0000	24=	188.259	477.800		5,679	12,334
27. Olimarao	162.839	0.3743	2	0.085	2.419		0	0
28. Gaferut	11.902	0.0274	20	0.043	none		0	0
29. West Fayu	107.143	0.2463	6	0.024	2.178		0	0
30. Pikelot	123.748	0.2845	4	0.036	none		0	0
31. East Fayu	21.200	0.0487	18	0.144	none		0	0

[a] From Bryan (1971).

[b] From Pisarik (1975) except for * from Bryan (1971).

over Elato and Lamotrek, and according to Kotzebue (1821) rule was once transferred from Lamotrek to Woleai (second most between). If power relations did fluctuate, Lamotrek's long-term advantage was structural with respect to Woleai—its adjacency to West Fayu and Pikelot—and perhaps physical with respect to Satawal.

The relation we propose between power and the control of communication is further exemplified by the position of Losap (third most between of all inhabited low islands) and by certain historical events in the eastern end of the network. J. L. Fischer (personal communication) suggests interesting connections between stratification, betweenness, and military conquest involving the islands of Nama (17), Losap (18), and, in the Lower Mortlocks, Namoluk (19), Etal (20), Lukunor (21), and Satawan (22):

> I was told that the only atoll in Truk district which had more than one settled island and a traditional paramount chief was Losap. For all the others each inhabited island was separate except for Ta (Tëë) in Satawan Atoll, which was attached to Satawan Island. However there are two other independent communities in that atoll (Kütü and Mwoch). Note that Losap is a crucial stop just before reaching Truk. Nama is closer but would not be hard to omit. People on Nama, Losap and Kütü all told me that Nama used to be inhabited by people related to the Losap people and that at some time in the past invaders from Kütü replaced them. Those invaders are the ancestors of the present people of Nama. I suspect that before this invasion Nama was also under the paramount chief of Losap Atoll (including two inhabited islands: Losap and Pis). Perhaps one reason for this invasion was that the people from the southern Mortlocks wanted their own way station to Truk.

When it controlled its immediate northern neighbour, Nama (most between of all inhabited low islands), Losap controlled not only all travel between Truk and the Lower Mortlocks but also travel between the latter islands and those to the west of Truk, between Lukunor and Puluwat for example.

(2) *Structure and environment.* The islands of Satawal and Olimarao illustrate the role of physical constraints in the exploitation of network position. Satawal ranks fifth in betweenness among inhabited low islands and its neighbourhood includes the strategically located islands of Lamotrek's neighbourhood, but as a raised island it is economically less stable than Lamotrek. According to Alkire (personal communication),

Although it is a higher and larger island than Lamotrek, this is a mixed blessing. As a raised island there are more serious problems with a constant water supply in times of drought; there are greater problems in cultivating taro that follow from this, and there is a longer recovery time for regaining agricultural production following storm damage. Today it supports a larger population than Lamotrek but in truly traditional times I doubt this would have been sustained (today it is more dependent for longer periods of time on greater food contributions of the Trust Territory Government, for example). Lacking a sheltering reef it also had a less reliable fish supply.

Olimarao is the second most between of all islands. Although it was inhabited at one time, its population was wiped out by a typhoon in the early nineteenth century. In his discussion of coral island ecosystems, Alkire (1978) proposes that in a population with normal sex and age ratios an island would have to support a minimum of 30–50 individuals in order to reproduce itself indefinitely. A smaller community might easily lose the requisite number of workers and reproductive members through natural disasters—storms, shipwrecks, epidemics, and droughts—or through warfare. Alkire estimates that Olimarao, which has a land area of only 0.08 square miles, is near the minimum size of islands required to support a permanent population. Native perception of what island biogeographers call 'extinction probability' (Williamson and Sabath 1984) may have discouraged subsequent resettlement of Olimarao in spite of its excellent network position—its distance as well as its betweenness centrality.

(3) *Network position and internal stratification.* In an application of the surplus theory of social stratification to Micronesian atolls modelled after Sahlins's (1958) study of Polynesian societies, Leonard Mason (1968) shows, for a small sample of societies in the Gilbert, Marshall, and Caroline Islands, that the economically most productive islands have the greatest elaboration of chieftainship. Productivity is inferred from land and lagoon area, rainfall, and presence and severity of droughts and typhoons. Features of chieftainship, arranged implicitly as a scalogram, include: the direction of communal sharing and work, rights to receive first-fruits, special respect, and luxury goods, power to apply physical force to orders, and separation from commoners as a class. Orans (1966) in his reinterpretation of Sahlins's analysis argues that stratification can be accounted for quite simply on the basis of

population size, and he adduces population figures given by Mason as further support for a demographic explanation.

Table 3.2 supports neither the demographic nor the surplus theory of stratification as applied to low islands in the Carolines (islands 2–15 and 17–25). Whether population ranges are estimated from the three sets of figures or inferred from the size and presumed productivity of the islands,[7] Lamotrek clearly falls in the middle range, yet it was apparently the most stratified low island in the Carolines. The comparison with Ulithi, one of four Carolinian societies in Mason's sample, is particularly interesting because that island is a quantum leap above Lamotrek in population size and presumed economic productivity, but chieftainship was certainly not more and probably less developed. On Ulithi the direction of communal work and sharing was in the hands of a council of elders headed by a chief who, depending on his personality, may have had influence but no power. District chiefs commanded respect and had some unspecified judicial authority. The paramount chief of the atoll was sacred and surrounded by numerous sex and food taboos, but his major political function involved relations with Yap rather than the control of internal affairs (Lessa 1966). Mason interprets the operative level of Ulithi politics as quite democratic, with a 'superficial overlay of political posts because of Ulithi's role in the historically derived chain of command extending eastward from Yap' (1968: 318).

From Damm and Sarfert (1935) it appears that chieftainship on Puluwat, which is much larger than Lamotrek but significantly smaller than Ulithi, was also fairly elaborate in traditional times. Puluwatese chiefs directed communal undertakings, received special respect, consumed special foods, exercised judicial functions, including the application of force in so far as this was not carried out by the clan, and received fines for theft. Puluwat ranks in the middle range on betweenness, a little less than half as between as Lamotrek, but high on distance centrality. Lamotrek has a better political neighbourhood in the sense of communication control, while Puluwat has a better economic neighbourhood in the sense of

[7] Assuming roughly comparable amounts of rainfall. According to Alkire (1978: 12), 'The two belts of heaviest rainfall are located 5 to 10 degrees latitude on each side of the equator in the Western Pacific. Most of the coral islands of the Carolines lie in the northern belt of heavy precipitation where the average annual rainfall exceeds 120 inches.' (See also Table 3.4.)

immediate access to resources. Puluwat received tribute from several adjacent islands and it had a reputation for raiding as well as trading—it was known as the 'scourge and terror of the Carolines' (Gladwin 1970). Our suggestion, therefore, is that in a system like the Carolines overseas exchange network, network position can lead to the elaboration of chieftainship in two ways. Betweenness, the potential for control of communication, encourages mobilization for warfare and conquest and thus a strong institution of chieftainship, as suggested by Clain's comment on Lamotrek chiefs coming to power in war. Distance centrality (closeness), the ease or efficiency of communication, favours success in trading, with its common corollary, raiding. Both alternatives increase the flow of wealth, especially exotic items, into a community for prestige-building activities. These can take the form of tribute payments, which not only symbolize hierarchical relations between groups, but also enhance the fundamental material expression of chiefly power and status in redistribution and special life-crisis rites and in obligatory displays of generosity.[8] This is at least suggested by reports that Puluwatese chiefs redistributed the tribute from adjacent islands to members of their own group; that Puluwatese weddings involved significant exchanges, including the imported turmeric and red spondylus shell necklaces (Damm and Sarfert 1935); that the corpses of men of high rank in the Carolines were painted with turmeric (Cantova 1728); and that chiefs on Ifaluk were expected to distribute gifts such as tobacco, another luxury good, on public and ceremonial occasions (Burrows and Spiro 1957).

In summary, the Carolines overseas exchange network simultaneously ensured the common survival and encouraged the differential economic and political development of its constituent communities. In contrast to familiar Melanesian networks, exchange was based on extended voyaging and political hierarchy. The analysis of this system requires a different interpretation of distance centrality and provides a new application of betweenness centrality, both supplemented by a consideration of local structure, the neighbourhood of a point in a graph. In the Mailu trade network in Melanesia (Irwin 1978), distance centrality measures the efficiency with which a single item, pottery, could be traded from a

[8] Redistribution and special life-crisis rites are major attributes of social stratification in Polynesian societies (Sahlins 1958).

given location to an entire region, whereas in the Carolines it measures the relative ease of access which each island had to all other islands to obtain a variety of goods. Unlike the radial structure of the Vitiaz Strait (Harding 1967), voyaging was not restricted to a single community (or small subset of communities), and unlike the chain or single path structure of the Huon Gulf (Hogbin 1947) direct exchange was not limited to pairs of adjacent communities. As in the *kula* ring, political hierarchy in these societies was associated with differential access to wealth as determined by network position (Brunton 1975; Hage, Harary, and James 1986), but unlike this and other Melanesian systems, hierarchical relations obtained between as well as within societies. Hierarchy appears to have been based on the potential for control of communication, as measured by betweenness centrality.

Our analysis has an important implication for speculations concerning social stratification on low islands in Oceania. In *Social Stratification in Polynesia*, Sahlins (1958) brackets together all the low islands in his sample—Ontong Java, the Tokelaus, and Pukapuka—as relatively unstratified in comparison to the moderate and highly stratified high islands. In his critique of Sahlins, Rappaport (1979) proposes an alternative theory which derives stratification not from the production of surplus but from the organizational requirements of specified types of ecosystems. According to Rappaport, high islands with large populations and heavy reliance on horticulture are threatened by environmental degradation in the form of erosion, soil depletion, and deforestation. Low islands with small populations and greater reliance on marine resources are threatened more by natural catastrophes including typhoons, tsunamis, and drought. Elaborate stratification on high islands is then interpreted as a 'managerial response' to the need of a large or expanding population to stabilize or increase food supplies in a fragile ecosystem subject to human control. On low islands the unpredictability and lack of control over natural calamity and over marine resources offer little adaptive advantage for stratification. It is interesting to note that none of the low islands in Sahlins's sample is part of a large overseas exchange system: Ontong Java and Pukapuka are relatively isolated, while the Tokelaus are only a small cluster of islands. If Lamotrek was as stratified as early accounts suggest, then clearly, network variables should be incorporated into explanations concerning the elaboration of stratification in island communities.[9]

Rooted Graphs and Island Settlement Patterns in Micronesia

A *rooted graph G* is one in which a specified point called the *root* is singled out. Figs. 3.7a and 3.7b show a rooted graph and a *doubly rooted* graph, one in which two points are distinguished from all others. Fig. 3.7c shows all the rooted trees with four points.

In constructing a classification of island networks based on graph theory it would be natural to ask whether the graphs are rooted. The most obvious root in a coral island network is a high island, because

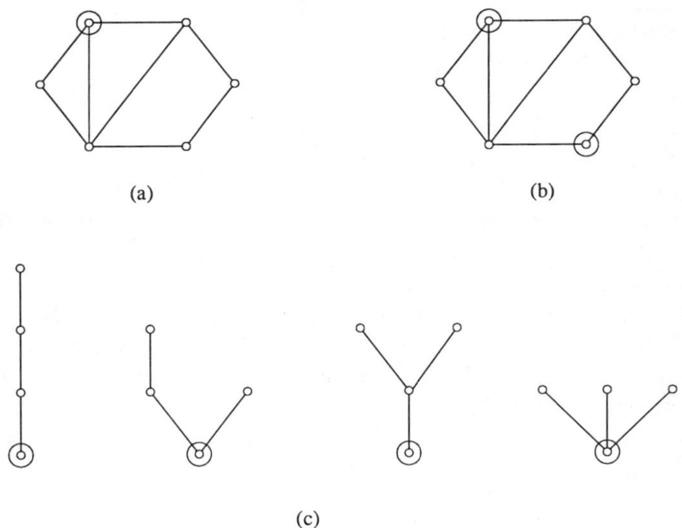

FIG. 3.7. Rooted graphs

[9] Network variables may affect other aspects of social organization, for example variations in descent structure, as Damas (1979) conjectures in the case of Pingelap, a relatively isolated island in the Eastern Carolines. In Polynesian studies, Friedman's (1981) 'regional systems analysis' of social stratification is based on network variables such as density, distance, and relative isolation.

the presence of such an island has major social, genetic, and ecological implications for surrounding low islands: it may be the real or imagined homeland of low islanders (Alkire 1984), or a 'centre of influence' (Goodenough 1986), and because it has such abundant resources it may ensure survival on poorly endowed, vulnerable low islands. We therefore propose the category of a 'rooted complex' in coral island taxonomy. We illustrate the utility of this classification by comparing settlement patterns in two island groups: the Western and Central Carolines, whose graph is rooted, and the Marshalls whose graph is not.

In a comparative work on the cultural ecology of coral islands, Alkire (1978) distinguishes four types of microenvironments based on the criteria of relative isolation and physiographic heterogeneity. In his classification, a 'fringing reef island' is either part of the reef of a high island or in such proximity to it as to be regarded as an offshore island. Examples are Pis which lies on the barrier reef of Truk, and Kayangel, Peleliu, and Angaur which lie off Babelthuap, Palau. Such islands are usually included under the social organization of adjacent high islands. An 'isolate' is a coral island which has itself as its environment and exchange partner, for example Kapingamarangi, the Polynesian outlier in Micronesia. A 'cluster' consists of two or more coral islands of similar economic potential which exchange with each other. Examples, from Polynesia, include Manihiki and Rakahanga in the Northern Cooks, Nukunonu, Atafu, and Fakaofo in the Tokelaus, and Raroia and Takume, Napuka and Tepoto, Takaroa and Takapoto, Hao and Amanu, and a large group that included Rangiroa, Makatea, Tikehau, Matahiva, Arutua, Apataki, Kaukura, and Niau, all in the Tuamotus. A 'complex' consists of a chain of islands which are dissimilar in size, height, or climatic characteristics and which are thus more than quantitatively interdependent. Two examples are located in the Marshalls and the Carolines. It seems to us, however, that there are major differences between the former, which consists entirely of low islands varying climatically along a wet–dry continuum, and the latter in which islands are distinguished topographically as high vs. low, with all that this distinction implies socially, economically, and demographically. In our proposed modification of Alkire's taxonomy, the former is a complex and the latter is a rooted complex.

The exchange complex of the Carolines is doubly rooted, with the high islands of Yap in the west and Truk in the east. One of the roots, Yap, was at the head of a hierarchical structure which undoubtedly sustained permanent settlement of virtually all low islands in the Western and Central Carolines. In native conception the political structure of the Yapese Empire consisted of a 'chain of authority' which originated at Yap and stretched eastward for some 1,100 kilometres, indirectly linking all the low islands as far as Truk. This was the communication structure down which orders for tribute flowed from Yap. Lessa's (1950) implicit graphical representation of this chain is shown in Fig. 3.8.

It is possible to depict a rooted tree in two entirely different ways. The first, as in Fig. 3.7c, puts a circle around the root point. The second way changes the tree into a digraph, which is not rooted, but in which the root point becomes the only transmitter and all arcs are oriented from the root. The latter depiction is given in Fig. 3.8, in which Yap is the root. The resulting digraph is an illustration of an oriented tree, which means that it is asymmetric.

Three observations may be made concerning the structural basis of the stability of the tribute relation. First, the tree structure guaranteed that there could never be conflicting orders, and therefore questions about legitimacy, because in a tree there is exactly one path from the transmitter root u to each other point. And, in fact, Lessa (1950) emphasizes that orders from Yap had to be transmitted exactly as shown in this tree if they were to be accepted. Secondly, the tree reinforced local status structures, and thereby the entire status structure, by making heads of local hierarchies cutpoints and thus communicators in the graph. The relation between precedence in communication and status can be seen in the way in which the order was relayed. It did not, for example, go from Ifaluk to its nearest neighbour Elato, but went from Ifaluk to Lamotrek and then back to Elato; similarly, the order went from Satawal to Puluwat and thence to the latter's immediate subordinates, Pulap, Pulusuk, and Namonuito. Thirdly, the tree structure connecting all the islands was reinforced by a tree structure connecting parts of each island. Thus when the paramount chief of Ulithi, the head of Lamathakh sib, received the order from Yap, he originated a fixed sequence of transmission from sib to sib on the islets of Ulithi atoll, as shown in Lessa's oriented tree reproduced in Fig. 3.9.

The substantive basis of the tribute relation is evident: the low

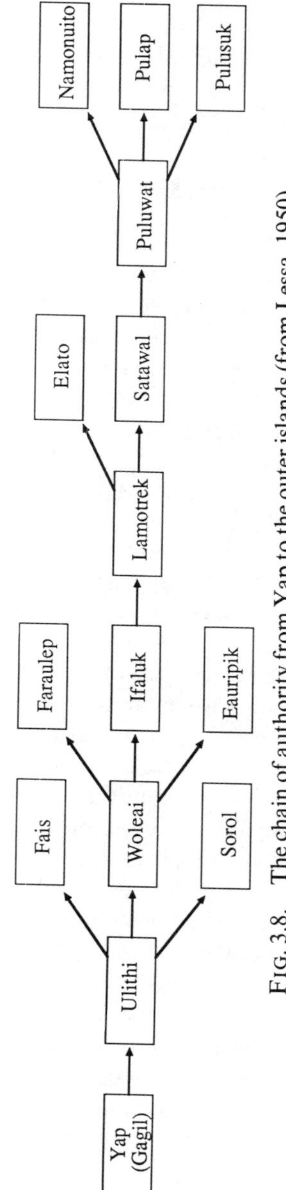

FIG. 3.8. The chain of authority from Yap to the outer islands (from Lessa 1950)

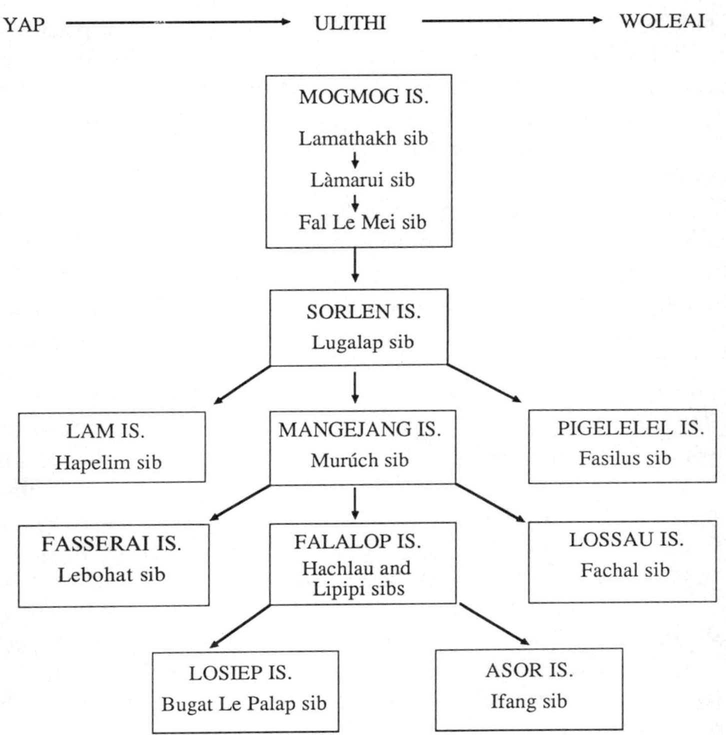

FIG. 3.9. The diffusion of orders for tribute within Ulithi (from Lessa 1950)

islanders gained economic necessities such as food and timber and luxuries such as turmeric and ochre, while the Yapese gained exotic goods with which to manipulate their own internal system of alliances. The role of exchange with Yap in ensuring the survival and enhancing the life of the low islanders is emphasized by Lessa:

> Should the low caste [low island] peoples kick over the traces they would stand to lose greatly desired sources of raw materials. Yap is a 'high' island and can supply timber, turmeric, yams, betel nut and other materials and foods not found on the atolls. Some of these are real necessities, while others help alleviate the monotony of their own diet and supplies. Informants on Ulithi specifically told me what a loss it would be not to be able to procure materials from Yap (1956: 71).

The Yapese Empire came to an end on its eastern periphery during the German administration of the Carolines (1899–1914), but closer to home it did not begin to deteriorate until the period of Japanese administration (1914–45) (Lessa 1950; Alkire 1965). *Sawei* gifts were offered into the 1930s and possibly the 1950s in several cases (W. Alkire, personal communication). The colonial powers in effect took over some of the economic functions of Yap *vis-à-vis* the low islands. Some trade with Truk continued into the modern era. The significance of the high islands of Yap and Truk in motivating and ensuring a tradition of permanent habitation of all but the tiniest islands of the Carolines can be judged by comparing settlement patterns in the Carolines and the Marshalls.

The Marshall Islands consist of 29 atolls and five coral islands, strung out in two parallel chains, Ralik and Ratak, running from south-east to north-west for over 900 kilometres (Map 3.1). The islands lie perpendicular to the Carolines and to the prevailing north-east trade winds and eastern swell. Voyaging was frequent within each chain. Because of the north–south orientation of the islands, navigation relied especially on distinctive patterns of wave refraction and reflection produced by the movement of swells as they meet particular islands (Davenport 1960).[10] There are no high islands in the Marshalls; instead the dominant ecological contrast is between the drier northern and wetter southern islands. Crops in the former were largely restricted to coconuts, pandanus, and arrowroot, while bread-fruit and taro grew in the latter. Northern and southern islands exchanged arrowroot flour for bread-fruit preserves. Pollock (1975) also mentions that turmeric grew in the north and was traded south. Political organization was confined to each chain or part of a chain. According to Alkire (1978), Ralik was divided into two districts: Namu and islands to the north, and Jabwot and islands to the south. The paramount chief of Ailinglapalap was said to 'own' the islands in both groups and periodically made a circuit of them to collect tribute. In Ratak, the chief of Maloelap was said to control all the islands to the north, but not Majuro and Arno in the south, at least not on a permanent basis.

The situation in the Marshalls is not really equivalent to that in the Carolines. The ecological complementarity, between the wet and

[10] As in the Carolines, star courses were also used (Erdland 1914).

dry islands, is less radical and the inter-island political structure was less stable, fluctuating with the fortunes of rival warring chiefs and dependent in part on military control. The tribute relation seems to have been mainly exploitative and based on a threat of physical coercion. The exchange system lacked the high island–low island richness and diversity of the Carolines. This contrast between the Carolines and Marshalls complexes suggests a major quantitative qualification of a recent theoretical model of island settlement patterns.

I. Williamson and M. Sabath (1984) have proposed a model of extinction probability for human settlement on oceanic islands based on the general theory of island biogeography (MacArthur and Wilson 1967). According to the authors, it is known from mathematical ecology (Pielou 1969) that, in general, extinction probability and population size are inversely related. Island populations, which are characteristically small and unstable, are particularly at risk. Extinction probability can be lowered by regular communication and exchange between islands, but there is a lower limit beneath which a population cannot sink if it is to reproduce itself indefinitely. Alkire, as mentioned above in connection with Olimarao Island in the Carolines, puts this number in the range of 30–50 individuals. There is evidence that islanders are aware of the relation between extinction and the instability of small populations, and that islands whose location is known may therefore not be settled if they are perceived as capable of supporting only very small populations.

The hypothesis that island settlement patterns are affected by small population instability is tested by examining the relationship between island settlement existence and island carrying capacity. The hypothesis of a threshold population size is supported when in a group of islands all those with a carrying capacity below a certain level are not settled while all those above that level are.

Williamson and Sabath specify four conditions for testing the hypothesis: (1) a culturally homogeneous and stable island group, (2) consisting of both settled and unsettled islands, (3) with variation in island population extinction probabilities determined only by island carrying capacities, and with all islands subject to comparable resource fluctuations; and (4) a method for quantifying estimates of carrying capacities. They consider that the Marshall Islands satisfy these conditions. The authors assume that 'contact costs' between all islands are equal.[11] The measurement of carrying capacity is given by the product of mean annual rainfall with the island land area, called mesophytic index (MI). Williamson and Sabath's Table 1 gives the land area in square miles, rainfall in millimetres, mesophytic index, and population size for 31 of the 34 islands in the Marshalls. A corrected and slightly modified version of their table is shown in Table 3.3.[12]

Table 3.3 shows that the largest MI for unsettled islands is 1,692 and the smallest MI for settled islands is 982, which puts the MI threshold for settlement and non-settlement between 982 and 1,692.

The Carolines are similar to the Marshalls, with one significant exception: the presence in the former of high islands (Yap and Truk). The effect of Yap on the low islands was twofold: it was a major source of material goods for the low islands, especially in times of crisis (severe storm damage), and through its hierarchical political structure Yap facilitated easy, frequent, and extensive exchange between the low islands themselves. The settlement patterns in the two groups provide a sort of natural experiment. Table 3.4 shows the land area, rainfall, and mesophytic index for 1973 for the low islands in the Carolines, including the two Polynesian outliers and those outside the Yapese sphere in the Truk and Ponape areas.[13]

The settlement threshold as defined by MI is much lower for the Carolines than for the Marshalls: the largest MI for the unsettled islands is 531 (Kuop) and the smallest MI for the settled islands is 254 (Eauripik). There are in fact six settled islands in the Carolines with MI values below the smallest MI value for settled islands in the Marshalls (982). These, in addition to Eauripik (254), are Faraulep

[11] Although this is not true it does not seem to affect their argument.

[12] The corrections consist of several MI computations and all the population figures to correspond with those given in Hainline (1965). The population figures for Bikini and Eniwetok, prior to their evacuation in 1946 and 1948, were 164 and 135 respectively. Williamson and Sabath (1982) excluded Mili Atoll, because its 1948 population size probably reflected the effects of damage from bombing during the Japanese occupation. They excluded Mejit because of a large discrepancy in the figures for land area given in Bryan (1971) and Fosberg (1956). Knox Atoll is considered part of Mili Atoll.

[13] Two islands not shown in Table 3.4 are Ant and Pakin (0.718 and 0.421 square miles respectively, with rainfall probably in the range of Mokil and Ngatik). According to J. L. Fischer (personal communication), both islands were 'more or less continuously occupied' but no population figures are available. In close proximity to the high island of Ponape, the communities on these islands were under the political control of two different chiefs of that island.

TABLE 3.3. *Williamson and Sabath's (1984) table of land areas, rainfall, mesophytic indices, and population sizes for Marshall atolls and islands*

Atoll or island	Land area (square miles)[a]	Rainfall (mm)[b]	Mesophytic index[c]	Population size (1948)[d]
Settled				
Eniwetok	2.26	1307	2954	0*
Bikini	2.32	1451	3366	0*
Rongelap	3.07	1527	4688	95
Wotho	1.67	1882	3143	40
Ujae	0.72	2292	1650	244
Lae	0.56	2532	1418	138
Kwajalein	6.33	2407	15236	1043
Namu	2.42	3165	7659	341
Ailinglapalap	5.67	3352	19006	705
Jaluit	4.38	3987	17463	960
Namorik	1.07	4736	5068	461
Ebon	2.22	5684	12618	747
Utirik	0.94	1515	1424	166
Ailuk	2.07	1816	3759	319
Likiep	3.96	2100	8316	682
Wotje	3.16	1986	6276	328
Maloelap	3.79	2589	9812	457
Aur	2.17	2790	6054	418
Majuro	3.54	3648	12914	1473
Arno	5.00	3753	18765	1071
Lib Island	0.36	2729	982	84
Unsettled				
Taongi	1.20	909	1091	
Bikar	0.19	1245	237	
Taka	0.22	1507	332	
Jemo Island	0.06	1852	111	
Erikub	0.59	2250	1328	
Rongerik	0.65	1456	946	
Ailinginae	1.08	1507	1628	
Jabwot Island	0.22	3165	696	
Kili Island	0.36	4700	1692	
Ujelang	0.67	1956	1311	

* See n. 12.

[a] Land area from Bryan (1971).

[b] Rainfall data from US Department of Commerce (1968), Fosberg (1956), Wiens (1962), and Williamson and Sabath (1982).

[c] Mesophytic index = rainfall × land area.

[d] Population figures from Hainline (1964) for 1948 census.

TABLE 3.4. *Land areas, rainfall, and mesophytic indices for Caroline atolls and islands*

Atoll or island	Land area (square miles)[a]	Rainfall (mm)[b]	Mesophytic index[c]
Settled			
Ulithi	1.799	2896	5210
Fais	1.083	2896	3136
Sorol	0.361	3048	1100
Woleai	1.749	3124	5464
Faraulep	0.163	3048	497
Eauripik	0.091	2794	254
Ifaluk	0.369	2540	937
Elato	0.203	2540	516
Lamotrek	0.379	2642	1001
Satawal	0.505	2642	1334
Pulusuk	1.083	3302	3576
Puluwat	1.313	2794	3669
Namonuito	1.710	2794	4778
Pulap	0.383	2794	1070
Nama	0.289	3302	954
Losap	0.396	3302	1308
Namoluk	0.322	3556	1145
Etal	0.731	3302	2414
Lukunor	1.090	3302	3599
Satawan	1.757	3302	5802
Murilo	0.497	2794	1389
Nomwin	0.716	2794	2001
Ngulu	0.165	3048	503
Ngatik	0.674	3048	2054
Pingelap	0.676	2794	1889
Mokil	0.478	2540	1214
Nukuoro	0.644	2540	1636
Kapingamarangi	0.521	2032	1059
Unsettled			
Olimarao	0.085	2642	225
Oroluk	0.192	2540	488
Kuop	0.190	2794	531
Gaferut	0.043	3048	131
West Fayu	0.024	3048	73
Pikelot	0.036	2794	101
East Fayu	0.144	2794	402

[a] Land area from Bryan (1971).

[b] Rainfall data from Pisarik (1975). *Note*: For islands not in Pisarik, rainfall estimates were made from rainfall of nearest neighbour at similar latitude.These islands are: Fais, Sorol, Satawal, Nama, Olimarao, Gaferut, West Fayu, Pikelot, East Fayu, Oroluk, and Pulusuk. Since according to Manchester (1951) rainfall in the Carolines is heaviest in a belt between 1° 30′ and 8° 30′ where it is usually more than 3048 mm, these guesses are probably not too far off.

[c] Mesophytic index = rainfall × land area.

(497), Elato (516), Ngulu (503), Ifaluk (937), and Nama (954). The first four islands, all of which have extremely low MI values, are in the Yapese Empire: Eauripik, Faraulep, and Elato are in the empire headed by Gagil District, Yap, and Ngulu is in a miniature empire consisting of itself and Goror District, Yap. The fifth island, Ifaluk, is also in the Gagil empire. The sixth island, Nama, is adjacent to Truk. We suggest that when high islands are integrated into a low island exchange system, it significantly reduces the threshold of carrying capacity for permanent settlement, and that there are therefore sound reasons for distinguishing between rooted and unrooted complexes in the cultural ecological taxonomy of coral islands.

The four largest unsettled islands each illustrate important points concerning the relation between settlement threshold and network position, and the definition of mesophytic index. Olimarao, as mentioned, was once inhabited, but not resettled after it was devastated by a storm in the nineteenth century. In spite of its excellent network position, its MI value is perhaps so low as to make settlement a matter of chance. We note that most of the Carolines are subject to frequent typhoons, whereas the Marshalls are not. Mason's (1968) study suggests that an index incorporating this additional variable would be more useful. It would make the lower settlement threshold in the Carolines, attributed to the presence of high islands, even more dramatic. Oroluk, which is over twice the size and has almost twice the MI value of Eauripik, was once inhabited but, like Olimarao, not resettled. It was used as a stopover in trips between Truk and Ponape. Not being part of a regular exchange network, it illustrates the potential fate of a relatively isolated small island, as compared, say, to the Polynesian outlier Nukuoro, which is also relatively isolated but much larger. Kuop (not shown on the map) could be regarded as an offshore island of Truk. According to J. L. Fischer (personal communication) there is no tradition of permanent settlement. Kuop is considered a possession of Uman Island in the Truk lagoon, whose people own tracts on it and use it for fishing. Finally, East Fayu has an MI value higher than Eauripik's but is not settled. Possibly this is because it lacks a lagoon, another variable not included in the mesophytic index.[14] It is traditionally regarded as a possession of Nomwin, whose people exploit it for turtles and turtle eggs.

[14] The relation between lagoon area and population size, however, is problematical (Hainline 1965).

We have shown the application of three graph theoretic models, centrality, betweenness, and the neighbourhood of a point, to the analysis of economic success, political stratification, and dominance in overseas trade networks, and we have shown the utility of a fourth model, rooted graphs, for network taxonomy and demographic studies. These models undoubtedly have application to other kinds of exchange and communication systems. Use of the first two often requires some of the matrix methods defined in the next chapter.

4
Matrix Analysis

What, then, is a table?

Jack Goody, *The Domestication of the Savage Mind*

In 'Mathematics and the Metaphysicians', Bertrand Russell observed that 'Obviousness is always the enemy of correctness. Hence we invent some new and difficult symbolism, in which nothing seems obvious. Then we set up certain rules for operating on the symbols, and the whole thing becomes mechanical' (1917: 77). Certainly in the case of large exchange structures it would be desirable to have purely mechanical procedures to ensure accurate descriptions and rapid, reliable computation. This can be accomplished by representing a graph as a matrix and applying elementary and easily programmed algebraic manipulations.

We define six basic matrix operations: the binary operations of addition, subtraction, multiplication, and the elementwise product, and the unary operations of transposition and raising to higher powers. We also discuss three matrices associated with a graph: the adjacency, reachability, and distance matrices. We then apply this symbolism to the analysis of social, economic, and political exchange networks from Melanesia and Micronesia. We first explicate what is meant by connectivity and centrality analysis in Melanesian archaeology, using Irwin's (1974, 1978) reconstruction of the Mailu network off the south-east coast of Papua New Guinea. Then we show how the distance matrix can be used to find the betweenness values of points in large graphs such as the Micronesian sea-lane structure analysed in Chapter 3. Next, we demonstrate how all three matrices (adjacency, reachability, and distance) can be used to read off various properties of exchange networks. Our example is the political structure of yam exchange in a Trobriand hamlet. Finally, we translate set theoretic operations into matrix operations to give the adjacency matrices of the intersection, union, and symmetric difference digraphs. We apply this technique to the analysis of preferential partnerships in an informal Papuan exchange network. Then we illustrate one application in the Grofman and Landa (1983) proto-coalition theory of trade network origins, using their example of the nearest neighbour relation in the *kula* ring. We interpret their theory as one of three alternative solutions to the minimum spanning tree problem, each of which is a potential model of the evolution of an exchange network.

Matrix Representation

The graph of an exchange network can be represented by a list of lines, a diagram, or a matrix. When a graph is small a diagram is superior to a list, because it provides immediate comprehension of a structure. Even small lists such as Bell's (1935) tables of relations between the five Tanga clans described in Chapter 2 are more easily interpreted pictorially. When a graph is large a matrix representation is superior to a diagram, because it permits algebraic manipulations that enable the elucidation and quantification of structural properties. The most basic and natural matrix representation of a graph is the adjacency matrix.

The *adjacency matrix of a labelled graph G* with p points, written $A = A(G) = [a_{ij}]$, is the $p \times p$ matrix in which $a_{ij} = 1$ if v_i is adjacent with v_j and $a_{ij} = 0$ otherwise. A *symmetric matrix A* satisfies $a_{ij} = a_{ji}$. A *binary matrix* has entries 0 and 1 only. Thus $A(G)$ is a symmetric binary matrix with zero diagonal. There is clearly a one-to-one correspondence between labelled graphs with p points and such binary matrices. Because of this correspondence, every graph theoretic concept is reflected in the adjacency matrix. Thus, for example, the row (or equivalently, the column) sums of $A(G)$ are the degrees of the points of G, as can be seen in Fig. 4.1, which shows a graph and its adjacency matrix. The numbers 1, 2, 3, 4, 5 to the left of the rows and above the columns indicate the points.

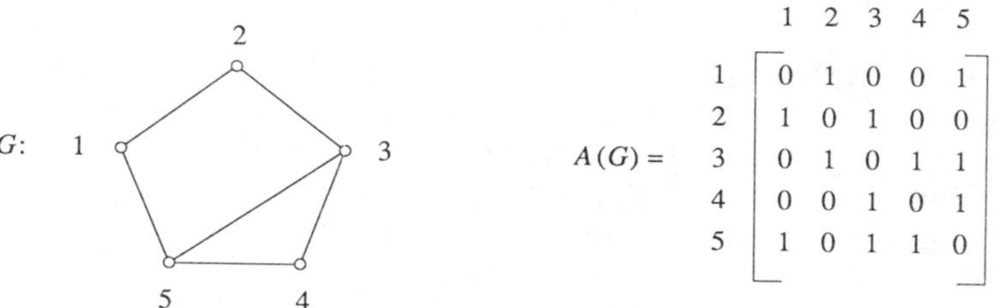

	1	2	3	4	5
1	0	1	0	0	1
2	1	0	1	0	0
3	0	1	0	1	1
4	0	0	1	0	1
5	1	0	1	1	0

FIG. 4.1. A graph and its adjacency matrix

The *adjacency matrix of a labelled digraph D* is defined similarly: $A = A(D) = [a_{ij}]$ has $a_{ij} = 1$ if arc $v_i v_j$ is in D and $a_{ij} = 0$ otherwise. Note that $A(D)$ is binary, but not necessarily symmetric. In $A(D)$ the row sums give the outdegrees and the column sums the indegrees of the points, as can be seen in Fig. 4.2.

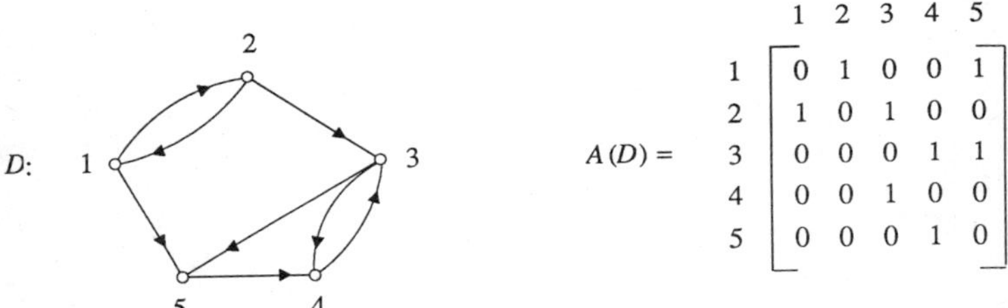

FIG. 4.2. A digraph and its adjacency matrix

Since there is a one-to-one correspondence between a labelled graph (or digraph) and its adjacency matrix, two graphs are certainly isomorphic if their adjacency matrices are equal. Two matrices $A = [a_{ij}]$ and $B = [b_{ij}]$ are *equal matrices* if $a_{ij} = b_{ij}$ for all i and j. This is illustrated in Fig. 4.3 for two 'different-looking' graphs G_1 and G_2 in which the points have been labelled in such a way as to show their isomorphism by means of equal adjacency matrices A_1 and A_2. (Note that when two graphs or digraphs are isomorphic their adjacency matrices need not be equal, because of labelling.)

We can use the graphs in Fig. 4.3 to show one way in which the patterning of an adjacency matrix reflects the structural properties of a graph. When a society has unrecognized dual divisions, its bipartite structure can be shown by partitioning its adjacency matrix in a certain way, as noted in Chapter 1.

Given an adjacency matrix A, we separate its set of rows into two subsets, the first r rows and the rest. We also separate its columns into the first s columns and the rest. Thus we obtain a 'partitioned matrix' in which A_{11} is $r \times s$, A_{12} is $r \times (p-s)$, and so forth:

$$A = \begin{bmatrix} A_{11} & A_{12} \\ A_{21} & A_{22} \end{bmatrix}$$

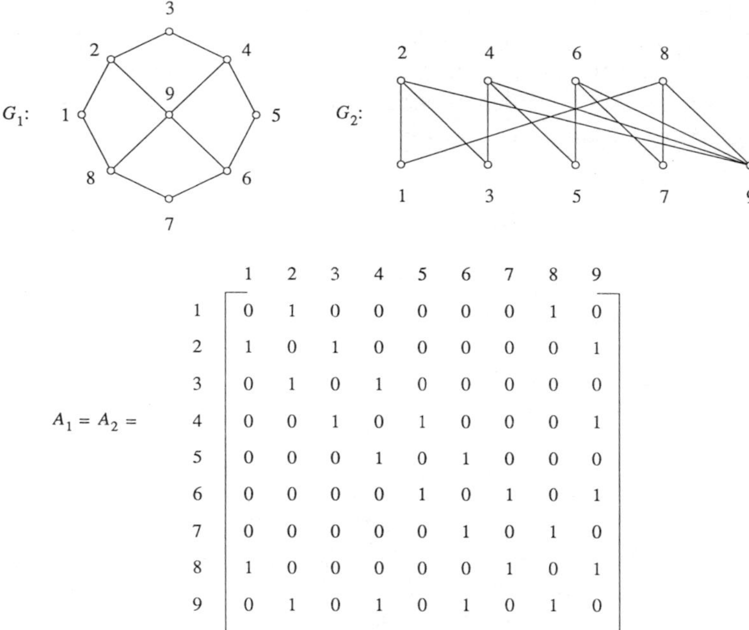

$$A_1 = A_2 = \begin{array}{c} \\ 1 \\ 2 \\ 3 \\ 4 \\ 5 \\ 6 \\ 7 \\ 8 \\ 9 \end{array} \begin{array}{c} \begin{array}{ccccccccc} 1 & 2 & 3 & 4 & 5 & 6 & 7 & 8 & 9 \end{array} \\ \left[\begin{array}{ccccccccc} 0 & 1 & 0 & 0 & 0 & 0 & 0 & 1 & 0 \\ 1 & 0 & 1 & 0 & 0 & 0 & 0 & 0 & 1 \\ 0 & 1 & 0 & 1 & 0 & 0 & 0 & 0 & 0 \\ 0 & 0 & 1 & 0 & 1 & 0 & 0 & 0 & 1 \\ 0 & 0 & 0 & 1 & 0 & 1 & 0 & 0 & 0 \\ 0 & 0 & 0 & 0 & 1 & 0 & 1 & 0 & 1 \\ 0 & 0 & 0 & 0 & 0 & 1 & 0 & 1 & 0 \\ 1 & 0 & 0 & 0 & 0 & 0 & 1 & 0 & 1 \\ 0 & 1 & 0 & 1 & 0 & 1 & 0 & 1 & 0 \end{array}\right] \end{array}$$

FIG. 4.3. Two isomorphic graphs and their common adjacency matrix

If the symmetric matrix A can be so partitioned with $r = s$ such that both A_{11} and A_{22} are square zero submatrices, then G is bipartite, as illustrated for the following presentation of the adjacency matrix A_1 of G_1 in Fig. 4.3, in which the rows and columns have been simultaneously reordered to obtain this form:

$$A = \begin{array}{c} \\ 1 \\ 3 \\ 5 \\ 7 \\ 9 \\ 2 \\ 4 \\ 6 \\ 8 \end{array} \begin{array}{c} \begin{array}{ccccccccc} 1 & 3 & 5 & 7 & 9 & 2 & 4 & 6 & 8 \end{array} \\ \left[\begin{array}{ccccc|cccc} 0 & 0 & 0 & 0 & 0 & 1 & 0 & 0 & 1 \\ 0 & 0 & 0 & 0 & 0 & 1 & 1 & 0 & 0 \\ 0 & 0 & 0 & 0 & 0 & 0 & 1 & 1 & 0 \\ 0 & 0 & 0 & 0 & 0 & 0 & 0 & 1 & 1 \\ 0 & 0 & 0 & 0 & 0 & 1 & 1 & 1 & 1 \\ \hline 1 & 1 & 0 & 0 & 1 & 0 & 0 & 0 & 0 \\ 0 & 1 & 1 & 0 & 1 & 0 & 0 & 0 & 0 \\ 0 & 0 & 1 & 1 & 1 & 0 & 0 & 0 & 0 \\ 1 & 0 & 0 & 1 & 1 & 0 & 0 & 0 & 0 \end{array}\right] \end{array}$$

Empirically interpreted, the matrix A shows that exchanging clans (or other social groups) 1, 3, 5, 7, 9 are in one moiety and clans 2, 4, 6, 8 are in the other.

There is another interesting related possibility. If intermarrying clans form two endogamous groups then the adjacency matrix of the marriage exchange graph can be decomposed. If A can be partitioned into four submatrices such that A_{11} and A_{22} are square and A_{12} and A_{21} consist entirely of zeros, then A is said to be *decomposed* by this partition, and the graph is disconnected, as illustrated in Fig. 4.4.

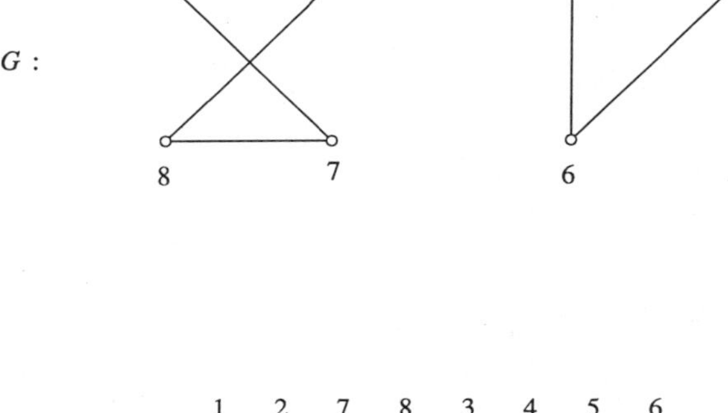

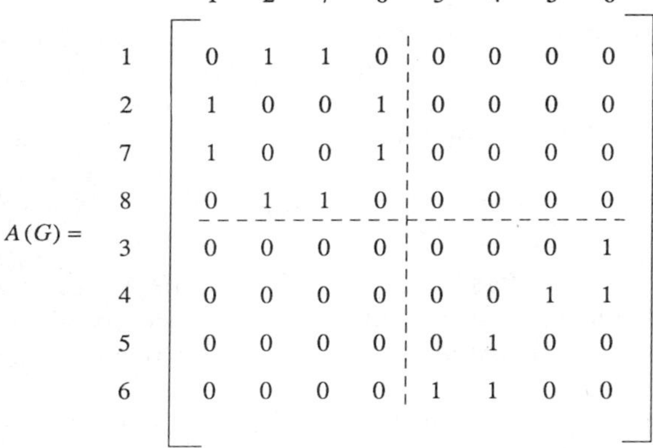

FIG. 4.4. The decomposed adjacency matrix of a disconnected graph

Matrix Operations

We consider two kinds of operations on matrices. In a *binary operation* we begin with two matrices A and B and combine them by some unspecified operation denoted by a very small circle to get a single matrix $A \circ B$. Then, as expected, a *unary operation*, denoted by an asterisk, takes just one matrix A and does something to it to produce another matrix A^*. We begin with four binary operations.

We first define the *sum*, $A+B$, the *difference*, $A-B$, and the *elementwise product*, $A \times B$, of two matrices A and B of the same size $m \times n$. For a matrix C, it is convenient to denote its general i, j entry by $(C)_{ij} = c_{ij}$, as in:

(4.1) $$(A+B)_{ij} = a_{ij}+b_{ij},$$

(4.2) $$(A-B)_{ij} = a_{ij}-b_{ij},$$

(4.3) $$(A \times B)_{ij} = a_{ij}\, b_{ij}.$$

These operations are illustrated by the following matrices:

$$A = \begin{bmatrix} 1 & 1 & 2 \\ 0 & 3 & 0 \end{bmatrix} \qquad B = \begin{bmatrix} 3 & 2 & 1 \\ 0 & 2 & 1 \end{bmatrix}$$

$$A+B = \begin{bmatrix} 4 & 3 & 3 \\ 0 & 5 & 1 \end{bmatrix} \quad A-B = \begin{bmatrix} -2 & -1 & 1 \\ 0 & 1 & -1 \end{bmatrix} \quad A \times B = \begin{bmatrix} 3 & 2 & 2 \\ 0 & 6 & 0 \end{bmatrix}$$

The operation of *multiplication* of two matrices A and B is defined only when the number of columns of A equals the number of rows of B. Thus consider A as $m \times n$ and B as $n \times r$. Their *product* $C = AB$ is $m \times r$ and is defined by the formula

(4.4) $$c_{ij} = a_{i1}\, b_{1j}+a_{i2}b_{2j}+\ldots+a_{in}\, b_{nj} = \sum_{k=1}^{n} a_{ik}b_{kj}.$$

Thus c_{ij} is obtained from the ith row of A and the jth column of B by what is called the 'scalar product' of these two vectors. Note that AB and BA are not necessarily equal, i.e. matrix multiplication is non-commutative.

In particular, the product of a square $n \times n$ matrix A with itself is always defined. Here A is called a *square matrix of order n*. The product AA is written A^2, and its i, j entry is $a_{ij}^{(2)}$:

$$(4.5) \qquad a_{ij}^{(2)} = a_{i1}a_{1j} + a_{i2}a_{2j} + \ldots + a_{in}\,a_{nj}.$$

Similarly, A^r denotes the rth *power* of matrix A, and its general entry is written $a_{ij}^{(r)}$. Clearly each power A^r is a unary operation on A. The next theorem (Harary 1969) tells the structural meaning of the entries of A^r.

Theorem 4.1. Given the adjacency matrix A of a digraph D, the i,j entry of its rth power A^r is the number of walks of length r from v_i to v_j.

The entries of A^2 and A^3 are illustrated in Fig. 4.5. Notice that the entries give the number of walks, not paths.

We now define the unary operation A' on matrix A, which corresponds to the converse of a relation or of a digraph, obtained by reversing all ordered pairs or arcs. The *transpose* A' of a matrix A is obtained from A by interchanging its rows and columns. Thus the i,j entry of A' is the same as the j,i entry of A. Symbolically, if $A = [a_{ij}]$ and $A' = [a'_{ij}]$, then $a'_{ij} = a_{ji}$. Formally, a matrix is called *symmetric* if it equals its own transpose, that is, $A' = A$. Fig. 4.6 shows a digraph D and its converse D' together with the adjacency matrix A and its transpose A'. Table 4.1 summarizes these six matrix operations.

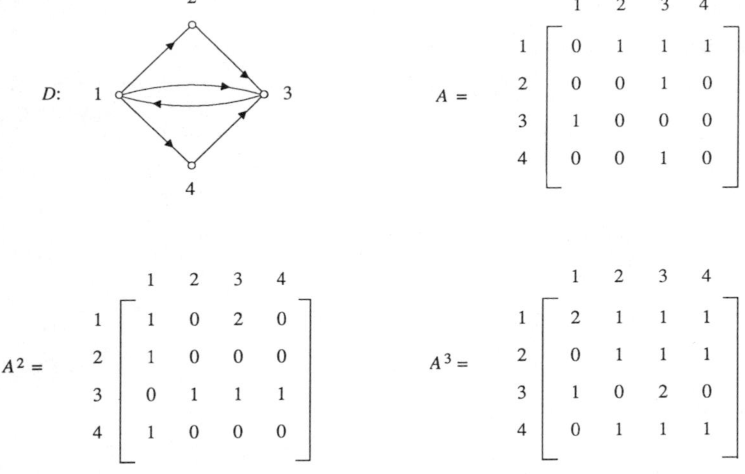

FIG. 4.5. The powers A^2 and A^3 of the adjacency matrix A of D

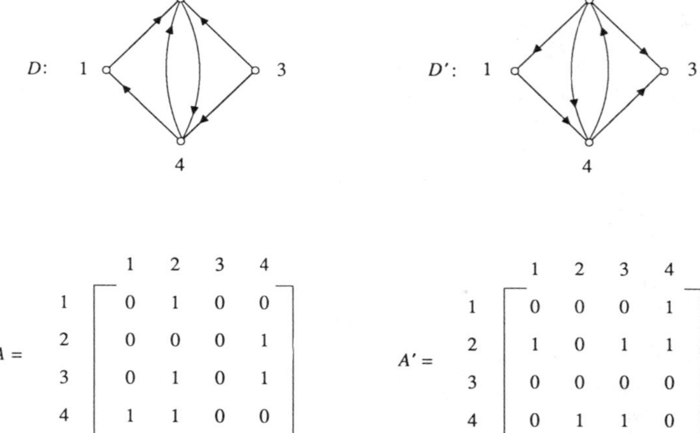

FIG. 4.6. The transpose A' of an adjacency matrix A

Two special matrices need to be defined. The *identity matrix* of order n is denoted by I_n, or more briefly I when its order is clear by context; its entries are 1 on the diagonal and 0 elsewhere. The reason for the name of this matrix is that for any square matrix C having

TABLE 4.1. *Matrix operations*

Size of A	Size of B	Operation	Notation	Size of result	i, j entry of result
$m \times n$	$m \times n$	sum	$A+B$	$m \times n$	$a_{ij}+b_{ij}$
$m \times n$	$m \times n$	difference	$A-B$	$m \times n$	$a_{ij}-b_{ij}$
$m \times n$	$m \times n$	elementwise product	$A \times B$	$m \times n$	$a_{ij}b_{ij}$
$m \times n$	$n \times r$	product	AB	$m \times r$	$\sum_{k=1}^{n} a_{ik}b_{kj}$
$m \times n$		transpose	A'	$n \times m$	a_{ji}
$n \times n$		power	A^r	$n \times n$	$a_{ij}^{(r)}$

the same order as I, the product $IC = C$, and also $CI = C$. Thus matrix I is the identity matrix with respect to multiplication. The *universal matrix* J is a matrix all of whose entries are 1. The matrix I is always square, whereas J is not necessarily square. Thus, for example,

$$I = I_4 = \begin{bmatrix} 1 & 0 & 0 & 0 \\ 0 & 1 & 0 & 0 \\ 0 & 0 & 1 & 0 \\ 0 & 0 & 0 & 1 \end{bmatrix}$$

$$J = J_{3,4} = \begin{bmatrix} 1 & 1 & 1 & 1 \\ 1 & 1 & 1 & 1 \\ 1 & 1 & 1 & 1 \end{bmatrix}$$

It is also necessary to define, for several uses of matrices, the arithmetic of Boolean algebra on the integers 0 and 1. With only one exception, addition and multiplication are exactly the same as for ordinary arithmetic. The exception is that $1+1 = 1$. It follows that a sequence of ordinary arithmetical operations on 0 and 1 that leads to any positive integer will yield the element 1 using Boolean arithmetic. If x and y are any two non-negative integers, we indicate their *Boolean sum* by the symbol $(x + y)\#$, so that the sum is either 0 or 1. Since $(1+1)\# = 1$, we write $2\# = 1$, $3\# = 1$, and so forth.

'Connectivity Analysis' in Melanesian Archaeology

In 1965 F. R. Pitts introduced a graph theoretic model into historical geography when he showed how intuitive notions about the advantages of central location could be quantified using a method he called 'connectivity analysis'. We will explain the method after necessary concepts are introduced. His study of the river-based trade network of 12th- and 13th-century Russia provided support for controversial situational as opposed to cultural and historical explanations for the dominance of Moscow, by demonstrating that it was among the most centrally located or best 'connected' of all cities in the region. The special value of the model for geography lies in structural comparisons of the same places in different transport nets and in different time periods. Clearly, there are implications for archaeology.

Irwin (1974, 1978) has shown how this same model can be used in conjunction with standard archaeological techniques to account for the pre-eminence of communities in prehistoric trade networks. In the Mailu area of south-east coastal Papua New Guinea over a period extending from the first millennium AD up to the end of the late 19th century, one community, Mailu Island, eventually came to dominate trade, especially the manufacture and distribution of pottery. By historic times, it was distinguished from neighbouring villages by a much larger population and greater social complexity. Pottery analysis reveals that in the early period, manufacture was functionally generalized (spread over all communities in the network), but that it gradually became more specialized and was eventually monopolized by Mailu Island. At the same time, graph analysis of archaeological sites in three successive time periods, Early, Mayri, and Mailu, shows that Mailu Island became progressively more central in the evolving network of trade relations joining all communities. Irwin rejects ecological explanations for the dominance of Mailu Island, for it was not uniquely in possession of good clay deposits, and demographic explanations as well, interpreting larger populations as a consequence rather than a cause of the community's success. He emphasizes instead Mailu Island's increasing locational advantage in a changing spatial configuration of settlements.

More recently Irwin (1983) has applied connectivity analysis to ethnographic problems, proposing central location as a cause of social stratification in the northern Trobriand Islands. If this model is to become a general tool of structural analysis in Oceanic archaeology and anthropology, it will require some clarification. Specifically, connectivity must be correctly defined and distinguished from centrality in graphs. The associated matrices of a graph must be identified and properly labelled. We need first of all to define and show the construction of the reachability and distance matrices. We then require the concept of planarity and planar graphs.

Recall that a point v_j is reachable from a point v_i if there is a path from v_i to v_j. The *reachability matrix* of a digraph D or a graph G is a matrix with entries $r_{ij} = 1$ if v_j is reachable from v_i and $r_{ij} = 0$ otherwise. Fig. 4.7 shows a digraph and its reachability matrix $R(D)$. It is conventional to stipulate that every point is reachable from itself, as shown by the diagonal entries, $r_{ii} = 1$.

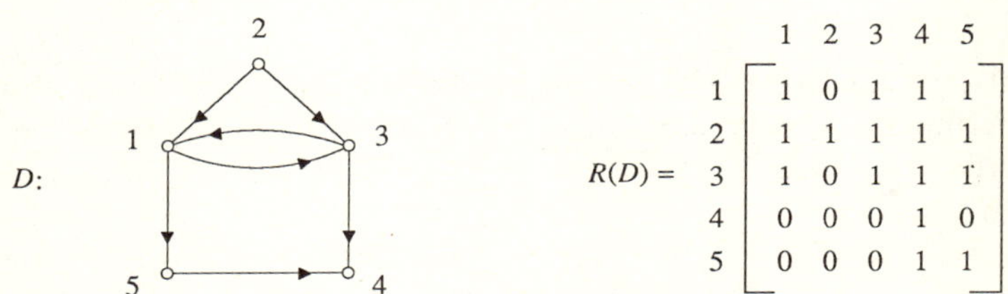

FIG. 4.7. The reachability matrix of a digraph

To obtain the reachability matrix R of a graph from its adjacency matrix A, we make use of Theorem 4.1. Since in constructing R we are concerned only with whether v_i can or cannot reach v_j and not with the number of ways it can be done, we make use of Boolean addition. By the binary matrix $A^2\#$ we shall mean the matrix obtained by applying Boolean arithmetic to compute the entries of the matrix A^2. Symbolically, $A^2\# = [a_{ij}^{(2)}\#]$, and in general by $A^n\#$ we shall mean the matrix obtained when A^n is computed using Boolean operations. The next statement is an immediate consequence of a combination of Theorem 4.1 with the definition of Boolean arithmetic.

Corollary 4.1a. The i, j entry $a_{ij}^{(n)}\#$ of $A^n\#$ is 1 if and only if there is at least one walk of length n in D from v_i to v_j.

This corollary is illustrated using the digraph in Fig. 4.5. The following matrices, in which the rows and columns have the same ordering as the points of that figure, should be compared with the matrices A^2 and A^3.

By *limited reachability* is meant reachability within a certain number n of steps. Thus R_2 denotes the reachability matrix for paths of length 2 or less, R_3 for paths of length at most 3, and so on. Let us denote by R the reachability matrix within $p-1$ steps, which is the length of any longest possible path in a graph or digraph with p points. (Note that the preceding theorem refers to walks, but any

$$A^2\# = \begin{bmatrix} 1 & 0 & 1 & 0 \\ 1 & 0 & 0 & 0 \\ 0 & 1 & 1 & 1 \\ 1 & 0 & 0 & 0 \end{bmatrix} \qquad A^3\# = \begin{bmatrix} 1 & 1 & 1 & 1 \\ 0 & 1 & 1 & 1 \\ 1 & 0 & 1 & 0 \\ 0 & 1 & 1 & 1 \end{bmatrix}$$

walk contains a path.) Then the reachability matrices R_n and R are determined by the next theorem, from Harary, Norman, and Cartwright (1965).

Theorem 4.2. For every positive integer n,

(4.6) $R_n = (I+A+A^2+\ldots+A^n)\# = (I+A)^n\#$, and

(4.7) $R = (I+A+A^2+\ldots+A^{p-1})\# = (I+A)^{p-1}\#$

In other words, we get the reachability matrix by adding, in Boolean fashion, the identity matrix, the adjacency matrix, and successively higher powers of the adjacency matrix up to power $p-1$. The matrices R_2 and R for the digraph in Fig. 4.5 are

$$R_2 = \begin{array}{c} \\ 1 \\ 2 \\ 3 \\ 4 \end{array} \begin{array}{c} \begin{array}{cccc} 1 & 2 & 3 & 4 \end{array} \\ \begin{bmatrix} 1 & 1 & 1 & 1 \\ 1 & 1 & 1 & 0 \\ 1 & 1 & 1 & 1 \\ 1 & 0 & 1 & 1 \end{bmatrix} \end{array} \qquad R = \begin{array}{c} \\ 1 \\ 2 \\ 3 \\ 4 \end{array} \begin{array}{c} \begin{array}{cccc} 1 & 2 & 3 & 4 \end{array} \\ \begin{bmatrix} 1 & 1 & 1 & 1 \\ 1 & 1 & 1 & 1 \\ 1 & 1 & 1 & 1 \\ 1 & 1 & 1 & 1 \end{bmatrix} \end{array}$$

Note that $R = J$ here. Clearly a digraph D is strong if and only if $R = J$.

The distance matrix is defined as follows. Recall that the distance from v_i to v_j, denoted by d_{ij}, is the length of a shortest path from v_i to v_j. If there is no path from v_i to v_j, then d_{ij} is considered to be infinite, symbolized by ∞. The *distance matrix* of a digraph D, denoted by $N(D)$, or of a graph G, denoted by $N(G)$, is the square matrix of order p whose entries are the distances d_{ij}. The next theorem from Harary *et al.* (1965) gives the entries of this matrix.

Theorem 4.3. Let $N = [d_{ij}]$ be the distance matrix of a given digraph D or graph G. Then,

(1) Every diagonal entry $d_{ii} = 0$,

(2) $d_{ij} = \infty$ if $r_{ij} = 0$, and

(3) Otherwise, d_{ij} is the smallest power n to which A must be raised so that $a_{ij}^{(n)} > 0$, that is, so that the i,j entry of $A^n\#$ is 1.

The procedure for constructing the distance matrix N from the adjacency matrix A is illustrated for the digraph in Fig. 4.8. First of all, enter zeros on the diagonal of $N(D)$, showing $d_{ii} = 0$. Next, enter 1 in $N(D)$ whenever $a_{ij} = 1$, thus showing the distance $d_{ij} = 1$. Taking higher powers of A, whenever $a_{ij}^{(n)}\# = 1$ and there is no prior i,j entry in $N(D)$, enter an n to show where $d_{ij} = n$. Finally, note that in $A^4\#$, wherever 1 occurs there is already an entry in $N(D)$. Hence

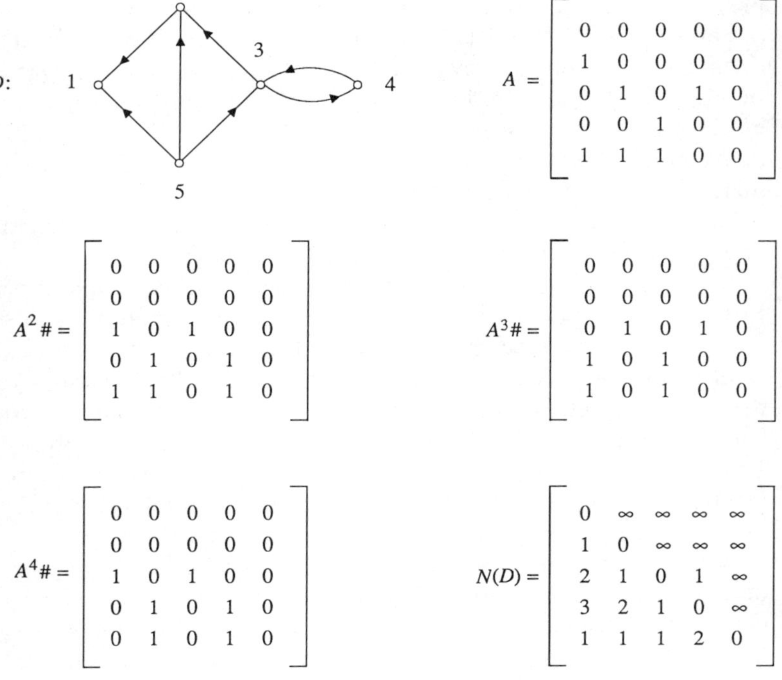

$$A = \begin{bmatrix} 0&0&0&0&0\\ 1&0&0&0&0\\ 0&1&0&1&0\\ 0&0&1&0&0\\ 1&1&1&0&0 \end{bmatrix}$$

$$A^2\# = \begin{bmatrix} 0&0&0&0&0\\ 0&0&0&0&0\\ 1&0&1&0&0\\ 0&1&0&1&0\\ 1&1&0&1&0 \end{bmatrix} \qquad A^3\# = \begin{bmatrix} 0&0&0&0&0\\ 0&0&0&0&0\\ 0&1&0&1&0\\ 1&0&1&0&0\\ 1&0&1&0&0 \end{bmatrix}$$

$$A^4\# = \begin{bmatrix} 0&0&0&0&0\\ 0&0&0&0&0\\ 1&0&1&0&0\\ 0&1&0&1&0\\ 0&1&0&1&0 \end{bmatrix} \qquad N(D) = \begin{bmatrix} 0&\infty&\infty&\infty&\infty\\ 1&0&\infty&\infty&\infty\\ 2&1&0&1&\infty\\ 3&2&1&0&\infty\\ 1&1&1&2&0 \end{bmatrix}$$

FIG. 4.8. The construction of the distance matrix

we enter ∞ in all remaining open locations, indicating all ordered pairs (v_i, v_j) for which there is no path from v_i to v_j in D.

We can now clarify the method of 'connectivity analysis'. Irwin's graph G of the Early Period Mailu trade network is shown in Fig. 4.9. The points represent settlements and the lines inferred trading relations between them. The graph is superimposed on the coastal area of New Guinea. Mailu Island is point 17.

We require a preliminary definition. The *diameter* $d(G)$ of a connected graph G is the length of any longest geodesic. Thus $d(G)$ is the maximum distance between any two points of G, i.e. it is the maximum entry in the distance matrix $N(G)$. The diameter of G in Fig. 4.9 is 4.

Following Pitts, Irwin calls the adjacency matrix A of a graph G the 'connectivity matrix'. He refers to walks as 'paths'. Using matrix operations, he defines two measures of 'centrality'. He calls the first one 'short path connectivity' and the second 'connectivity centrality'. The first of these is precisely distance centrality as presented in Chapter 3. The second is more complicated to explain.

For a given graph G, let A be its adjacency matrix and let d be its diameter. He forms the power matrix A^d, which has as its i, j entry, by Theorem 4.1, the number of walks of diameter length d from v_i to v_j. This is shown *in exemplo* in Fig. 4.9. In this illustration the diameter is clearly $d = 4$, so we show a fragment of A^4. The distance matrix $N(G)$ is also given in full for this example.

He then defines the 'connectivity score' of a point as its row sum in A^d expressed as a percentage of the 'total connectivity' of the matrix A^d, the total of the row sums. This definition calls for two comments. First, the word 'path', which in graph theory means that all points and lines in a sequence are distinct, must be replaced by the word 'walk' which allows for repetition of points and lines. Secondly, the word connectivity must be eliminated, for it has nothing to do with either the number of walks or the centrality of a graph. Unfortunately this measure based on the number of walks whose length is the diameter of G, borrowed from geography, does not appear to have theoretical or empirical significance.

Because the terms connectivity and connectedness have such diverse uses in network analysis (Barnes 1969) and are so consistently misused, and because they will eventually prove to have legitimate applications, we shall define them precisely as they are used in graph theory. The *connectivity* $\kappa = \kappa(G)$ of a graph G is the

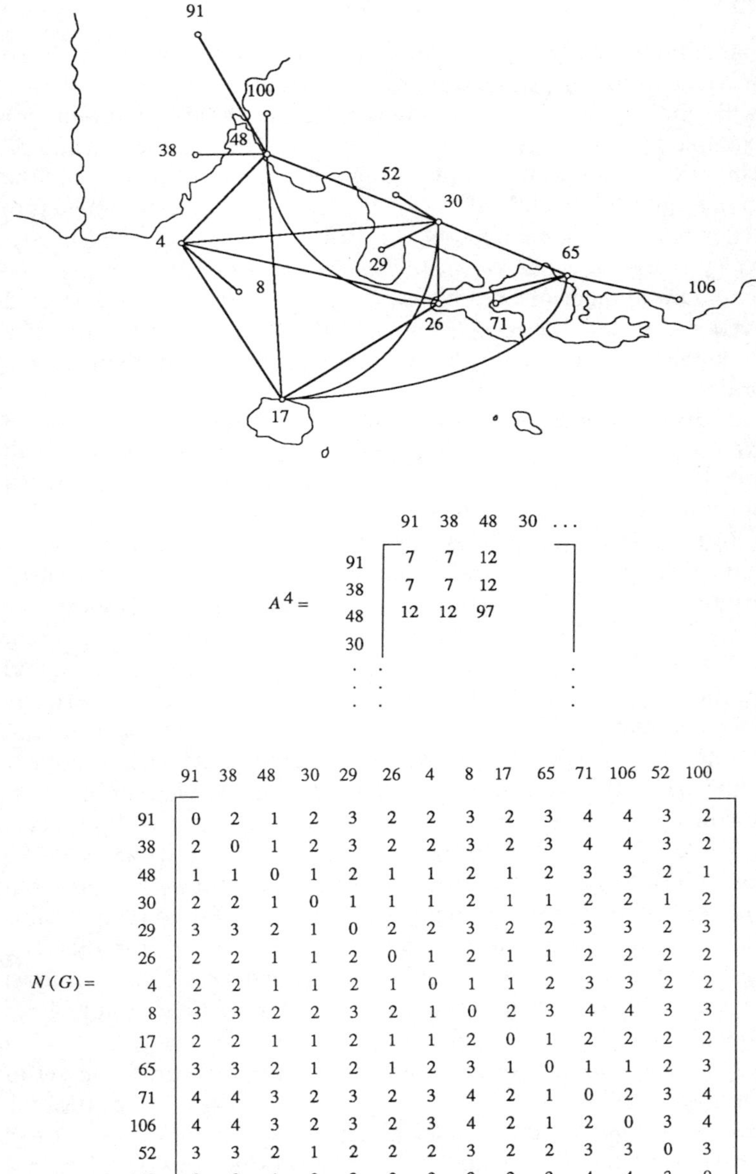

$A^4 =$

	91	38	48	30	...
91	7	7	12		
38	7	7	12		
48	12	12	97		
30					
⋮					⋮

$N(G) =$

	91	38	48	30	29	26	4	8	17	65	71	106	52	100
91	0	2	1	2	3	2	2	3	2	3	4	4	3	2
38	2	0	1	2	3	2	2	3	2	3	4	4	3	2
48	1	1	0	1	2	1	1	2	1	2	3	3	2	1
30	2	2	1	0	1	1	1	2	1	1	2	2	1	2
29	3	3	2	1	0	2	2	3	2	2	3	3	2	3
26	2	2	1	1	2	0	1	2	1	1	2	2	2	2
4	2	2	1	1	2	1	0	1	1	2	3	3	2	2
8	3	3	2	2	3	2	1	0	2	3	4	4	3	3
17	2	2	1	1	2	1	1	2	0	1	2	2	2	2
65	3	3	2	1	2	1	2	3	1	0	1	1	2	3
71	4	4	3	2	3	2	3	4	2	1	0	2	3	4
106	4	4	3	2	3	2	3	4	2	1	2	0	3	4
52	3	3	2	1	2	2	2	3	2	2	3	3	0	3
100	2	2	1	2	3	2	2	3	2	3	4	4	3	0

FIG. 4.9. The graph and associated matrices of the Early Period Mailu network (from Irwin 1978)

minimum number of points whose removal results in a disconnected or trivial graph. Thus the connectivity of a disconnected graph is 0, while the connectivity of a connected graph with a cutpoint, for example the Mailu graph in Fig. 4.9, is 1. The complete graph K_p cannot be disconnected by removing any number of points, but the trivial graph results after removing $p-1$ points; therefore $\kappa(K_p) = p-1$. Sometimes κ is called the *point connectivity*. Analogously, the *line connectivity* $\lambda = \lambda(G)$ of a graph G is the minimum number of lines whose removal results in a disconnected or trivial graph. Thus $\lambda(K_1) = 0$, and the line connectivity of a disconnected graph is 0, whereas that of a connected graph with a bridge is 1.

A graph G is *n-connected* if $\kappa(G) \geqslant n$, and *n-line connected* if $\lambda(G) \geqslant n$. We note that a non-trivial graph is 1-connected if and only if it is connected, and that it is 2-connected if and only if it is a block having more than one line.

In the analysis of the Carolines trade network in Chapter 3, we defined the distance centrality (status) of a point as the sum of its distances to all other points in G. This number is given by the row sums of $N(G)$. This is Irwin's first measure of centrality, except that he calls it 'short path connectivity' instead of distance centrality, and labels $N(G)$ as $A(P)$, the 'short path matrix'. This definition perfectly suits his purpose, for he specifies that central location in the Mailu graph means the 'location from which a commodity such as pottery could be, and in fact was, traded from a single place to a whole region with greatest efficiency' (1978: 308).

Table 4.2 shows the distance centrality of all settlements in the Early Period graph. Mailu Island at this time is not most central but only second in a group of highly ranked places. Fig. 4.10 and Table 4.3 show Irwin's graph and the centrality measure for the Late Period, in which Mailu Island has emerged as clearly the most central of all settlements. We note that both of these graphs have connectivity 1, as each is connected and has a cutpoint.

One final concept contained in Irwin's graph theoretic model of prehistoric networks requires clarification. Irwin designates the graphs in Figs. 4.9 and 4.10 as 'planar', but in fact neither of these graphs is planar, as we shall see. The following considerations are from 'topological graph theory'.

A graph G is said to be *embedded* in a surface S when it is drawn on S so that no two lines intersect. A graph is *planar* if it can be

TABLE 4.2. *Distance centrality of Mailu settlements in the Early Period*

Settlement	Σd_{ij}	Rank
91	33	6=
38	33	6=
48	21	2=
30	19	1
29	31	5=
26	21	2=
4	23	3
8	35	7
17	21	2=
65	25	4
71	37	8=
106	37	8=
52	31	5=
100	33	6=

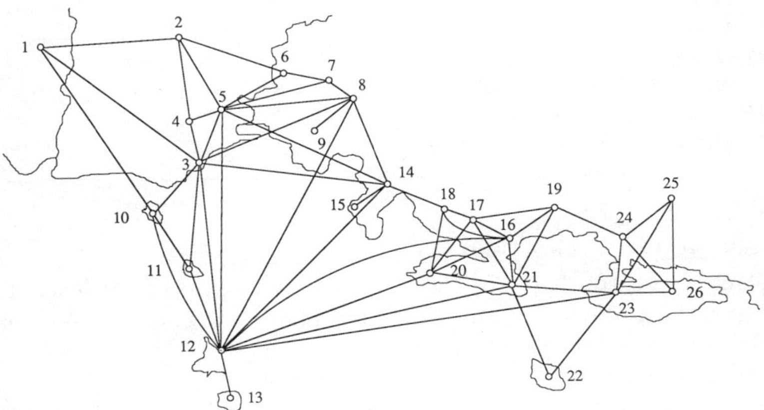

FIG. 4.10. The graph of the Late Period Mailu network (from Irwin 1974)

TABLE 4.3. *Distance centrality of Mailu settlements in the Late Period*

Settlement	Σd_{ij}	Rank
1 Magore	68	14
2 Block	67	13=
3 Kurere	48	2=
4 Asioro	67	13=
5 Woworo	48	2=
6 Oraoro	69	15=
7 Selai	69	15=
8 Derebai	50	4=
9 hamlet	74	18
10 Loupom	57	8
11 Laluoro	59	9
12 Mailu	39	1
13 Goigoi	63	11
14 Bauoro	49	3
15 hamlet	73	17=
16 Borebo	53	6
17 Masioro	66	12
18 Taioro	62	10
19 Sabiribo	67	13=
20 Boganaoro	55	7
21 Pedili	50	4=
22 Eunoro	70	16
23 Gauoro	52	5
24 Nobunoga	69	15=
25 Gema	73	17=
26 Geaoro	73	17=

embedded in a plane; a *plane graph* has already been embedded in a plane. The graph G_1 in Fig. 4.11 is planar although as drawn it is not plane; the graph G_2 is both planar and plane, and is isomorphic to G_1, as both of them are forms of the complete bipartite graph $K_{2,3}$; the graph G_3, which is the bigraph $K_{3,3}$ is not planar.

The following statement is the 'easy half' of the celebrated theorem of Kuratowski (1930) which gave a criterion for a graph to

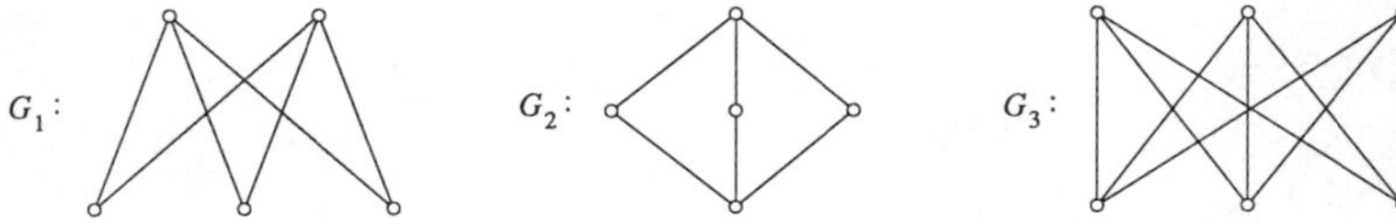

FIG. 4.11. Planar, plane, and non-planar graphs

be planar. It establishes immediately that the Early Period Mailu graph in Fig. 4.9 is non-planar:

Theorem 4.4. The graphs K_5 and $K_{3,3}$ are non-planar, and so is any graph which contains a subdivision of either of these as subgraphs.

Technically speaking, a graph is planar if and only if it contains no subgraph which is a subdivision of K_5 or $K_{3,3}$. A *subdivision* of a graph G is obtained from G by admitting the insertion of new points of degree 2 into its lines. The graph in Fig. 4.9 contains K_5 as a subgraph on the point set {4, 17, 26, 30, 48}. The graph in Fig. 4.10 contains a subdivision of K_5 defined by the point set {16, 17, 18, 20, 21}, in which the path with points {21, 12, 14, 18} can be regarded as the line {21, 18} subdivided by the points 12 and 14. This graph also contains a subgraph K_5 (not subdivided) on the points {3, 5, 8,12, 14}.

In some applications of graph theory, planarity is an important consideration. As an example from the physical sciences, the printed circuits of an electrical network must have a planar representation, otherwise the intersection of the wires could produce distortion due to mutual induction effects. The complete bigraph $K_{3,3}$ is often interpreted as the 'utilities problem'. In this bigraph, the points of one colour set represent three different consecutive houses on the same street, the points of the other colour set three different utilities, and the lines are connections between houses and utilities. The problem is, how can the houses be joined to the utilities so as to avoid potential conflicts arising from the intersection of their connections? No matter how the graph of this system is drawn, at least one pair of connections must intersect, as illustrated in Fig. 4.12. Thus 1 is its 'crossing number'. Formally, the *crossing number* $v(G)$ of a graph G is the minimum possible number of pairwise intersections of its lines when G is drawn in the plane. Obviously $v(G) = 0$ if and only if G is planar. The crossing number of the Early Period Mailu graph in Fig. 4.9. is 1. It is quite possible

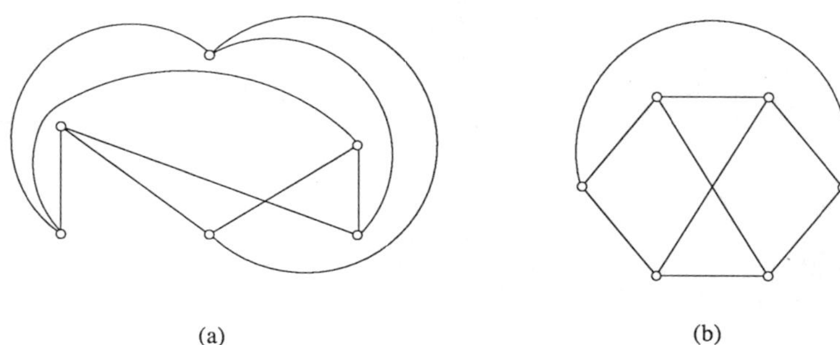

FIG. 4.12. $K_{3,3}$ drawn with one line intersection, in two isomorphic drawings

that planarity enters into the construction as well as the reality of networks in archaeology and anthropology.[1] Irwin's (1983) largely hypothetical graph of land routes linking Trobriand villages, for example, appears to be planar.

Geodetic Structure

To find the relative values of the distance centrality of each island in the Caroline network described in Chapter 3 and thus its relative closeness to resources, it is necessary to find the length of a geodesic between it and every other island. Geodetic length is given by the row entries and centrality values by the row sums of the distance matrix. To find the relative betweenness value of each island and thus its potential for the control of communication, it is necessary to enumerate all the geodesics between every pair of islands. The set of all such geodesics, also called the *geodetic subgraph* $D_g(v_i, v_j)$, can be found from the rows and columns of the distance matrix.

Recall that betweenness value refers to the frequency of occurrence of each point on the geodesics between all pairs of points in a graph. To find the rank of each point it is necessary to find the geodetic subgraph of each pair of points in G. We can define the procedure as follows, using the point pair 8, 16 (Ifaluk, Truk) from the graph in Fig. 3.2. The next theorem, based in part on Flament

[1] Planarity in graphs is also of crucial importance in computer science, chemistry, geography, and operational research.

(1963), is from Harary *et al.* (1965) but paraphrased for graphs instead of digraphs.

Theorem 4.5. Let v_i and v_j be joined by a path in a graph H, that is, v_i and v_j are in the same connected component of H. The geodetic subgraph H_g of v_i and v_j consists of all the points v_k such that $d_{ik} + d_{kj} = d_{ij}$, and all the lines $v_r v_s$ of H such that both $d_{ir} + 1 = d_{is}$ and $d_{rj} = d_{sj} + 1$.

The sums of the entries in the ith row of $N(G)$, which show d_{ik}, and the jth column, which show d_{kj}, disclose the points v_k in the geodetic subgraph $D_g(v_i, v_j)$. To illustrate for $D_g(8, 16)$, we take the eighth row and the sixteenth column of the distance matrix of the Caroline Islands subgraph in Fig. 3.2:

k	1	2	3	4	5	6	7	8	9	10	11	12	13	14	15	16	17	18
$d_{8,k}$	3	3	2	2	1	1	1	0	1	2	3	4	4	4	4	5	6	7
$d_{k,16}$	7	7	6	6	5	5	6	5	4	3	2	1	1	1	1	0	1	2
$d_{8,k} + d_{k,16}$	10	10	8	8	6	6	7	5	5	5	5	5	5	5	5	5	7	9

k	19	20	21	22	23	24	25	26	27	28	29	30	31
$d_{8,k}$	8	9	10	10	6	6	4	4	1	2	2	3	5
$d_{k,16}$	3	4	5	5	1	1	8	8	4	4	3	2	1
$d_{8,k} + d_{k,16}$	11	13	15	15	7	7	12	12	5	6	5	5	6

Since $d_{8,16} = 5$, points 8, 9, 10, 11, 12, 13, 14, 15, 16, 27, 29, and 30 are in $D_g(8, 16)$.

When the points v_k are arranged from left to right in order of their increasing distance from v_i and when a line is added from a point to its immediate right neighbour if the distance between them is 1, the result is the geodetic subgraph $D_g(v_i, v_j)$. Fig. 4.13 shows $D_g(8, 16)$ of the point pair 8, 16. These two points are on the left and right, and all other points lying on 8–16 geodesics appear between them in the order in which they occur in the graph G of Fig. 3.2.

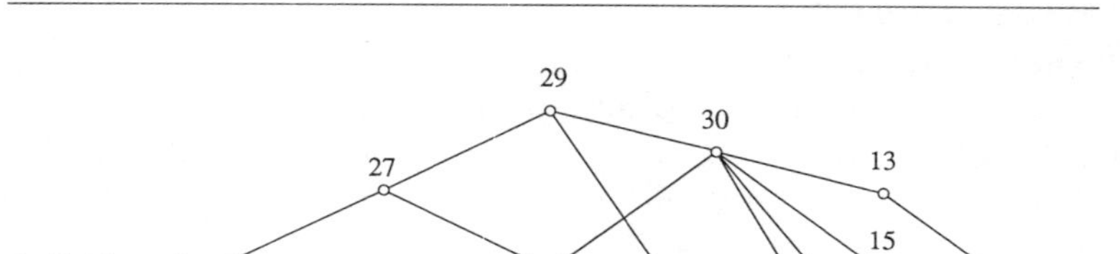

FIG. 4.13. The geodetic subgraph of the point pair 8, 16 (Ifaluk, Truk) in the graph of Fig. 3.2

The Raw and the Cooked: Hierarchy and Intimacy in Melanesian Exchange Networks

Commonly in Oceania, social relations are mediated by food exchanges based on a symbolic opposition between the raw and the cooked. In Tikopian marriage exchanges, the chief of the wife's clan gives raw food to the kinship group of the husband, which reciprocates with gifts of cooked food (Firth 1939). In Palauan ritual, raw food (*mengeseb*) and cooked food (*ongat*) are distinguished in gifts to a girl's matrilineage at first child ceremonies, while the new mother is herself 'cooked' by pots of boiling taro in a steam hut as a final preparation for her re-entry into public life (D. Smith 1983).[2] In Dobuan mourning exchanges, the village *susu* (matrilineal clan) of the surviving spouse makes a large gift of uncooked yams to the *susu* of the dead and receives a minor gift of cooked food from the latter, and if the dead is a man, from the children of the dead (Fortune 1932). In Arapesh kinship classification, ego's relatives are distinguished behaviourally on the basis of exchanges of raw and cooked food (Mead 1947; see Chapter

[2] In the Trobriands, a *kula* canoe, an object made by men but viewed as a woman (Tambiah 1983), is ritually 'cooked' before launching (Malinowski 1922: 140). One might also cite, as instances of symbolic culinary mediation, *tapu* removal of a new-born child among the Maori (Johansen 1954) and punishments for murder, theft, and adultery in Samoa (Hage 1979b).

6 below). The raw/cooked opposition which defines such egocentric kinship structures also defines collective network structures. These may be based on either formal asymmetrical or informal symmetrical relations, as exemplified by Trobriand yam exchange and Orokaiva taro exchange respectively. We will illustrate how the description of such networks can be improved by explicit set theoretic operations applied to the adjacency matrices of exchange digraphs, and how analytical generalization can be checked by deductions from entries in the adjacency, reachability, and distance matrices.

In the Trobriand Islands yams are the staple food and at the basis of exchange relations connected with wealth and power. The obligation of a man to provide yams for his sister's husband, including in particular one who is a chief, is well known from the classic works of Malinowski (1929, 1935), although the complexities of this relation went unrecognized until the modern study by A. B. Weiner (1976). According to Weiner the differentiation of yams into the categories of the cooked and the raw is correlated with the distinction between subsistence and exchange gardens. In the former, yams are immediately and unceremoniously converted into food, while in the latter they are formally harvested and kept in a raw state to be used for exchange. One of the most important types of exchange relations is that in which a younger man—a son, sister's son, or younger brother—provides raw yams (strictly speaking, labour in a yam exchange garden) to an older man—a father, maternal uncle, or elder brother—in return for valuables. The older man gains additional control over resources, and the younger man, through the acquisition of axes, shell ornaments, pots, etc., gains access to magical knowledge and participation in the *kula*. The number of exchange gardens a man makes varies, but it is said that a man must make an exchange garden for the manager of the hamlet in which he resides in order to obtain rights to garden land and the right to build a house foundation. In return for a presentation of raw yams (or labour) the manager gives cooked food, and if the exchange garden is especially large, valuables, thus inspiring men of the hamlet to increase their efforts.

Fig. 11 of *Women of Value, Men of Renown* presents what might be called a 'Weiner diagram', a kind of picture list of ordered pairs of exchange relations between men of a hamlet. The essential features of this relation are shown in Fig. 4.14, in which each of her 'little

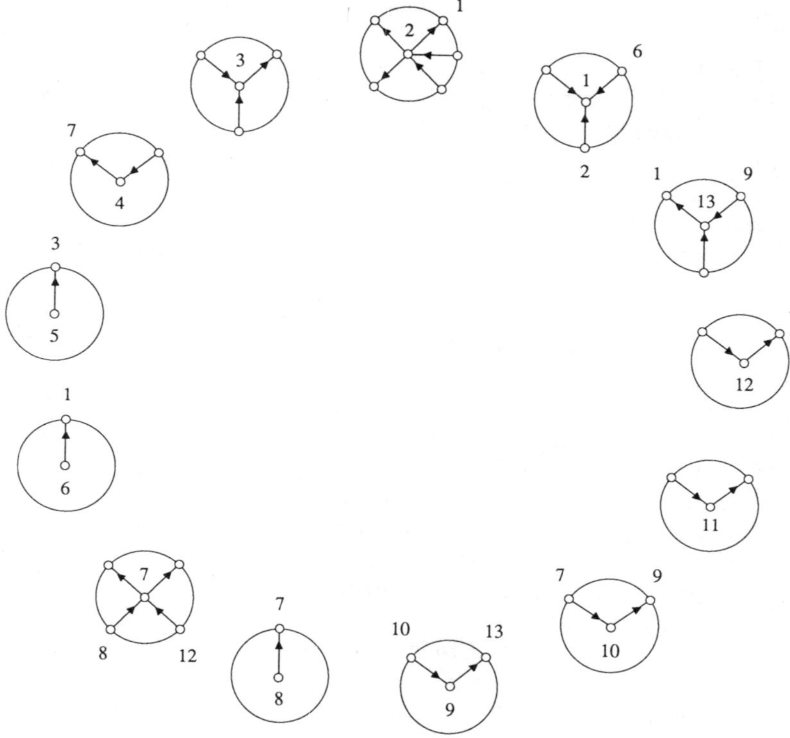

FIG. 4.14. A Weiner diagram of Trobriand yam exchange

pictures' is drawn as a digraph. The numbers represent men of the hamlet, and the arrows a yam exchange garden made by one man for another. Number 1 is the hamlet manager. Unnumbered men are presumably outside of the hamlet.[3]

Weiner's diagram is entitled 'Circulation of yam exchange gardens in one hamlet'. However, the pattern of exchange is not cyclical as the title and layout suggest, but hierarchical instead. Weiner describes this hierarchical relation in three different statements:

[3] Although it is not numbered, Weiner's diagram appears to show an arrow reaching from 13 to 1. A diagram such as this one is not a very systematic presentation. It sometimes shows i giving to j and j receiving from i, but at other times it shows only one of these relations.

(1) Every man who lives in a hamlet (whether or not it is his own *dala*'s [matrilineal subclan's] land) must prepare a yam garden for the hamlet manager (1976: 42).
(2) Whether men reside on the land of their own *dala* or on that of someone else's *dala*, they must validate the right to live on that land by producing a yam exchange garden for either the hamlet manager or a linking intermediary male (1976: 146).
(3) Schematically (see Fig. [4.14]) a hamlet is composed of individual clusters in which younger men are linked to one older man. Each man who is head of a cluster is linked directly to the hamlet manager (1976: 147).

In order to see whether these statements are actually true, one would either have to connect the dots in the diagram or else put it into matrix form. We shall use the latter method, as a convenient illustration of how structural properties of large exchange graphs may be found from their associated adjacency, reachability, and distance matrices. In this particular case one could imagine how the method would facilitate the comparative study of a large number of hamlet exchange digraphs.

In the adjacency matrix $A(D)$ of the yam exchange garden digraph, each male resident of the hamlet has a row and a column, and the i, j entries are 1 if i makes a garden for j and 0 if not. By definition this is a *sparse matrix*, that is, it has a high proportion of zeros. From $A(D)$ we conclude that it is not true that every man makes a garden for the hamlet manager, because not all the entries in 1's column are 1.

In the reachability matrix $R(D)$, which is less sparse than $A(D)$, the i, j entries are 1 if i makes a garden for j or if i makes a garden for some k who makes a garden for j, etc. From $R(D)$ we conclude that it is not true that all men produce a yam exchange garden for either the hamlet manager or a linking intermediary male, because not all entries in 1's column are 1.

In the distance matrix $N(D)$ the i,j entries are 1 if i gardens for j, 2 if i gardens for k who gardens for j, and so on, with entries of 3, 4, . . ., n depending on the number of intermediary links. Note that each ∞ entry of $N(D)$ is associated with a 0 entry of $R(D)$. From $N(D)$ we conclude that it is not true that all heads of clusters (older men who are immediately linked to one or more younger men) are directly linked to the hamlet manager, because not all finite entries are 0, 1, or 2.

$$A(D) = \begin{array}{c} \\ 1\\2\\3\\4\\5\\6\\7\\8\\9\\10\\11\\12\\13 \end{array} \begin{array}{c} \begin{array}{ccccccccccccc} 1 & 2 & 3 & 4 & 5 & 6 & 7 & 8 & 9 & 10 & 11 & 12 & 13 \end{array} \\ \left[\begin{array}{ccccccccccccc}
0 & 0 & 0 & 0 & 0 & 0 & 0 & 0 & 0 & 0 & 0 & 0 & 0 \\
1 & 0 & 0 & 0 & 0 & 0 & 0 & 0 & 0 & 0 & 0 & 0 & 0 \\
0 & 0 & 0 & 0 & 0 & 0 & 0 & 0 & 0 & 0 & 0 & 0 & 0 \\
0 & 0 & 0 & 0 & 0 & 0 & 1 & 0 & 0 & 0 & 0 & 0 & 0 \\
0 & 0 & 1 & 0 & 0 & 0 & 0 & 0 & 0 & 0 & 0 & 0 & 0 \\
1 & 0 & 0 & 0 & 0 & 0 & 0 & 0 & 0 & 0 & 0 & 0 & 0 \\
0 & 0 & 0 & 0 & 0 & 0 & 0 & 0 & 0 & 1 & 0 & 0 & 0 \\
0 & 0 & 0 & 0 & 0 & 0 & 1 & 0 & 0 & 0 & 0 & 0 & 0 \\
0 & 0 & 0 & 0 & 0 & 0 & 0 & 0 & 0 & 0 & 0 & 0 & 1 \\
0 & 0 & 0 & 0 & 0 & 0 & 0 & 0 & 1 & 0 & 0 & 0 & 0 \\
0 & 0 & 0 & 0 & 0 & 0 & 0 & 0 & 0 & 0 & 0 & 0 & 0 \\
0 & 0 & 0 & 0 & 0 & 0 & 1 & 0 & 0 & 0 & 0 & 0 & 0 \\
1 & 0 & 0 & 0 & 0 & 0 & 0 & 0 & 0 & 0 & 0 & 0 & 0
\end{array}\right] \end{array}$$

Although matrix analysis does not confirm Weiner's assertions about the structure of the yam exchange garden relation, it does suggest an interesting comparative possibility for social structures of this type.

In a digraph D, the *reachable set* $R(v)$ of a point v is the collection of points reachable from v. In the reachability matrix $R(D)$ this set is given by the entries of 1 in v's row. Trobriand society is divided into hamlets each headed by a manager. In political affairs the power of a manager presumably depends on the control he is able to exert over his own hamlet. If the converse of the subordinate gardening relation is interpreted as a superordinate influence relation, then the hamlet manager's row entries in the transpose R' of R would indicate the extent of his control over the men of the hamlet. More powerful hamlet managers should have larger reachable sets.

$R(D) =$

	1	2	3	4	5	6	7	8	9	10	11	12	13
1	1	0	0	0	0	0	0	0	0	0	0	0	0
2	1	1	0	0	0	0	0	0	0	0	0	0	0
3	0	0	1	0	0	0	0	0	0	0	0	0	0
4	1	0	0	1	0	0	1	0	1	1	0	0	1
5	0	0	1	0	1	0	0	0	0	0	0	0	0
6	1	0	0	0	0	1	0	0	0	0	0	0	0
7	1	0	0	0	0	0	1	0	1	1	0	0	1
8	1	0	0	0	0	0	1	1	1	1	0	0	1
9	1	0	0	0	0	0	0	0	1	0	0	0	1
10	1	0	0	0	0	0	0	0	1	1	0	0	1
11	0	0	0	0	0	0	0	0	0	0	1	0	0
12	1	0	0	0	0	0	1	0	1	1	0	1	1
13	1	0	0	0	0	0	0	0	0	0	0	0	1

Among the Orokaiva in the Northern District of Papua the staple crop is taro. Although this plant cannot be stored as long as yams and thus lacks equivalent investment potential, there is none the less a fundamental cultural distinction between gifts of raw and cooked taro. According to Schwimmer (1973) raw taro is given by men in feasts to start or transform a social relationship. Large gifts of raw taro symbolize a feast giver's strength and superiority at the same time as they express his desire for amicable relations. Small but frequent gifts of cooked taro, on the other hand, are given by women on behalf of their households as an expression of a desire to maintain intimate relations. The frequency of taro gifts is an indication of the degree of intimacy between households. In a large village it would obviously be materially impossible for every household to maintain intimate relations with all other households;

$N(D) =$

	1	2	3	4	5	6	7	8	9	10	11	12	13
1	0	∞	∞	∞	∞	∞	∞	∞	∞	∞	∞	∞	∞
2	1	0	∞	∞	∞	∞	∞	∞	∞	∞	∞	∞	∞
3	∞	∞	0	∞	∞	∞	∞	∞	∞	∞	∞	∞	∞
4	5	∞	∞	0	∞	∞	1	∞	3	2	∞	∞	4
5	∞	∞	1	∞	0	∞	∞	∞	∞	∞	∞	∞	∞
6	1	∞	∞	∞	∞	0	∞	∞	∞	∞	∞	∞	∞
7	4	∞	∞	∞	∞	∞	0	∞	2	1	∞	∞	3
8	5	∞	∞	∞	∞	∞	1	0	3	2	∞	∞	4
9	2	∞	∞	∞	∞	∞	∞	∞	0	∞	∞	∞	1
10	3	∞	∞	∞	∞	∞	∞	∞	1	0	∞	∞	2
11	∞	∞	∞	∞	∞	∞	∞	∞	∞	∞	0	∞	∞
12	5	∞	∞	∞	∞	∞	1	∞	3	2	∞	0	4
13	1	∞	∞	∞	∞	∞	∞	∞	∞	∞	∞	∞	0

instead, each household has a restricted set of intimates or preferential exchange partners. This restriction on the number of partners leads Schwimmer to pose a structural question: how can a limited number of dyadic pairs combine to create a 'circuit of mediation'? In the village of Sivepe studied by Schwimmer,

> The simplest way in which 22 households can be joined in intimate association by a circuit of mediation would be if there were intimacy between households 1 and 2, 2 and 3, 3 and 4, and so on, the chain being closed by an intimate association between households 22 and 1. Households 1 and 3 might not maintain an intimate association, but they would share

intimacy with 2 and this would serve to mediate relationships between them. In theory such circuits could be unlimited in length, though the system would be rather cumbersome and expensive if extended too far (1973: 113).

Schwimmer regards his circuit of mediation as a mathematical model resembling a Hamiltonian cycle in graph theory. As in a digraph D, a *Hamiltonian cycle of a graph G* is a spanning cycle, one which contains all the points of G, and a *Hamiltonian graph* contains a Hamiltonian cycle, for example a complete graph. So Schwimmer's designation of his model is correct. This, however, seems much too stringent a condition for a structure of mediated intimacy: it severely and unrealistically restricts the particular partnership choices of each household, which would undoubtedly operate in at least partial disregard of global considerations. Common clanship also inflects the structure of exchange. We therefore suggest a path instead of a circuit of mediation. This would require only that each household be able to reach all others through a succession of mediators, and that its graph therefore be connected. In such a graph there may or may not be Hamiltonian cycles.

There are two tasks. The first is to define the graph G generated by preferential exchange partners, and the second is to determine whether G is connected. On the basis of a detailed quantitative analysis of taro exchanges, Schwimmer discriminates for each of the 22 households in the village a first, second, and third preferential partner, as summarized and listed in order in Table 4.4. The letters in parentheses following each household give its clan membership, Sorovi (V), Seho (E), or Jegase (J).

Schwimmer then uses these preferential partnerships to construct a graph of intimate links between households. His method is implicitly set theoretic, based on the operations of the union and intersection of digraphs. Because the procedure is only implicit and probably for that reason contains ambiguities, and because it has general application in defining the threshold or existence of an exchange relation, we shall make it explicit by stating a theorem (from Copi 1954) on set theoretic operations on the adjacency matrices of digraphs.

Theorem 4.6. Let D_1 and D_2 be two digraphs having the same set of labelled points, with adjacency matrices A_1 and A_2. Then the adjacency matrices of the intersection, union, and symmetric difference digraphs are as follows:

$$A(D_1 \cap D_2) = A_1 \times A_2,$$
$$A(D_1 \cup D_2) = (A_1 + A_2)\#,$$
$$A(D_1 \oplus D_2) = (A_1 + A_2)\# - (A_1 \times A_2).$$

The theorem is illustrated for two digraphs D_1 and D_2 in Fig. 4.15.

The list in Table 4.4 contains three separate digraphs: D_1, D_2, and D_3, consisting of the first, second, and third partners chosen by each household. According to Schwimmer, first and second partners are in a privileged position in comparison to others, averaging 62 per cent of all taro transactions. He therefore distinguishes a 'closest

TABLE 4.4. *Preferential taro exchange partners in Sivepe (from Schwimmer 1973)*

Household no. (clan)	First partner	Second partner	Third partner
1 (V)	17	2	5
2 (V)	1	3	17
3 (V)	2	17	4
4 (V)	5	3	7
5 (J)	4	6	21
6 (J)	7	5	4
7 (J)	6	19	4
8 (J)	9	10	7
9 (E)	8	10	13
10 (E)	8	9	11
11 (E)	12	20	18
12 (E)	13	14	11
13 (E)	14	12	9
14 (E)	13	12	15
15 (V)	14	16	12
16 (E)	17	15	7
17 (V)	1	16	18
18 (J)	22	17	11
19 (J)	7	22	12
20 (J)	21	11	5
21 (J)	20	5	11
22 (J)	18	19	17

Note: Household 18A in Schwimmer's Table vi/2 is relabelled as household 22 in Table 4.4.

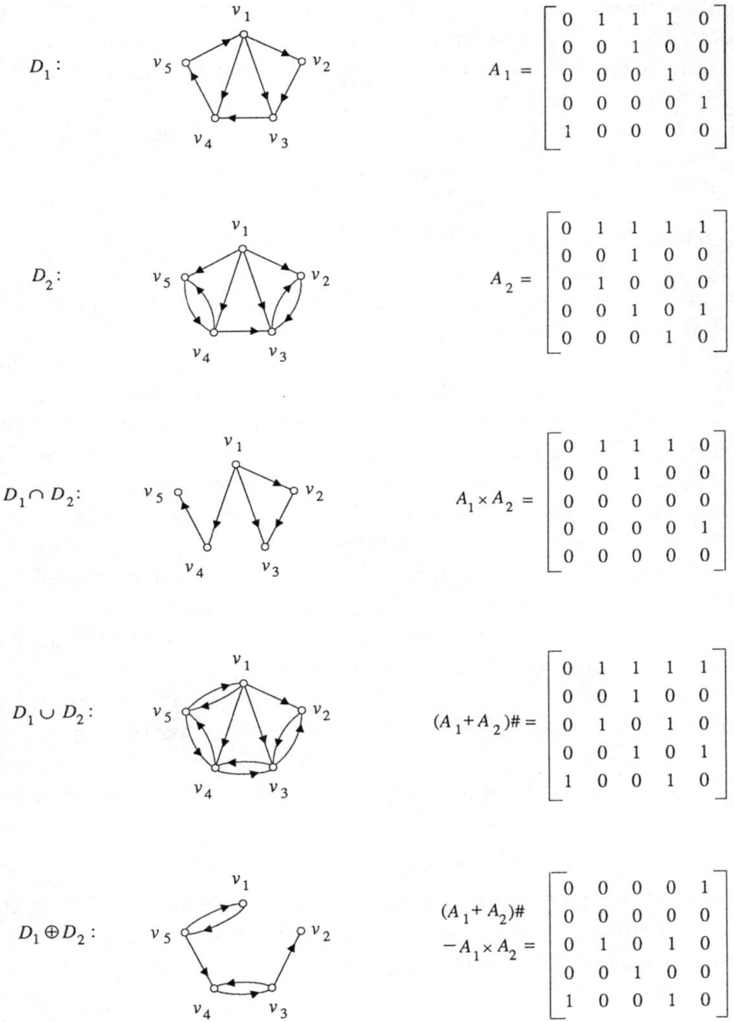

FIG. 4.15. Illustration of the adjacency matrices of the intersection, union, and symmetric difference of two labelled digraphs

mutual relationship' from others: 'Households A and B may be said to stand in a "closest mutual relationship" . . ., if B is a first or second partner to A and A is a first or second partner to B' (1973: 129). Representing the digraphs D_1 and D_2 by their adjacency matrices A_1 and A_2 and using the operation of the transpose, the definition of this exchange graph is

$$A(D_1 \cup D_2) \cap A(D_1' \cup D_2') = (A_1 + A_2)\# \times (A_1' + A_2')\#,$$

which is the matrix (with zeros omitted) on p. 147.

In order to tell whether this graph is connected, that is, whether all pairs of households can reach each other by a path of mediators, we form the reachability matrix $R(G)$. As not every entry in it is 1, $R(G) \neq J$, so the graph G is disconnected. Instead, G has five components, shown by the partitioning of R, each of which is dominated by common clan membership as shown by the clan symbols (V), (E), (J) to the left of the row numbers of $R(G)$, p. 148.

Using partitioned matrices, we can write $R(G)$ as on p. 149.

Schwimmer then relaxes his definition of the taro exchange network by including third preferential partnerships: those for which 'none, or only one of [the above] conditions are satisfied . . .' (1973: 129). This definition is ambiguous as stated and does not appear to have any consistent application. It includes all cases in which B is a non-reciprocated first or second choice of A, but it does not consistently include or exclude cases in which A and B are reciprocal or non-reciprocal third choices. If we disregard the distinction between first, second, and third choices and require only a symmetrical choice of any kind, the definition of the new graph is

$$A(D_1 \cup D_2 \cup D_3) \cap A(D_1' \cup D_2' \cup D_3') =$$
$$(A_1 + A_2 + A_3)\# \times (A_1' + A_2' + A_3')\#.$$

The result (not shown) is a universal reachability matrix, $R = J$, the matrix consisting entirely of 1s, and so the graph is connected. All households can reach each other by a path of reciprocal links.

Nearest Neighbours and the Evolution of Trade Networks

In many networks, exchange is strongly influenced by the nearest neighbour relation. For example, preferential partners in the Orokaiva system just described often live in adjacent dwellings. Recently in archaeology and economic anthropology the nearest

	1	2	3	4	5	6	7	8	9	10	11	12	13	14	15	16	17	18	19	20	21	22
1		1															1					
2	1		1																			
3		1																				
4					1																	
5				1		1																
6					1		1															
7						1													1			
8									1	1												
9								1		1												
10								1	1													
11																				1		
12													1	1								
13												1		1								
14												1	1									
15																1						
16															1		1					
17	1															1						
18																						1
19							1															1
20											1										1	
21																				1		
22																		1	1			

		1	2	3	15	16	17	4	5	6	7	18	19	22	8	9	10	12	13	14	11	20	21
(V)	1	1	1	1	1	1	1	0	0	0	0	0	0	0	0	0	0	0	0	0	0	0	0
(V)	2	1	1	1	1	1	1	0	0	0	0	0	0	0	0	0	0	0	0	0	0	0	0
(V)	3	1	1	1	1	1	1	0	0	0	0	0	0	0	0	0	0	0	0	0	0	0	0
(V)	15	1	1	1	1	1	1	0	0	0	0	0	0	0	0	0	0	0	0	0	0	0	0
(E)	16	1	1	1	1	1	1	0	0	0	0	0	0	0	0	0	0	0	0	0	0	0	0
(V)	17	1	1	1	1	1	1	0	0	0	0	0	0	0	0	0	0	0	0	0	0	0	0
(V)	4	0	0	0	0	0	0	1	1	1	1	1	1	1	0	0	0	0	0	0	0	0	0
(J)	5	0	0	0	0	0	0	1	1	1	1	1	1	1	0	0	0	0	0	0	0	0	0
(J)	6	0	0	0	0	0	0	1	1	1	1	1	1	1	0	0	0	0	0	0	0	0	0
(J)	7	0	0	0	0	0	0	1	1	1	1	1	1	1	0	0	0	0	0	0	0	0	0
(J)	18	0	0	0	0	0	0	1	1	1	1	1	1	1	0	0	0	0	0	0	0	0	0
(J)	19	0	0	0	0	0	0	1	1	1	1	1	1	1	0	0	0	0	0	0	0	0	0
(J)	22	0	0	0	0	0	0	1	1	1	1	1	1	1	0	0	0	0	0	0	0	0	0
(J)	8	0	0	0	0	0	0	0	0	0	0	0	0	0	1	1	1	0	0	0	0	0	0
(E)	9	0	0	0	0	0	0	0	0	0	0	0	0	0	1	1	1	0	0	0	0	0	0
(E)	10	0	0	0	0	0	0	0	0	0	0	0	0	0	1	1	1	0	0	0	0	0	0
(E)	12	0	0	0	0	0	0	0	0	0	0	0	0	0	0	0	0	1	1	1	0	0	0
(E)	13	0	0	0	0	0	0	0	0	0	0	0	0	0	0	0	0	1	1	1	0	0	0
(E)	14	0	0	0	0	0	0	0	0	0	0	0	0	0	0	0	0	1	1	1	0	0	0
(E)	11	0	0	0	0	0	0	0	0	0	0	0	0	0	0	0	0	0	0	0	1	1	1
(J)	20	0	0	0	0	0	0	0	0	0	0	0	0	0	0	0	0	0	0	0	1	1	1
(J)	21	0	0	0	0	0	0	0	0	0	0	0	0	0	0	0	0	0	0	0	1	1	1

neighbour relation has been used to construct hypothetical networks of community interaction and models of network origin. The proximal point method, as it is called in archaeology, consists of joining each member of a set of sites or settlements to its nearest neighbour (NN), its second NN, its third NN, etc., to obtain a graph of the entire system. Terrell (1977), reasoning from gravity models in

$$R(G) = \begin{bmatrix} J_6 & 0 & 0 & 0 & 0 \\ 0 & J_7 & 0 & 0 & 0 \\ 0 & 0 & J_3 & 0 & 0 \\ 0 & 0 & 0 & J_3 & 0 \\ 0 & 0 & 0 & 0 & J_3 \end{bmatrix}$$

geography in which human interaction is analysed as a function of the relative size and nearness of population units, constructs a proximal point network graph of islands and parts of islands in the Solomons as a means for studying probable direction of trade, migration, and marriage. Irwin (1983) uses this technique to demonstrate the effects of distance and direction on links between communities of the *kula* ring.

Nearest neighbour models are eminently graph theoretic. For clarity of definition and consistency of application it is worth making this explicit. It is customary to assume that all distances are different, to avoid having to deal with ties. The nearest neighbour relation is non-symmetric: even if X is the nearest neighbour of Y, it need not follow that Y is the nearest neighbour of X. Depending on one's purpose, an undirected graph G based on the nearest neighbour relation can be constructed either from $D \cap D'$, or in matrix terms $A \times A'$, in which case i and j must be mutually nearest neighbours, or from $D \cup D'$ whose adjacency matrix is $(A + A')\#$, in which case they need not be. Irwin's application of a proximal point graph is not consistently based on either definition and is therefore ambiguous.

In an important contribution to the study of trade network origins, B. Grofman and J. T. Landa (1983) use the nearest neighbour relation to model hypothetical developmental sequences in network formation among spatially separated traders. Their approach differs from global strategies that attempt to specify probable trade routes through the application of optimal routing algorithms, by positing instead that trade routes arise through pairwise independent choices. It differs from proximal point methods by specifying that choice based on the nearest neighbour relation is symmetrical and expanding.

The Grofman and Landa model, called sequential proto-coalition formation, is based on the idea of overlapping memberships between sets of trading communities. It is applied as follows. Begin with the adjacency matrix A of a nearest neighbour digraph D. If points i and j are mutual nearest neighbours, as shown by the entries of 1 in $A \times A'$, they become permanent trading partners and form a 'proto-coalition'. In the first stage all proto-coalitions will be dyads or singletons. By dyads in this first stage are meant edges (lines) of a graph, and by singletons, isolated points. Ties cannot occur, because of the empirically valid assumption that all distances are different.

In the second stage new proto-coalitions are generated. These new subgraphs, being obtained from independent edges and isolated points by adding new edges, must still have connected components which are paths of length 2 or 3. Whenever proto-coalitions I and J from the first stage are nearest neighbours of each other as determined by the distance between individual traders i and j in I and J, they form a new proto-coalition K. This process is continued until all traders are part of a single coalition, that is, until the graph of the trading relation is connected; see Fig. 4.16a, which is similar to but simpler than Map 3 of Grofman and Landa. (The trading communities are 1–5 from left to right.)

The numbered edges in Fig. 4.16b show that at the first stage in Fig. 4.16a we have two edges and one isolate, and at the second stage we have the edge on the left and the 2-step path on the right with edges marked 1 and 2. In this particular example the third stage adds edge 3 in Fig. 4.16b and completes a single coalition consisting of a single path. In general, however, at the third and later stages it is not necessary to have components which are paths, as we shall see below.

Grofman and Landa apply this model to the developmental sequence of links leading to the emergence of the *kula* ring. Their application takes account of the roughly circular or elliptical arrangement of *kula* communities, and it makes two crucial empirical assumptions. The first is that every community is a middleman between communities on each side of it and thus has something that they want. This assumption seems fully justified. To give a single example from Malinowski, the Amphlett Islanders give

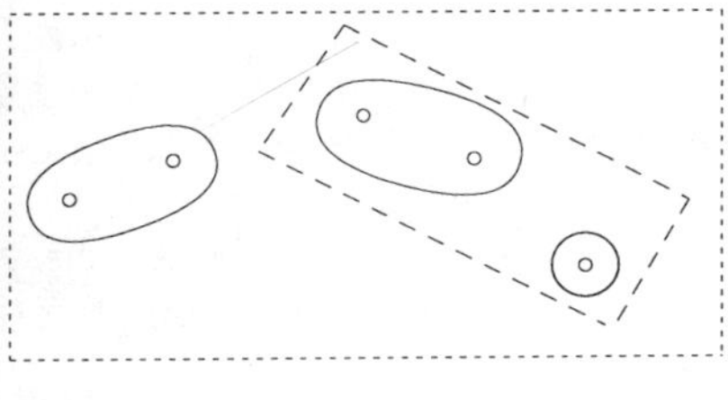

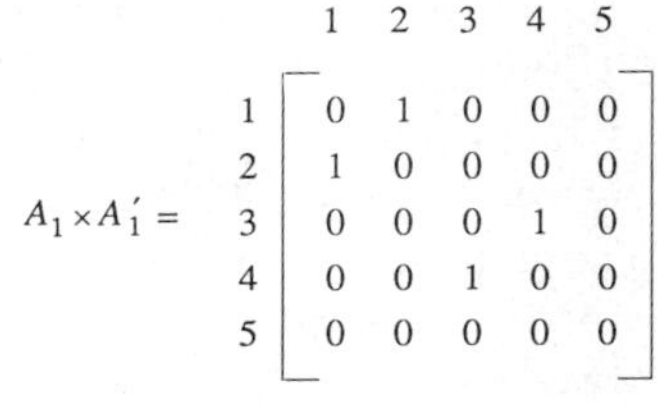

$$A_1 \times A_1' = \begin{array}{c} \\ 1 \\ 2 \\ 3 \\ 4 \\ 5 \end{array} \begin{array}{c} \begin{array}{ccccc} 1 & 2 & 3 & 4 & 5 \end{array} \\ \begin{bmatrix} 0 & 1 & 0 & 0 & 0 \\ 1 & 0 & 0 & 0 & 0 \\ 0 & 0 & 0 & 1 & 0 \\ 0 & 0 & 1 & 0 & 0 \\ 0 & 0 & 0 & 0 & 0 \end{bmatrix} \end{array}$$

(a)

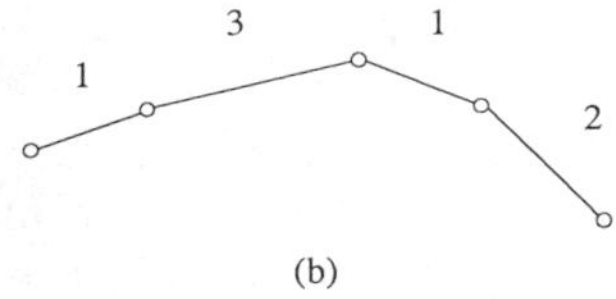

(b)

FIG. 4.16. A graph and matrix to illustrate the formation of proto-coalitions: solid lines = first stage; broken lines = second stage; dotted lines = third stage (adapted from Grofman and Landa 1983)

pots, tortoise shell ear-rings, red ochre, and pumice stone, all locally obtained, to their immediate northern neighbours in the Trobriand Islands in exchange for sago, pigs, yams, and taro, but they also give products obtained from their immediate southern neighbours in Dobu such as seeds for necklaces, rattan, and red parrot feathers, which Trobrianders pass on to get stone for tools, which they obtain indirectly from Woodlark Island in the east. The second assumption is that there is an 'inner' (*Ur-*) *kula* ring. This is likely, although its exact membership is a matter of some conjecture.

Referring back to Malinowski's graph reproduced in Fig. 1.5, Grofman and Landa consider eleven communities as members of an inner, original *kula* ring. These are: Vakuta, Sinaketa, Kiriwina, Kitava, Iwa and Gawa in the Marshall Bennett Islands, Woodlark (Kaurai), Tubetube, SE Dobu, Dobu, and NW Dobu. The proto-coalitions in the first stage are the two edges (Sinaketa, Kiriwina) and (NW Dobu, Dobu), with all others singletons. In the second stage Vakuta is joined to the first of these edges through Sinaketa, and SE Dobu to the second through Dobu. By the seventh stage all communities are in one coalition, with a path joining Tubetube, SE Dobu, Dobu, NW Dobu, Vakuta, Sinaketa, Kiriwina, Kitava, Iwa, Gawa, Woodlark. The graph consists of one path and is thus connected, so the process stops. To get a ring (cycle) Tubetube is joined to Woodlark. One can similarly construct a cycle for the illustration of Fig. 4.16a, by adding a new edge joining the two end points of the path in Fig. 4.16b.

Grofman and Landa's sequential proto-coalition model must necessarily succeed in connecting all members of the *kula* ring. The previously used so-called symmetrical nearest neighbour model is merely their first stage, which we have seen yields independent edges and isolated points. Most of the specific links in Grofman and Landa's graph are contained in Malinowski's graph, but not all. Our main empirical objection to this application is that the designation of the communities in the inner ring is arbitrary. For example, it includes Kiriwina, which Malinowski (1922), reasoning from the distribution of *kula* myths, considered a later entry into the ring, and it excludes the Amphletts, which Malinowski considered an original participant.[4] Geographically, Kiriwina is also outer. It also omits the two communities of Wawela and Okayaulo, not shown on Malinowski's map or on Uberoi's (1971) map, the one used by Grofman and Landa, but described in Malinowski's text, which were between Sinaketa and Kitava and joined to each as *kula* partners.

We therefore propose a modification of the proto-coalition model which makes no assumptions about an inner and outer ring. The diagram in Fig. 4.17, adapted from Irwin (1983), shows the approximate boundaries of each *kula* community, consisting of an island, a part of an island, or a group of islands. It treats Gawa and Iwa as a single point (the Marshall Bennetts) and it includes the communities of Wawela and Okayaulo. The lines joining these communities are those produced by the proto-coalition model.

It can be seen that, contrary to what Grofman and Landa seem to imply, the proto-coalition model will not always generate a ring, that is, a cycle obtained from a path. If in Fig. 4.17 Misima and Laughlan are joined by a line the result will be a cycle in graph G, but G will not consist of a single cycle. The proto-coalition graph will be

[4] Landa (1983: 142) states that 'it is only this Inner Kula Ring which is described in great detail by Malinowski'. But in fact Malinowski devotes an entire chapter of *Argonauts* to the Amphletts.

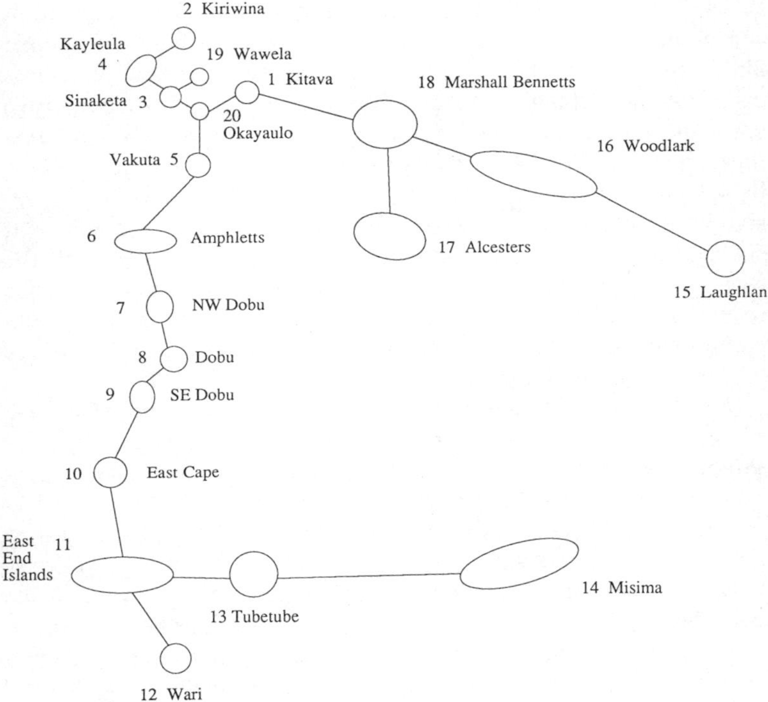

FIG. 4.17. A modified proto-coalition model of the original *kula* ring

a tree, but not necessarily one consisting of a single path: it may 'branch' like *G* in Fig. 4.17. We note that Grofman and Landa's own graph should actually branch. According to their mileage chart of distances between *kula* communities, Kitava should be joined to Sinaketa not Kiriwina, thus producing a graph in which Sinaketa has degree 3, being adjacent with Kitava, Vakuta, and Kiriwina.[5]

Evolutionary Trees in Exchange Networks

Although Grofman and Landa's proto-coalition model will not generate the *kula* ring, it will undoubtedly prove applicable to other

[5] But Malinowski's (1922) diagram of the *kula* ring in *Argonauts* and Grofman and Landa's map show Kitava closer to Kiriwina than to Sinaketa.

types of exchange and communication networks. Their algorithm can actually be regarded as an independent discovery of a third alternative solution to the minimum spanning tree problem, as it is called in operational research. This problem arises in numerous contexts. Historically one of them involved the question of how much per month to charge a large company with many branches for its telephone network. The company correctly argued that they should not be billed for the total distance in a complete graph, as it suffices to connect its various branches by joining them with the structure of a spanning tree of minimum total length. Similar reasoning can be used in modelling the evolution of exchange and communication networks either when the network consists of a tree or when it originated as one.

We illustrate the alternative solutions for the example of five islands (u, v, w, x, y) shown in Fig. 4.18a, with relative distances as shown, by constructing a minimum spanning tree using three different methods. These could refer to three different historical processes. All three algorithms can be called 'greedy', because they bite off the most desirable pieces (shortest lines) of the network.

Kruskal's (1956) algorithm begins by drawing line 1 joining the two islands which have the shortest possible distance between them, i.e. they are the nearest of the nearest neighbour pairs. This is shown as line 1 in Fig. 4.18b, which joins islands u and v of Fig. 4.18a. We now have four components; three are the still isolated islands w, x, y and the fourth is the line uv. The second step again joins two components which have the shortest possible distance between them. This turns out to be x and y, shown as line 2 in Fig. 4.18b. We now have three components: the isolate w and the two lines uv and xy. Repeating this process we join v and w as line 3. Finally, and similarly, line 4 is xv.

Prim's (1957) algorithm, published just one year later, begins with an arbitrary island, say u without loss of generality. (In practice one might choose an island with specified physical or social characteristics.) Prim begins with the shortest line (minimum distance) joining u to another island, in this case v. Then the second step continues to construct a connected pattern by joining either u or v to one of the still isolated islands by the shortest possible line, which is vw shown as line 2 in Fig. 4.18c. Lines 3 and 4, vx and xy, then complete a minimum spanning tree which is precisely the same

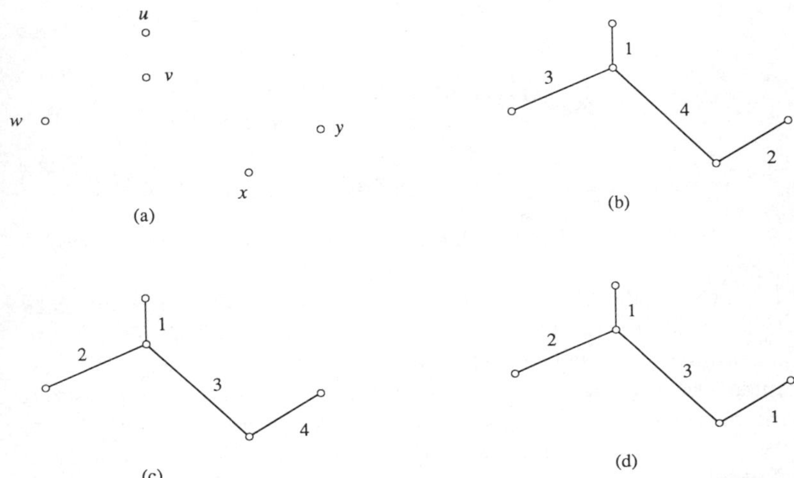

FIG. 4.18. Illustration of the minimum spanning tree algorithms of Kruskal, Prim, and Grofman and Landa

tree as given in Fig. 4.18b by Kruskal's algorithm, although the order of selecting its lines is different.

Grofman and Landa's (1983) algorithm illustrates a third way to construct the same minimum spanning tree. At their first stage they get both the lines *uv* and *xy*, as these are the two pairs of mutually nearest neighbours. Their second and third stages are as shown in Fig. 4.18d.

Recent historical research has unearthed the fact that Boruvka (1926a, b) already anticipated the algorithms of Kruskal and Prim with his discovery of the third algorithm. More remarkable still is Graham and Hell's (1985) report that preliminary work on the minimum spanning tree problem was done by the physical anthropologist Czekanowski (1909, 1911) in research on classification.

In the next chapter we continue our analysis of the *kula* ring, modelling it as a numerical network and using matrix methods to simulate the flow of valuables.

5

Markov Chains

> The saying, that a chain is not stronger than its weakest link does not, let us hope, apply to Ethnology.
>
> Bronislaw Malinowski, *Argonauts of the Western Pacific*

In the classification of ceremonial exchange in Highland New Guinea we proposed the model of a non-numerical network to accommodate structures with multiple exchange relations. We now introduce a special case of a numerical network known as a Markov chain to simulate the flow of wealth through an exchange system. We use a chain to estimate the distribution of valuables in different communities of the *kula* ring for the purpose of showing the relation between access to wealth and the development of political hierarchy. In the Carolines the closeness of one island to all other islands and its position between all pairs of islands are structural determinants of economic and political dominance. In the *kula* ring we conjecture that the position of a community in the circulation of wealth is a structural determinant of internal stratification.

Wealth and Hierarchy in the *Kula* Ring

In the recent updating of the Malinowski–Fortune ethnography of the *kula* ring (Leach and Leach 1983), two uncertainties remain concerning the operation of this system of exchange: (1) How are the valuables distributed over the individual *kula* communities? (2) Does the *kula* ring exist, that is, how unified are the rules governing the exchange of valuables? Both questions, the one theoretical and the other ethnographic, are basic to evolutionary theories of chieftainship in the Massim. We proceed as follows. We first provide a structural model for simulating the flow and predicting the distribution of valuables in the *kula* ring, and, by implication, in any similar network. Instead of only modelling the *kula* ring by an undirected graph and trying to guess the distribution of valuables by inspecting individual lines in it, we treat it as a regular Markov chain and deduce this distribution from the entries of its limiting probability vector. Then, we use these results to generalize Brunton's (1975) theory that derives chieftainship in the *kula* ring from restricted access to valuables. Our suggestion is that either of the extreme conditions of scarcity or abundance of valuables could facilitate the emergence of political hierarchy, depending on the associated ecological circumstances of a society and the particular manner in which it defines *kula* exchange. We conclude with an evolutionary speculation on the *Ur-kula* ring.

The Kula *Ring*

The following ethnographic facts are relevant to our analysis. As described by Malinowski the *kula* ring consists of some 20 communities joined in a 'circuit' off the eastern tip of New Guinea. A *kula* community, consisting of an island, a part of an island, or a group of islands, is defined as '. . . a village or a number of villages, who go out together on big overseas expeditions, and who act as a body in the *kula* transactions, perform their magic in common, have common leaders, and have the same outer and inner social sphere, within which they exchange their valuables' (Malinowski 1922: 103). Pairs of communities periodically exchange two complementary classes of shell valuables (*vaygu'a*): armshells (*mwali*) and necklaces (*soulava*), conceived of as 'female' and 'male' respectively.[1] The necklaces flow clockwise and the armshells counter-clockwise around the ring, and the rule of exchange is that a necklace can only be given for an armshell and conversely.[2] A completed exchange is a 'marriage'. The exchanges consist of an opening gift (*vaga*) and a restoration gift (*yotile*) given in a delayed format: A visits B to receive armshells (or necklaces), and later in the year B visits A to receive necklaces (or armshells).[3] The valuables are roughly ranked,

[1] Reversals of this symbolism are mentioned in Leach and Leach (1983).

[2] Malinowski mentions that at one time boars' tusks accompanied the armshells.

[3] The classification of gifts on the basis of gender and their exchange in opposite directions is common in Oceania. In Samoa, for example, 'female property' (*toga*)—fine mats—is exchanged for 'male property' (*'oloa*)—raw food, canoes, houses, cloth, and other items—between the 'side of the bride' and the 'side of the groom' and between chiefs (*ali'i*) and talking chiefs (*tulāfale*) (Turner 1861, 1884; Mauss 1950; Mead 1969 [1930]; Shore 1982). Landa (1983) in her analysis of the *kula* ring, and Shore in his analysis of Samoan social organization, emphasize the stabilizing effects of such complementary exchanges. Landa bases her interpretation on the 'economics of signalling' and Shore bases his on Bateson's (1958) model of 'complementary schismogenesis'.

but there is no haggling; every pair of exchange partners operates on the basis of their own sense of equivalence. In this, *kula* exchange is sharply distinguished from barter (*gimwali*). In the Trobriands, at least, economic exchange does not enter into ceremonial exchange, but runs parallel to it. The valuables cannot be held permanently but must circulate endlessly around the ring. Temporary possession, however, is 'exhilarating, comforting, soothing in itself' (Malinowski 1922: 512), and a source of immense individual prestige. In the Trobriands participation is monopolized by chiefs; in Dobu (Fortune 1932) it is egalitarian.

The papers in the Cambridge symposium (Leach and Leach 1983) reveal a number of subtleties not mentioned in early accounts of *kula* exchange. Thus certain valuables, *kitoms*, are said to be owned, and may be removed or reintroduced into the flow at the owner's discretion. According to J. W. Leach (1983) this concept probably varies in definition from community to community. There are also *kula* paths, called *keda* or 'chains', along which valuables may ideally flow. Such paths, however, are multiple, fragile, and intersecting (Campbell 1983). We shall ignore these refinements and concentrate instead on the global properties of the system: the annual exchange of thousands of valuables (Firth 1983 estimates 3,000 of each type as a very minimal figure) between thousands of individuals (Malinowski 1922: 85).

Fig. 5.1 shows a graph G of the *kula* ring drawn in roughly geographical fashion. Each point of G represents a *kula* community and each line an exchange relation. It can be seen immediately, as noted in Chapter 1, that the *kula* ring consists of a number of different cycles along which the valuables flow.

Scarcity and Hierarchy

R. Brunton (1975) rejects explanations of chieftainship in the northern Trobriands, particularly Kiriwina (point 2 in the graph G of Fig. 5.1), based on population density (Uberoi 1971; Powell 1969) and agricultural productivity (Harris 1971), pointing out that chieftainship is absent in Dobu (points 7, 8, 9) where land is fertile, and 'hardly operative' in Vakuta (5) in the southern Trobriands, where population density is high. He proposes instead a theory based on a combination of network constraints, political logic, and geographical accident.

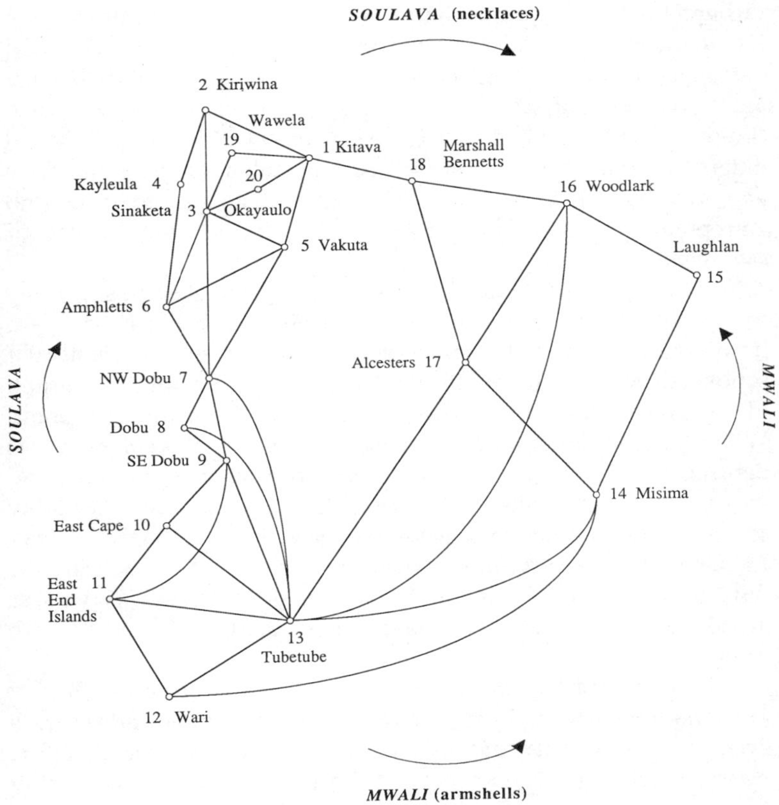

FIG. 5.1. The graph G of the *kula* ring

He focuses on the subgraph of G consisting of the points 1, 2, 3, 4, 5, 6, 7, 19, 20 together with all the lines between them. He observes that Kiriwina (2) is in a 'precarious position' in this structure, by which he means that Kiriwina can be bypassed in the flow of valuables (armshells) going from Kitava (1) to NW Dobu (7). This is the 'double ramification' consisting of 'long and short routes' between Kitava and Dobu noted by Malinowski, but ignored by some later commentators. Kiriwina is in fact only one of four recipients of valuables passing through Kitava.

Brunton conjectures that Kiriwina entered the *kula* ring late through intermarriage with Kitava or Sinaketa and Vakuta (men of Kiriwina would receive valuables in exchange for the food payments, *urigubu*, given to their sisters' husbands), and that the monopolization of scarce and already prestigious 'new resources' (Bailey 1969) by some men of Kiriwina provided the basis for social differentiation. Since Kiriwina lacked exportable resources and skills and was thus 'closed off' to other trading relations, there was no entrepreneurial basis for challenges to an emerging order of stratification.

The theory is unexpected but plausible, ingenious even, according to one of its critics (Irwin 1983).[4] Brunton's graphical analysis, however, from which all else follows, requires comment. In general it is not sufficient to estimate the relative position of a point in a graph like G simply on the basis of its adjacency from another point. Whatever one community receives in a round of *kula* exchange depends on what its neighbours receive from their neighbours, and so on around the ring. One would have to consider the entire structure of such a network, including the values assigned to its lines. In this particular case there is also the problem of considering the flow in the opposite direction, that is, of necklaces as well as armshells. There need not have been equal amounts of each category.

J. Persson (1983) proposes a general theory relating the flow of *kula* valuables to political organization, assuming, however, a cyclical relation between scarcity and abundance and therefore between hierarchical and egalitarian social structures. In support of his theory he compares the position of the Amphletts (6) and Tubetube (13). Neither is economically self-sufficient, but must specialize (in pottery manufacture and trade) to support their populations. According to Persson,

> The survival of these societies—perhaps even their birth and development—may be ascribed to the parallel and simultaneous exchange in the *kula* of valuables and articles of necessity. Since both of these groups must at the same time attract both flows, it is possible, I think, to derive their egalitarian tribal structure from a constant overabundance of *kula* items (1983: 44).

This is a puzzling comparison in the light of Malinowski's

[4] Irwin points out that Kiriwina does have one important exportable skill: the polishing of stone axe blanks received (indirectly) from Woodlark.

description of Tubetube as a prosperous commercial centre and 'point of convergence' for *kula* articles, and of the Amphletts as a community known for being 'stingy and unfair in all *kula* transactions, and as having no real sense of generosity and hospitality' (1922: 47). This description also does not accord with Macintyre's (1983b) report of chieftainship on Tubetube, as we shall see.

Unlike Brunton and Persson, Fortune (1932: 206) assumes an even distribution of valuables over the *kula* ring: 'It must be understood that the entire circuit does not revolve in a regular procession from one district to the next, then to the next, and so on, every year. Each year's overseas expeditions cease with an approximately equal number of armshells and spondylus shell necklaces everywhere.' Firth (1983: 96) imagines the valuables 'distributed, as it were in nodules, through the network'.

A Markov Chain Analysis

In order to estimate the distribution of both classes of valuables in all communities of the *kula* ring, we propose the structural model of a finite stationary Markov chain. It will be easiest to begin with our assumptions concerning the circulation of *kula* valuables, and then illustrate the model by a graph representing a miniature *kula* ring. The first three assumptions are the same as those made by Firth (1983) in his careful scrutiny of Malinowski and Fortune on quantitative aspects of *kula* exchange.

(1) Expeditions between *kula* communities take place on a regular seasonal basis. Both Malinowski (1922) and Fortune (1932) mention fixed dates and annual expeditions.

(2) The flows of the two classes of valuables are decoupled. As Firth notes, in spite of *kula* ideology of one-to-one matching equivalence of armshells and necklaces, it seems quite unlikely that the numbers of each were ever exactly equal. Mechanisms for flexibility in matching would include the delayed exchange format, intermediate or holding gifts (*basi*) given until an equivalent return can be made, fluctuating partnerships, and broad notions of equivalence as well as the ancillary non-*kula* gifts and the conversion of *kula* into non-*kula* items, as suggested by Firth.

(3) Each community retains some of its stock of valuables. Although Fortune wrote that communities exported all of their

armshells and necklaces every year, it seems likely that some were withheld for a time for internal exchange, for reasons of the 'sharp dealing' Fortune describes, or for attempts to maximize the exchange of especially valuable items.

(4) When one community has two or more other communities as partners, exchange with each is equiprobable. Other assumptions could perhaps be made, but the results suggest that this may not be an unreasonable one at least as a first approximation.

The following definitions are necessary. A *probability matrix* is a square matrix with non-negative entries whose row sums are all 1.

A *Markov chain*, or more briefly a *chain*, is a directed network N (with directed loops permitted) in which the value of each arc is a positive number, and the sum of the values of the arcs *from* each point is exactly 1. Thus the matrix of a chain, called its *transition matrix*, is a probability matrix, as illustrated in Fig. 5.2. (The values shown in Fig. 5.2 have been chosen arbitrarily, for purposes of illustration only.)

The points of a chain are called its *states*. The value of an arc x is the conditional probability that if the present state is fx, the first point of x, then the next state will be sx, the second point of x. The theory of Markov chains studies sequences of events together with a given distribution of initial probabilities that are used only to determine the initial state. Thus a chain has the single property that the probability of the next event in a sequence depends only on the present event and not on the preceding ones.

A *finite Markov chain* is simply a chain in which the number of states is finite. In the general definition of a Markov chain, the

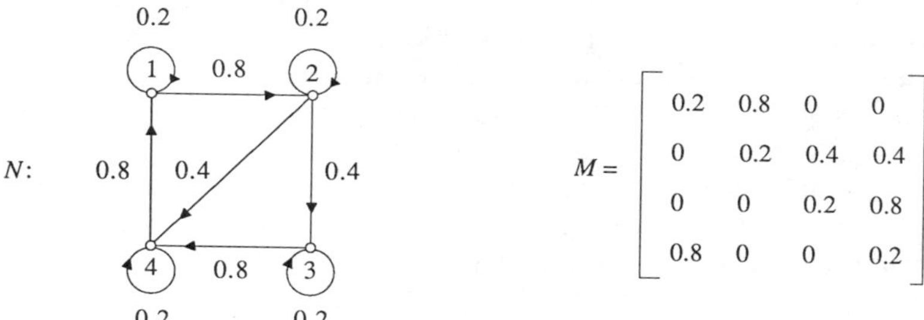

FIG. 5.2. A Markov chain N and its transition matrix M

probability value assigned to each arc is permitted to change with the passage of time. A *stationary chain* is one in which these probabilities are constant over time. We shall only be concerned with finite stationary Markov chains.

We can now simulate a *kula* ring. For purposes of exposition we shall first consider a model with only four points. Later we shall introduce a similar but larger model with 20 points corresponding to the exchanging communities in the *kula* ring. Let the network N in Fig. 5.2 describe the flow of one class of valuables, say necklaces, in a *kula* ring. Then state v_i means a particular valuable is in community v_i, and the arc $v_i v_j$ means the probability that a valuable in v_i will be given to v_j in a round of exchange—interpreted here as the proportion of the set of valuables now in v_i given by v_i to v_j.

A $1 \times n$ matrix is called a *vector of length n*. A *probability vector* $P = [p_1, p_2, \ldots, p_s]$ has non-negative entries and the sum of its entries is 1. A probability vector may be used to indicate the likelihood that each point is the first state to occur in the chain. In this case it is known as the *initial probability vector* of the chain. Let us represent the initial distribution of *kula* valuables in the four communities by $P_0 = [p_1, p_2, p_3, p_4] = [0.2, 0.4, 0.2, 0.2]$. From the product of P_0 and the matrix M we obtain the vector P_1 which gives the proportions of valuables after one round of exchange. The calculation $P_0M = P_1$ is as follows:

$$[0.2, 0.4, 0.2, 0.2] \begin{bmatrix} 0.2 & 0.8 & 0 & 0 \\ 0 & 0.2 & 0.4 & 0.4 \\ 0 & 0 & 0.2 & 0.8 \\ 0.8 & 0 & 0 & 0.2 \end{bmatrix} = [0.2, 0.24, 0.2, 0.36].$$

Assuming that the transition matrix remains constant, we can form the product P_1M to obtain P_2, the distribution of valuables after two rounds of exchange, and so on. If this process is continued, eventually P_{n+1} will be about the same as P_n, which means that the distribution of valuables after many rounds of exchange will remain about the same (no matter what the initial distribution might have been!). That is because this is a 'regular chain'. Two theorems from Harary *et al.* (1965) give a formal characterization of such a chain and a method of computation.

Theorem 5.1. Let M be the transition matrix of a given chain. Then in M^n the i, j entry is the nth transition probability from v_i to v_j.

In the matrix M in Fig. 5.2 the i, j entry of M^2 is the probability of a valuable going from v_i to v_j in exactly two steps. For example, the entry of m_{12} in M^2 is 0.32.

In a *regular chain* some power n of its transition matrix M is all positive, that is, M^n has no zero entries.

Theorem 5.2. For a regular chain, the powers of its transition matrix M approach a limit. This limit has all rows alike, and each is equal to the unique probability vector P for which $PM = P$.

The entries in M^8 of the transition matrix in Fig. 5.2, for example, are

$$\mathrm{M}^8 = \begin{bmatrix} 0.311 & 0.264 & 0.132 & 0.294 \\ 0.294 & 0.311 & 0.132 & 0.264 \\ 0.263 & 0.324 & 0.149 & 0.264 \\ 0.264 & 0.263 & 0.162 & 0.311 \end{bmatrix}$$

Theorem 5.2 states that the rows of the limit matrix are given by the probability vector for which $PM = P$. Since we know M, we can find the limit matrix by solving $PM = P$ for P. We know that $P = [p_1, p_2, \ldots, p_s]$ and that M is an $s \times s$ matrix. Therefore we have the following equations:

$$\begin{gathered} p_1 m_{11} + p_2 m_{21} + \ldots + p_s m_{s1} = p_1, \\ p_1 m_{12} + p_2 m_{22} + \ldots + p_s m_{s2} = p_2, \\ \vdots \\ p_1 m_{1s} + p_2 m_{2s} + \ldots + p_s m_{ss} = p_s, \end{gathered}$$

and since P is a probability vector, we have

$$p_1 + p_2 + \ldots + p_s = 1.$$

This last equation together with any $s - 1$ of the preceding ones gives us a system of s equations in s unknowns. By solving these simultaneously (it does not matter which $s - 1$ equations are chosen), we obtain the vector P. Applied to the matrix M of the chain in Fig.

5.2, we obtain $P = [0.286, 0.286, 0.143, 0.286]$ which is the final distribution of valuables in the communities 1, 2, 3, 4. If Brunton is right, community 3 with its scarcer resources has more potential for developing monopolies and chieftainship than the others.

For our illustration of a Markov chain we used a simple model with only four points, and with probabilities chosen arbitrarily. Let us now apply this procedure to a model representing the *kula* ring, with probabilities derived from the four assumptions listed above.

In Fig. 5.3 the digraph D represents a Markov chain model of the flow of armshells. The converse digraph D' (not shown), in which all the arc directions in D are reversed, represents the flow of necklaces. In D (and D') the arcs from v_i to v_j are assigned equal probabilities,

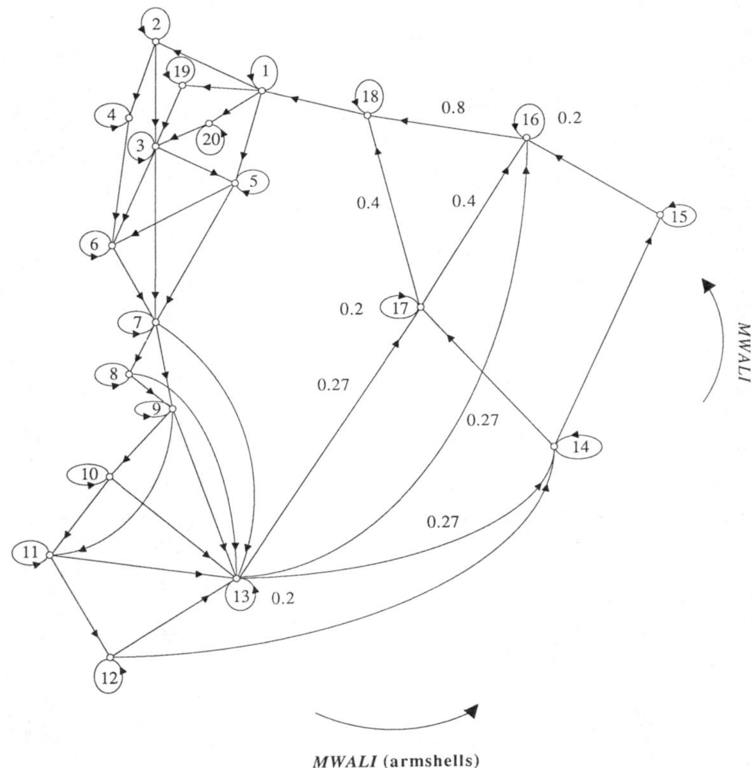

FIG. 5.3. A network N of the flow of *kula* valuables (armshells)

as illustrated for points 13, 16, 17, and there are loops at each point standing for some small supply of valuables that each community is presumed to withhold in a round of *kula* exchange. Since this is a regular chain, the entries of the unique probability vector P for which $PM = P$ give the proportion of valuables in each community. These proportions and the relative rank of each community are shown in Table 5.1.

The following observations may be made on Table 5.1.

First of all, the results are in general agreement with Malinowski's characterizations of several different *kula* communities and thus inspire confidence. Kayleula (4), which according to the model is one of the pair of communities with the fewest armshells, and poorest of all in necklaces, is described as 'slightly anomalous for they make

TABLE 5.1. *The predicted distribution of valuables in* kula *ring communities*

Kula community	Proportion of armshells	Rank	Proportion of necklaces	Rank
1 Kitava	0.1037	1=	0.0960	1=
2 Kiriwina	0.0259	10=	0.0320	7=
3 Sinaketa	0.0648	4	0.0640	3=
4 Kayleula	0.0130	13=	0.0107	10
5 Vakuta	0.0475	7	0.0427	5
6 Amphletts	0.0583	5	0.0320	7=
7 NW Dobu	0.1037	1=	0.0960	1=
8 Dobu	0.0346	9	0.0480	4=
9 SE Dobu	0.0518	6=	0.0720	2=
10 East Cape	0.0173	12	0.0360	6=
11 East End Islands	0.0259	10=	0.0480	4=
12 Wari	0.0130	13=	0.0360	6=
13 Tubetube	0.0972	2	0.0720	2=
14 Misima	0.0389	8	0.0480	4=
15 Laughlan	0.0194	11	0.0160	9
16 Woodlark	0.0778	3	0.0480	4=
17 Alcesters	0.0518	6=	0.0640	3=
18 Marshall Bennetts	0.1037	1=	0.0960	1=
19 Wawela	0.0259	10=	0.0213	8=
20 Okayaulo	0.0259	10=	0.0213	8=

kula only on a small scale . . .' (Malinowski 1922: 476). Kayleula also has the reputation of being 'hard' in the *kula*, that is, of passing on valuables only reluctantly.[5] On the other hand Kitava (1) and NW Dobu (7), which rank first, and SE Dobu (9), which ranks second in necklaces and sixth in armshells, are the opposite of 'hard', that is, 'good' in *kula* exchange (1922: 360). Tubetube (13) in the southern sector ranks second in Table 5.1 and is described as a 'place of concentration' of *kula* valuables. So is Woodlark (16), which ranks third in armshells.

Secondly, the valuables are not evenly distributed around the *kula* ring as Fortune's statement implies, but irregularly as Firth's 'nodule' metaphor suggests. Furthermore, the relative proportions of armshells and necklaces do not always match. The Amphletts, for example, are necklace poor while Tubetube is armshell rich. This may have some bearing on the Trobrianders' perception of the Amphlett Islanders, from whom they receive necklaces, as 'stingy and mean' in *kula* transactions, and it suggests that the main southern point of 'leakage' of armshells out of the *kula* ring is at Tubetube rather than Wari.[6] Such quantitative imbalances (which J. W. Leach (1983) notes for modern times) may provide research leads into the non-*kula* use of one or the other class of valuables in particular communities, and into symbolic ascriptions based on sex or on the 'tightness' and 'looseness' in the flow of each class through different *kula* communities.

Thirdly, Kiriwina (2) is indeed predicted to have few valuables, as Brunton supposes, not only armshells, but necklaces as well. Malinowski emphasizes a decline in the institution of chieftainship moving south from Kiriwina to Sinaketa (3) to Vakuta (5). There is not, however, a parallel increase in access to valuables, of either kind, for of these three communities Sinaketa has most. This does not affect Brunton's main argument but rather his criticism of Uberoi's (1971) theory, which predicts greater elaboration of rank with better *kula* position, all else being equal. In Brunton's reading of Uberoi, all else is equal, and chieftainship should be most developed in Vakuta, which Brunton supposes to be in the strongest *kula* position of all Trobriand communities.

Finally, the most interesting result concerns the distribution of

[5] Róheim (1950) notes not only anal but also oral and genital symbolism, making the *kula* a total erotogenic phenomenon.

[6] Malinowski mentions both places.

valuables at Kiriwina (2), Tubetube (13), and the Amphletts (6). In Kiriwina, chieftainship is based on formal hereditary authority associated with *guyau* as opposed to commoner (matrilineal) subclans. Chiefly status is marked by special insignia and taboos, and power is based on the economic entailment of polygyny—the large food payments (*urigubu*) made by wives' brothers to sisters' husbands—and on the monopolization of *kula* exchange. In Tubetube, it now appears, there was also some form of chieftainship based on pre-eminence in *kula* exchange. In describing the current pastor of the Christian mission, Macintyre says:

He is a direct descendant of the chiefly (*guyau*) family on Tubetube and is also the most important *kune* (*kula*) trader on the island. In the past one family was acknowledged as hereditary *guyau* and people are still conscious of the rights this position conferred, such as the wearing of boar's tusk necklaces (*dona*) at feasts and the right to rattle one's limestick when getting lime from a pot. The pastor's prestige is grounded in a traditional leadership role, with his skill as an orator, his extensive knowledge of traditional customs, and most particularly his great skill in *kune*, in many respects more important than his position as pastor (Macintyre 1983b: 370–1).

In this connection Seligman (1910: 453) speaks of 'men of wealth' on Tubetube, of 'rich men, who had in their time bought many canoes and traded, i.e., travelled, extensively . . .' And he mentions the use of shell ornaments in payment for canoes and other imported goods (1910: 536). In Tubetube, *kula* and economic exchange do not run parallel as Malinowski emphasized for the Trobriands, but are evidently fully integrated. According to Macintyre (1983b: 374) a man 'may *kune* simply with *mwali* and *bagi* or he may integrate the exchange of valuables (maintaining the flow in the correct direction) with the acquisition of a pig or a canoe'. She gives an example from an informant worth quoting in full:

I went to my partner on Murua [Woodlark] and asked him for a canoe. On this trip I threw him a *mwalikau* [armshell of highest-rank category] as *logita* [opening or sustaining gift in *kula* relationship; same as *vaga* and *basi* in northern *kula* area]. One year later I returned with a canoe full of yams and pots, the meat of four pigs, and two high-ranking armshells. He took these as his *kitomwa* [personal *kula* shells free of extant obligation to other transactors]. The next year when the canoe was almost finished, I returned with two more armshells, two live pigs, and some yams and pots. That then was enough. I returned a couple of months later with enough men to sail back both canoes [the one brought and the one purchased]. I collected the canoe and my partner gave me the necklace which was the return for my first armshell given as *logita* (Macintyre 1983b: 374).

A canoe so obtained could then be exchanged with other communities such as Dobu.

Chieftainship in Tubetube seems to have emerged under exactly the opposite conditions as in Kiriwina: abundant rather than scarce valuables, used in economic exchange rather than disjoined from it. Whereas Kiriwina lacked exportable resources or skills, Tubetube made its living as a manufacturer of pots and nose ornaments and as a middleman in trade. As Seligman emphasized, Tubetube was not economically self-sufficient and had to import food and virtually everything else. The favourable position of Tubetube in the *kula* ring can be appreciated by comparing it to the Amphletts (6), whose ecological characteristics are otherwise similar—insufficient food supply and consequent economic specialization. Here, according to Malinowski (1922: 47), 'there are no chiefs'. Whereas Tubetube is a 'crucial point of the *kula*', a 'place of concentration of valuables' and an economic wheeler-dealer, the Amphletts have access to only a modest supply of armshells and necklaces and these are only partially integrated with economic exchange—in the demand for solicitory gifts or *pokala* for valuables. It is interesting to compare the casualness with which economic and *kula* exchange are integrated on Tubetube with the situation in the Amphletts. In Trobriand perception,

The Gumasila [Amphletts], their *kula* is very hard; they are mean, they are retentive. They would like to take hold of one *soulava*, of two, of three big ones, of four perhaps. A man would *pokala* them, he would *pokapokala*; if he is a kinsman he will get a *soulava*. The Kayleula only, and the Gumasila are mean. The Dobu, the Du'a'u, the Kitava are good (Malinowski 1922: 360).

Malinowski comments: 'This means that a man in Gumasila would let a number of necklaces accumulate in his possession; would require plenty of food as *pokala*—a characteristic reduplication describes the insistence and perseverance in *pokala*—and even then he would give a necklace to a kinsman only.' The Amphletts rank near the bottom, seventh, in the possession of *soulava* (necklaces).

So far as Kiriwina and Tubetube are concerned, political hierarchy can occur when valuables are scarce, external trade is minimal, and agricultural productivity is high, or when just the opposite extreme conditions obtain. Just as in the Caroline Islands (Chapter 3), stratified communities in overseas exchange networks are those with favourable locations.

Alternative *Kula* Rings

Irwin (1983) in his study of the *kula* ring raises two problems in network analysis: the indeterminacy of certain points and lines, and the weights assigned to the lines of a graph. There are in fact several different versions of the *kula* ring. Ours is like Brunton's in so far as it includes points 19 and 20 between 1 and 3, since Malinowski emphasized the absence of any direct connection between Kitava and Sinaketa. There are three main versions of the *kula* ring.

A second version of the *kula* ring given in Malinowski's book excludes two places whose participation he was unsure of: points 10 and 15 (together with their incident lines).

Belshaw's (1955) version, which is somewhat schematic, does not appear to include points 11, 12, 14, and 15.

Brunton's version is based on Belshaw, Fortune, Malinowski, and Lauer (1970). It eliminates points 11, 12, 14, 15 and adds points 21 (Bonvouloir) and 22 (Panamoti), both in the southern part of the *kula* ring, together with the lines (13, 21), (16, 22), (17, 21), (21, 22). It does not show the lines (13, 16) and (13, 17).

All of these versions when modelled as a Markov chain basically agree with the results in Table 5.1, as can be seen from Tables A2.1, A2.2, and A2.3 in Appendix 2. The Belshaw and Brunton versions show that Tubetube is in an even better position than in the graph in Fig. 5.1.

In his hypothetical gravity model of the *kula* ring, Irwin uses a 'proximal point method' which postulates the existence of lines in a network on the basis of nearest neighbour relations, distinguishing first, second, and third nearest neighbours. Our Markov chain model assigns equal probabilities to each of the arcs from v_i to all adjacent points v_j. We regard this as a first approximation which could perhaps be improved upon by some informed guesswork on the part of Massim area and network specialists. One might for example assign probabilities on the basis of the distance between communities, their dependence on trade, level of sailing technology, and significant economic specializations such as pottery and sailing canoes. Irwin's (1983) proximal point model of the *kula* ring, and his network model, which assigns six different weights to links between *kula* communities, suggest a series of simulation experiments.

A Speculation on the *Ur-Kula* Ring

The type of chain underlying the flow of valuables in a network has implications for social differentiation. The *kula* ring, for example, is a regular chain in its present form, with stratified and unstratified communities, but it may not have been so originally.

Depending on the structure of a network which defines a chain, various properties of the chain can be determined which are independent of the magnitudes of the probabilities assigned to its lines. For example in Fig. 5.4 the points v_0 and v_5 are receivers (points with zero outdegree and positive indegree) and are the only such points. The only limiting (eventual) outcomes of the chain are these two states. An *absorbing state* of a chain is a receiver of its network. An *absorbing chain* has an absorbing state. By contrast, an *ergodic chain* is a strong chain (its digraph is strongly connected) and thus has no absorbing state. Of course not every chain is absorbing or ergodic.

A regular chain, as already defined, has some power n of its transition matrix M positive, that is, M^n has only positive entries. Interpreted graphically, this means that for any two points v_i and v_j in the network, there is a walk from v_i to v_j of length n (and hence a path of length at most n). The underlying digraph of a regular chain is therefore strong, and so every regular chain is ergodic. But the converse does not hold: not every ergodic chain is regular. The chain consisting of a single cycle is ergodic but not regular. For example, Fig. 5.5 shows the three different powers of the transition matrix of a cyclic triple. Since $M^3 = I$, we find $M^4 = M$, $M^5 = M^2$, and every power of M is one of M, M^2, or I. Thus no power of M has all positive entries and this ergodic chain is not regular.

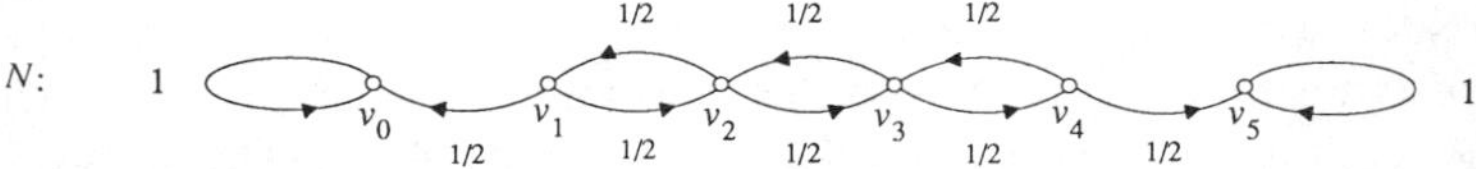

FIG. 5.4. An absorbing chain

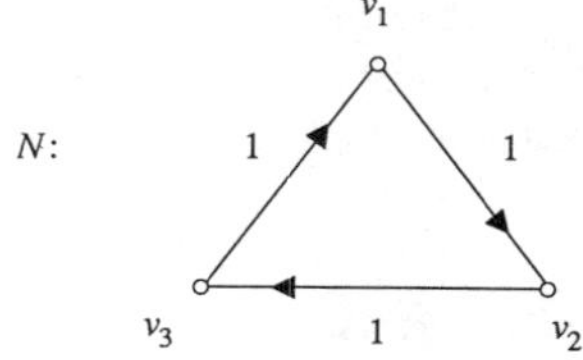

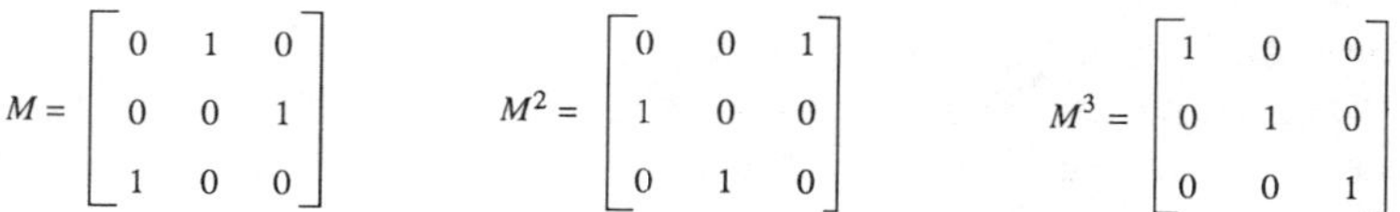

$$M = \begin{bmatrix} 0 & 1 & 0 \\ 0 & 0 & 1 \\ 1 & 0 & 0 \end{bmatrix} \qquad M^2 = \begin{bmatrix} 0 & 0 & 1 \\ 1 & 0 & 0 \\ 0 & 1 & 0 \end{bmatrix} \qquad M^3 = \begin{bmatrix} 1 & 0 & 0 \\ 0 & 1 & 0 \\ 0 & 0 & 1 \end{bmatrix}$$

FIG. 5.5. An ergodic non-regular chain

In *Argonauts of the Western Pacific*, Malinowski states that *kula* myths are localized at points on a semicircle consisting of Woodlark (16), the Marshall Bennetts (18), Kitava (1), Vakuta (5), the Amphletts (6), and Dobu (7, 8, 9). Since such myths are absent from the northern Trobriand communities, he suggests that they may have entered the *kula* ring late. This offers some support for Brunton's theory and provides for a conjecture of our own. By adding one more point, 13, to this set and joining each community with its neighbour on its right and its neighbour on its left, the result is the single cycle shown in Fig. 5.6, which is a subgraph of G in Fig. 5.1.

Assume that this is the graph of the original *kula* ring, and assume that in this case the digraphs D and D' representing the flow of armshells and necklaces have no loops, as each community passes on everything to the unique community to which it is adjacent. A chain consisting of a single cycle is not a regular Markov chain, because no power of its transition matrix has all entries positive, as illustrated by the cyclic triple in Fig. 5.5. In this system, if all communities have equal amounts of valuables to begin with, then nothing will change in successive rounds of exchange. The distribution will be even as Fortune supposed. If the communities have unequal amounts then the relation between scarcity and abundance will be cyclical. In this case an equilibrium will be reached—the distribution of valuables

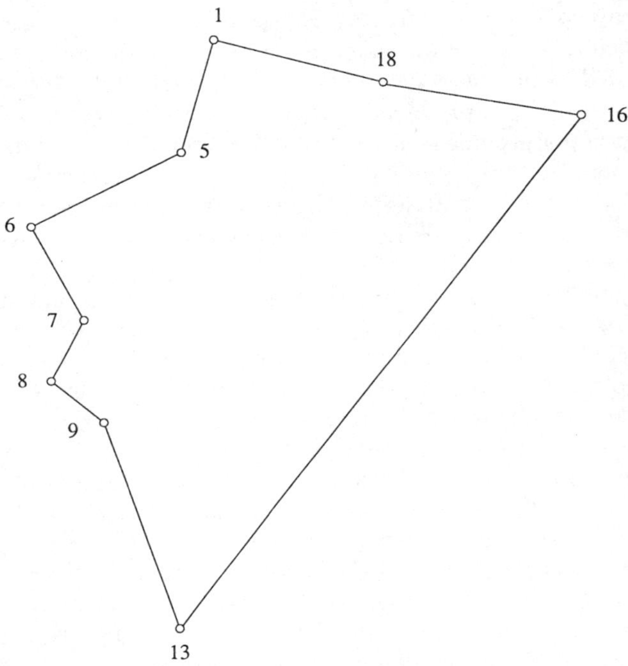

FIG. 5.6. The graph of an inferred original *kula* ring

will stabilize—only as the system, to use Malinowski's very graphical metaphor, 'ramifies'. The internal complexity of particular societies will result from the external complexity of their common exchange network.

On Models of Social, Trade, and Transport Networks

Our analysis of exchange systems in Chapters 3, 4, and 5 has important connections with graph theoretic analyses of transport networks in geography (Haggett 1967; Haggett and Chorley 1969; Taaffe and Gauthier 1973) and social networks in anthropology (Mitchell 1969; Barnes 1972). Like geographers, e.g. Pitts (1965) and Brookfield and Hart (1971), we have used distance centrality in graphs to study dominance in trade networks, and like anthropologists, e.g. Garbett (1980) and Kapferer (1969), we have used concepts of reachability in graphs and digraphs to study mediated communication and status in formal and informal social networks. We have already discussed certain aspects of social network analysis in *SMA*, including Kapferer's work and its matrix treatment, and the digraph model used by Garbett.[7] Here we wish to make a general point about the potential for cross-fertilization between adjacent fields. Since exchange theorists (including anthropologists and archaeologists), social networkers, and transport geographers are all concerned with communication networks and all employ graph theoretic models, we expect that each could benefit from studying the others' work. In this book, for example, the betweenness model of centrality (Freeman 1979) which we applied to the analysis of middlemen activity in trade networks is, as Garbett (1980) has suggested and as we have previously demonstrated (Hage and Harary 1981a, 1983a), applicable to the analysis of influence and power in social networks, and it should be applicable to various forms of transport networks as well. Similarly, our clarification of centrality and connectivity in trade networks is directed to archaeologists (Irwin 1974, 1983) but it obviously applies to the geographers (Pitts 1965; Haggett 1967; and Haggett and Chorley 1969) from whom these concepts were borrowed. Anthropologists who read further in geography will find a variety of potentially applicable models in addition to median centrality, for example 'branching networks', 'circuit networks', and 'barrier networks', which are based on trees, connected graphs with cycles, and polygonal graphs respectively (Haggett and Chorley 1969). They will also find suggestive applications of Markov chains to analyses of human migration and flows of goods and services in regional networks (Harvey 1967). Geographers who read this book may find uses for topological trees in addition to betweenness centrality.

To conclude our discussion of trade networks, we expect that questions will arise concerning the effects of topography, history, and system level on applications of graph theoretic models. In less

[7] Garbett uses level assignments in digraphs to analyse exchange and rank in the Indian caste system. This model was defined by Harary *et al.* (1965) and applied to anthropology by Hage (1976a). Kapferer's analysis of reachability and multiplex relations and Doreian's (1974) matrix method of analysis are discussed in chapter 5 of *SMA*.

'perfect' landscapes than the Caroline Islands, network structures may be affected by variations in coastline, altitude, and terrain. Brookfield and Hart observe with respect to Melanesia that

> In thickly populated country inland, the network of transfers is enormously complex, and in I. M. Hughes' words (pers. comm.) 'the network looks like a Löschian landscape modified by topography'. Where there are mountain passes, barriers of empty country or empty sea, the separate links of the network become elongated and funnelled into corridors. Strategically placed communities may then become nodes through which many strands are drawn, and may be able to establish an oligopolistic control over the traffic, by not permitting the passage of persons past them without leave, and thus assuming the role of obligatory middlemen by becoming the terminal points of contact fields from either side. They are then able to make considerable profits by means of 'exchange' transaction with separated peoples at either edge of their own contact field (Brookfield and Hart 1971: 317).

Let us note first of all that topographical distortions of network patterns may be a problem for models deriving from central place theory (Christaller 1966; Lösch 1954), but not for centrality models from graph theory, which make no assumptions about topographically idealized networks.[8] Topographical problems *are* inevitable in archaeology, where informed guesses must be made concerning the existence of links in a trade network. In archaeology and anthropology topographical effects on the ease, frequency, and volume of trade may make it difficult or impossible to convert graphs into network models. One solution, exemplified by Irwin's (1974, 1983) work, is to construct a number of models each of which emphasizes a particular aspect of the evidence in order to arrive at an optimal network. We should point out here that while a graph is less realistic than a network it may none the less be sufficient for the task at hand, as illustrated by the work of Pitts (1965) and Brookfield and Hart (1971).

Some trade networks may have long and complex histories. They are none the less amenable to graph theoretic analysis, which seeks only to determine the role of structural constraints in network development. As Pitts emphasized in his analysis of medieval Russian trade routes, the point is not to insist exclusively on situational factors or on social, political, and cultural factors in

[8] Central place theory as used in regional analysis, e.g. C. A. Smith (1976), and centrality in graphs are entirely distinct models and should not be confused.

accounting for the dominance of a community in a network, but rather to assess the importance of the former: 'Clearly, we need a way of filtering out the centrality factor, the "situational determinant" so that the variation in the remaining factors may be studied more closely. One way of doing just this is to call upon the resources of graph theory' (Pitts 1965: 16).

Finally, and very briefly, many if not most trade networks are subsystems contained within larger more global systems. In some cases boundaries may be drawn on the basis of the frequency of communication, e.g. 'regular vs. sporadic contacts', or on the basis of a cultural demarcation such as 'all participants in the *kula* ring'. In many cases, however, subsystems have significant interactions. In the *kula* ring, for example, communities such as Tubetube had important trade relations with non-*kula* communities, which implies something like a 'greater Massim trade network' (see the papers in Leach and Leach 1983). The analysis of global systems is clearly a next step in the study of Oceanic trade networks. A potentially useful starting-point is the Cartwright and Harary (1977) model of system–environment relations, which focuses precisely on communities which lie on the boundaries joining interacting subsystems.

6

Combination and Enumeration

Le structuralisme consiste à considérer par principe toute réalité comme un *cas*, dont il faut trouver la règle de déclinaison.

Jean Pouillon, 'L'hôte disparu et les tiers incommodes'

In a review of the state of kinship studies, occasioned by the publication of Scheffler's (1978) book on Australian systems, W. Shapiro (1982) dismisses Lévi-Strauss's theory of marriage alliance as 'philosophic posturing', citing in particular what he believes to be the analytical superficiality and empirical disconfirmation of the configuration of relations known as the 'atom of kinship':

The analysis is glib in the extreme: a 'plus' sign represents dyads which are 'free and familiar', a 'minus' sign dyads 'characterized by hostility, antagonism, or reserve . . .' (Lévi-Strauss 1963: 44). This all-too-simple notational system, lifted from Radcliffe-Brown's student Warner (1931: 180), is used to assert the proposition MB/ZS : B/Z : : F/S : H/W (ibid.: 42).[1]

In fact, the whole scheme is ethnographic balderdash (Shapiro 1975, 1976), and the asserted analogy has been disconfirmed cross-culturally (Ryder and Blackman 1970). Its continued employment in ethnographic analyses . . . reflects Lévi-Strauss's prestige, not the plausibility (much less the truth) of his propositions (Shapiro 1982: 261).

This hasty critique, which is not unique to Shapiro,[2] is unfortunate because it ignores three general implications of Lévi-Strauss's formulation for kinship analysis.

First, the + and − signs imply that structural analysis is properly

[1] The following conventional kin type notation is used in this chapter: F = father, M = mother, B = brother, Z = sister, D = daughter, S = son, H = husband, W = wife, with MB = mother's brother, ZS = sister's son, etc.

[2] Robin Fox has a very different assessment. In comparing descent and alliance theories, he emphasizes the irreducibility of the atom of kinship in exchange: '[Lévi-Strauss] argues that, for example, in times of strife when systems break down, they revert to this atomic situation. A man's special relation to his mother's brother—as opposed to his father's brother—is built in from the start. It was the MB who released the M in marriage and so produced the son. The relation of the S then to FB and MB is not symmetrical—the maternal uncle is a very special person. Medieval songs evidently celebrate this relationship at length. We are in anthropology only just beginning to explore the consequences of such thinking for the study of kinship' (1983 [1967]: 236–7).

concerned with culturally defined significant oppositions between kinship relations. These signs were introduced not as a formal, notational system but only as a convenient shorthand or mnemonic; provisionally defined, they do not in general refer to any specific empirical content. The cross-cultural study by Ryder and Blackman, which Shapiro cites, neither confirms nor disconfirms any propositions about relations in the atom of kinship, precisely because it ignores culturally defined contrasts and interprets the + and − signs as resultants or mysteriously defined averages of heterogeneous and culturally variable relations. This point was made in Hage (1976a).

Secondly, the formula P : Q :: R : S implies that individual kinship relations should be studied not in isolation or seriatim but from the standpoint of their position in a system of relations. A demonstration of the value of such an analysis, inspired in fact by Lévi-Strauss's atom of kinship model, is given in Biersack's (1982) article on parallel and cross-kinship relations in Tonga, which appeared in the same issue of the *Journal of the Polynesian Society* as Shapiro's review. Biersack's 'parallel and cross matrices', as she calls them, are, as we will show, part of a still larger system of kinship relations.

Thirdly, the interpretation of different atoms of kinship as transformations of each other implies that analysis might properly begin with the logical enumeration of a set of structures of related type—deductively rather than inductively. Anthropologists have sometimes recognized the analytical utility of such enumeration, as illustrated by Lowie's (1928), Kirchoff's (1932), and Murdock's (1949) typologies of kinship classification, Needham's (1971b) outline of descent modes, Fox's (1967) analysis of descent and residence interactions, and Lévi-Strauss's (1949) and Héritier's (1981) analysis of alliance structures. Leach's (1961) topologically inspired analysis of patterns of incorporation and alliance in kinship systems, presented in the first Malinowski Memorial Lecture, deserves special mention here. In the physical sciences, chemistry being a conspicuous example, enumeration is a standard procedure of classification and discovery. There is reason to believe that it will have similar value in anthropology. Indeed, there are already two impressive examples of such research in Oceanic studies: Epling, Kirk, and Boyd's (1973) and Marshall's (1984) structural analyses of sibling classification.

In order to develop these ideas we require (1) a language to define the different types of oppositions implicit in kinship relations, (2) models to characterize the coherent combination of relations, and (3) explicit techniques of structural enumeration. We illustrate these requirements using ethnographic and theoretical material from Melanesia and Polynesia.

Exchange Dualities

In the case of exchange relations one can begin a typology by considering three oppositions: the presence vs. the absence of exchange; the direction of exchange; and the positive or negative values associated with exchange. The graph theoretic model for these oppositions is structural duality (Harary 1957).

A duality principle has two properties: (1) the dual of the dual of a statement is the original statement; and (2) the dual of a true statement is true. We shall utilize several kinds of duality. The generic notation for a duality operation is *. We say that operation * acts on some configuration C and gives another configuration C^*. For all configurations C, the duality equation holds: $C^{**} = C$. For every type of graph, undirected, directed, and signed, there is a dual operation that changes it to give another graph with the same set of points and that when applied twice results in the original graph. This is *structural duality*.

Existential duality is the model for exchange relations which contrast by presence vs. absence. *Existential duality* is based on the operation of taking the complement of a graph (or digraph). The *complement* of a graph G is that graph $\bar{G}$ having the same set of points as G, but in which two points are adjacent if and only if they are not adjacent in G. Thus the lines that are present in $\bar{G}$ are precisely those that are absent in G. The complements of the graphs G_1 and G_2 in the top half of Fig. 6.1. are the graphs $\bar{G}_1$ and $\bar{G}_2$ in the lower half. We note that the (unlabelled) graph G_2 is *self-complementary* in that $\bar{G}_2$ is isomorphic to G_2.

A simple example of existential duality in an exchange relation occurs in the Arapesh origin myth, 'The bringing of yams by Sharok, a cassowary' (Mead 1940). Commonly, in origin myths, the precultural world is represented in various ways as the opposite of the cultural one (Lévi-Strauss 1978). In the Arapesh myth, men

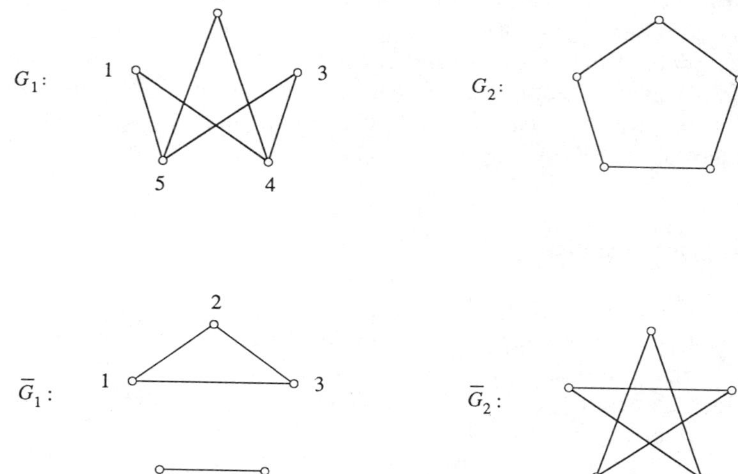

FIG. 6.1. Two graphs G_1 and G_2 and their complements $\overline{G}_1$ and $\overline{G}_2$

originally lived alone and ate cooked woodchips, referred to as 'rotten food', instead of (metaphorically equivalent) boiled yams normal in marriage. A cassowary woman named Sharok came and replaced the chips with yams. She showed men how to cook them and how to make yam gardens. The men harvested the gardens, painted the yams, and began exchanging them in *abullu* feasts (feasts given by men who have especially successful crops). The passage from nature to culture marked by the inauguration of reciprocal feasting is represented by the graphs G and $\bar{G}$ in Fig. 6.2. This is the primordial case of the complement of a graph, as there are just two points.

Directional duality is the model for exchanges which contrast by direction. *Directional duality* is based on the operation of taking the converse of a digraph. The *converse* D' of a digraph D is that digraph

FIG. 6.2. Reciprocity = the origin of culture in Arapesh myth

FIG. 6.3. A digraph D and its converse D'

with the same points as D in which the arc (directed line) uv occurs if and only if the arc vu is in D, as illustrated in Fig. 6.3.

In the *kula* ring the digraph D represents the flow of one class of valuables (armshells or necklaces) between communities, and its converse D' represents the flow of the other class (necklaces or armshells). The complement of the graph of exchanging communities of the *kula* ring, on the other hand, is the graph in which there is a line joining every pair of non-exchanging communities, giving the un-*kula* ring. (Compare Lewis Carroll's 'unbirthday'.) A kinship example of converse flows of valuables is shown in F. E. Williams's (1940) diagram of marriage exchanges in Elema society, reproduced in Fig. 6.4. After reciprocal prestations between relatives of the bride and groom, individual exchanges begin, at first between brothers-in-law, and then, upon the birth of a child, between mother's brother and sister's son, with shell ornaments and pigs (or portions of pigs) flowing in opposite directions. The exchanges between mother's brother and sister's son

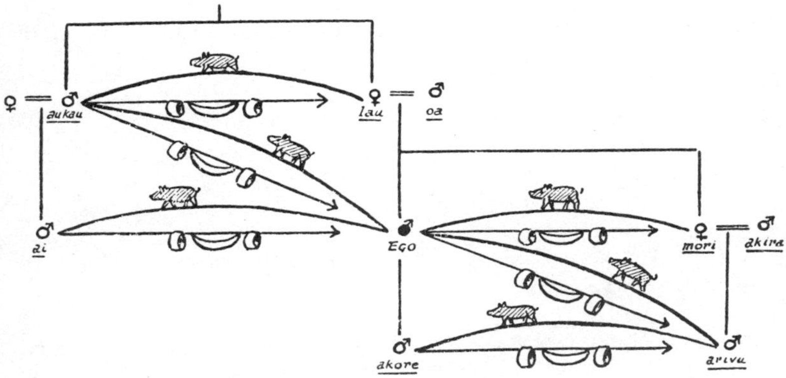

FIG. 6.4. Williams's (1940) diagram showing the directionally dual circulation of shell ornaments and pigs in Elema kinship relations

are made on the occasion of a succession of ceremonies the latter passes through: ear-piercing, nose-piercing, initiation to the bullroarer, etc. In the event of the former's death he is replaced by his son. The directionally dual exchange of different classes of valuables, undertaken at marriage and continuing into the next generation, connected with significant events in the life cycle, is a common pattern in New Guinea societies.

Antithetical duality is based on the operation of negation of a signed graph or digraph. The *negation* S^- of a signed graph or digraph S is obtained from S by changing the sign of each line or arc of S. Clearly, the negation of the negation is S itself. Fig. 6.5 shows the four signed triangles, displayed as two pairs of antithetical duals.[3]

In psychology, interpretations of antithetical duality include oppositions such as like vs. dislike, pleasant vs. unpleasant (Harary 1957), power vs. negative power (French 1956), and, in some folk models of balance, intimacy and familiarity vs. distance and restraint (Hage 1976b). In exchange relations, oppositions include competition vs. co-operation, as exemplified by the 'food friends' vs. 'food enemies' signed graph in Fig. 2.20 in Chapter 2, and compulsory vs. free and spontaneous gifts as exemplified in the next section.

An Atom of Kinship: Arapesh

The atom of kinship as conceived by Lévi-Strauss (1963c) is a small

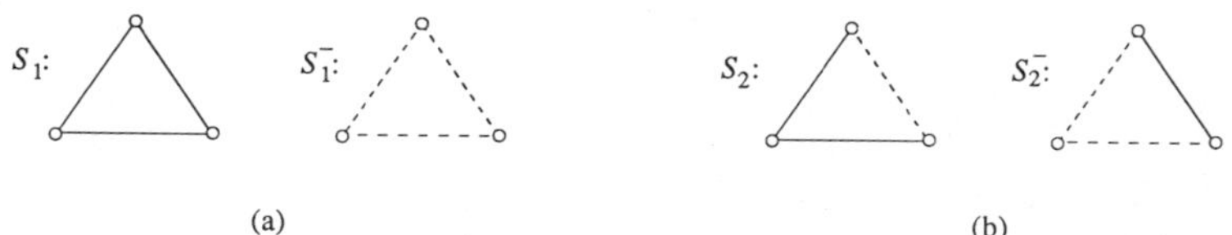

FIG. 6.5. Antithetical duality in the four signed triangles, shown as two pairs of duals (a) and (b)

[3] Although S^+ was used in Chapter 2 for the spanning subgraph of S containing only its positive lines, and S^- now indicates the signed graph obtained from S by changing every line sign, there should be no confusion, as we shall have no occasion to mention (other than here) the subgraph of S obtained by removing all its positive lines.

group consisting of a husband, a wife, an offspring, and, because of the exchange implication of the incest taboo, a representative of the group which has given the woman to the man. In the simplest case the fourth member is the wife's brother, the maternal uncle from the offspring's point of view. In complex atoms other relations may be added, as in the Mundugumor case, or the position of wife-giver may be occupied by someone else, mother's mother's brother for example, as among the Lele (Lévi-Strauss 1976).

The atom of kinship consists of six relations, although Lévi-Strauss is explicitly concerned with only four: brother/sister, husband/wife, father/son, and mother's brother/sister's son, hereafter labelled when convenient as B/Z, H/W, F/S, and MB/ZS. These relations are opposed as 'positive' and 'negative', although the specific referents of these signs are not defined in advance but are culturally determined. In those structures which on theoretical grounds are regarded as 'common' or 'frequent' the distribution of the positive and negative values is constrained by a rule such that both intra- and inter-generationally there is both a positive and a negative sign, that is, in each of the pairs of relations B/Z, H/W and F/S, MB/ZS. In the 'rare' or 'impossible' forms—those which, depending on their specific content, might lead to the synchronic or diachronic breakdown of the group (Lévi-Strauss 1963d)—each of the relations B/Z and H/W has the same sign and is opposed to the pair F/S and MB/ZS which has the opposite sign. Since there are four relations whose signs can vary independently, there are $2^4 = 16$ logically possible atoms, but only four of these have the specified combination of positive and negative signs. These four are shown in the top row of Fig. 6.6, with the two impossible cases in the bottom row.[4]

In the cross-cultural study cited by Shapiro, Ryder and Blackman (1970), using a sample of 27 societies, found no support for Lévi-Strauss's assertion concerning the expected combination of positive and negative relations. This is not surprising, since apart from the false assumption that the atom of kinship is everywhere defined by the same set of relations and has the same functional significance, their definition of positive and negative relations is so

[4] For the non-anthropological reader, the conventional kinship diagrams in Fig. 6.6 are interpreted as follows: the triangle and circle stand for male and female, the equal sign for marriage, the vertical line for descent, and the horizontal line for siblingship.

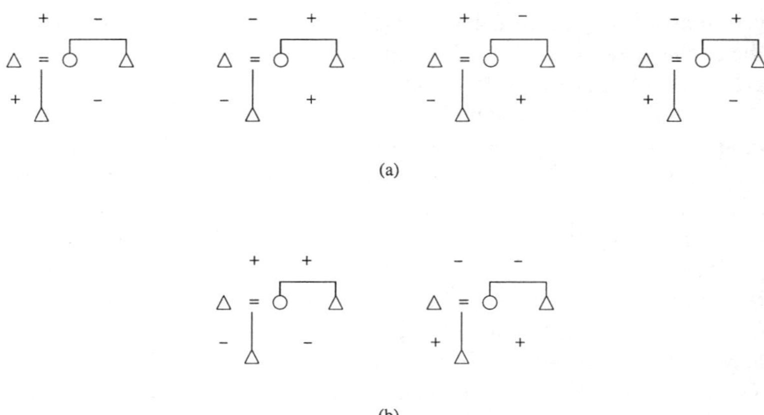

FIG. 6.6. The (a) possible and (b) rare or impossible atoms of kinship according to Lévi-Strauss (1963d)

encyclopaedic as to obliterate all significant oppositions. Thus, according to Ryder and Blackman:

> The positive attitudes of freedom and familiarity can be encompassed within the more general categories of solidarity and equality. The former category includes loyalty, devotion, cooperation in economic and social affairs, lack of reserve or enmity, public and/or private intimacy, reciprocity and mutuality, companionship, proximity, corporateness, and common interest. Equality denotes similarity of social status and free, familiar, and informal relations.
>
> Hostility and antagonism include relationships of open enmity, conflict, competition, and opposition between jural persons. The negative attitude of reserve involves structured formality, detached social relations, and restraint in interaction. In addition, negative attitudes are incorporated in superordination/subordination relationships. Such relationships imply inequality of status and jural and magico-religious authority, on the one hand, and deference or submission on the other (Ryder and Blackman 1970: 100).

For purposes of structural analysis this coding scheme, which in some unknown way averages all these features, is quite useless. The problem is not to consider all possible aspects of a kinship relation but rather to isolate and analyse specific and significant culturally defined contrasts. We can illustrate the result obtained by a culturally focused analysis with a single ethnographic example. The very first Oceanic society in Ryder and Blackman's sample[5] is Mountain Arapesh, which they code as the first 'impossible' structure shown in Fig. 6.6: the H/W and B/Z relations are positive and the F/S and MB/ZS relations are negative. In this particular case the coding difficulties are compounded by the authors' undue reliance on the popular account of Arapesh kinship in Mead's (1963 [1935]) *Sex and Temperament*, thus ignoring the more substantial one given in Mead's (1947) *The Mountain Arapesh III—Socioeconomic life*. (This would be equivalent to using Mead's (1928) *Coming of Age in Samoa* instead of her (1969 [1930]) *Social Organization of Manu'a* as the primary ethnographic source for Samoan culture.) Arapesh is a significant choice with respect to Melanesian ethnography because, in this monograph, Mead showed the correlation between exchange relations and kinship classification, and thus anticipated an entire trend of research exemplified by the work of Wagner (1967) on the Daribi of Highland New Guinea, Clay (1975) on the Mandak of New Ireland, and McDowell (1976, 1980) on the Bun of the Yuat River in the Sepik region of Lowland New Guinea.[6]

In response to a critique by Luc de Heusch (1958, 1971), Lévi-Strauss specifies that an atom of kinship, whatever its form, must satisfy the three originally stipulated conditions:

> (1) that an elementary kinship structure [atom of kinship] rest on an alliance relation to the same extent as on consanguine relations; (2) that the content of the avuncular relation is independent from the rule of descent; and (3) in the midst of this structure, that the attitudes which are opposed to one another (and which can thus, to simplify, be called positive or negative) form a balanced whole (Lévi-Strauss 1976: 94).

The Arapesh atom of kinship appears to satisfy these conditions. The relevant ethnographic facts may be summarized as follows. First of all, Arapesh marriage is based on and formulated as a relation of alliance. The favoured mode of marriage is the exchange of sisters, real, but probably more commonly classificatory. At the clan level, marriages can evidently be repeated, as suggested by Mead's statement: 'The men-folk of the two clans, already bound together by several ties, will urge a further tie' (1963 [1935]: 82). At

[5] Exclusive of Australia.

[6] In his lectures on Melanesia in 1978–9, Lévi-Strauss (1984) emphasized the originality and importance of Mead's work on exchange and kinship relations, with particular reference to her monograph on the Admiralty Islands (Mead 1934).

the lineage level, marriages are dispersed. Mead summarizes marriage prohibitions as follows:

> Formally, the children of parents who use brother and sister, cross-cousin, or two-generations-apart-child-of-cross-cousin terms to each other are not allowed to marry, nor may a man marry a woman whom he calls either aunt, daughter, or niece, nor may a woman marry a man whom she calls father, or mother's brother, or nephew (Mead 1947: 199).

In Rubel and Rosman's (1978) translation of these English terms into the Omaha terminology of the Arapesh, the result is a prohibition on the renewal of a marriage tie between ego's lineage and any lineage from which it has received or to which it has given a woman for a period of three generations. Mead notes the residential inflection of these rules as well as their significance in creating broad affinal networks:

> But in actually deciding upon eligibility for marriage, the children of related men are allowed to marry, if the mothers have come from far distant places, because the marriage will not then too greatly constrict the life of the next generation by cutting down on the number of new affinal relatives. This attitude is quite explicit, and the possible intermarriage of young children is discussed in these terms (Mead 1947: 199).

At the level of the local or family group, marriage is based on substantial and continuing payments of food and valuables from the groom's to the bride's family, beginning with infant betrothal and continuing through adolescence and the birth of children. The payments consist of rings and shell valuables and raw meat. They are compulsory but informally reciprocated: 'these interchanges of valuables and conspicuous payments of meat are the details most often referred to, they are the outer and visible signs that this is a true marriage of long planning and long standing' (Mead 1963 [1935]: 97).

Secondly, the avuncular relation has no necessary connection with the rule of descent, which is patrilineal. In Arapesh conception, the father/son tie is a 'feeding tie' while the uncle/nephew tie is a 'blood tie'. Mead distinguishes two unnamed but conceptually opposed sets of kinship relations in Arapesh: 'Class One for ties based upon ties of direct descent and same sex links, and Class Two for ties based upon opposite sex links' (1947: 190). The first group includes (among others) parents and children, same sex siblings, and spouse's same sex siblings; the second group includes mother's brother and sister's children, opposite sex siblings, and spouse's opposite sex siblings. According to Mead, this classification assumes that husband and wife are equated: a man regards his wife's brother as an opposite sex sibling and uses this kinship term for him, and he regards his wife's sister as a same sex sibling. A woman regards her husband's opposite and same sex siblings in the same way.

Class One relationships are always conceived of as based on seniority. Senior relatives are charged with helping, guiding, and nurturing junior relatives. Siblings always use elder and younger kinship terms, and every sibling relation, no matter how minimal the age difference, has the 'shadow' of the parent–child relation. In the Arapesh theory of conception the child's blood comes from its mother but it is fed by its father during the early period of gestation and after its birth. The father establishes his claim to the child by paying the mother's brother for its blood and by continuing to feed it: 'The father makes the body of the child.'

The father/son relation is the prototype of Class One relations, and this includes the relation of a man to his wife. According to Mead,

> The young husband 'makes the body of his wife' just as the father 'makes the body of his child'. When the girl is grown she belongs to the patrilineal household into which she is betrothed more thoroughly than do her husband's sisters who have been grown in the households of their betrotheds (Mead 1947: 196).

One's own sister is grown by others, and she thus becomes a 'partial stranger', identified with her husband and his group (Mead 1947: 201).

Class Two relationships, in contrast to those of Class One, 'carry the special aura of the mother's brother tie' (Mead 1947: 194). The mother's brother is specifically owed for the child's blood and, as a representative of the mother's group, for the mother herself. According to Mead, 'If the *formal demands of the system* are examined, the mother's brother is defined as one to whom the sister's child is always owing something, a one-sided relationship which does not carry any specially pleasant implication' (1947: 198; emphasis added). The debtor–creditor aspect of this relationship, and of all Class Two relationships in contrast to those of Class One, is defined by an indigenous binary classification of payments for services, symbolized by the opposition between the cooked and the raw which Mead describes as follows:

> 1. Simple payments, typically of cooked food garnished with meat, made to Class One relatives or persons who temporarily assume a Class One status, in return for small services rendered or about to be rendered, either informally as in building a fence, or more formally as to the midwife, the sponsor in an *abullu* ceremony, etc.
>
> 2. Payments to Class Two relatives, typically rings and uncooked meat, accompanied by a small feast, in response to situations in which the Class Two relative was conceived as having a right to demand these payments—betrothal, first menstruation, childbirth, scarification of a girl, initiation of a boy, death, battle wounds, or loss of dignity through insult, the giving of an *abullu*. Repayments of these (Mead 1947: 226).

Prototypical of Class One payments are those given by the men of a hamlet to the women who have helped with a feast: 'When the feast is over, the men will give a special small feast, garnished with meat, to their own wives, their mothers and sons' wives. And this is done in no spirit of accounting, but out of courteous consideration for the labors of the women' (Mead 1947: 227).

The Arapesh distinction between casual, spontaneous exchanges and formal compulsory ones which have the character of inescapable debts establishes that the relations MB/ZS, B/Z, F/S, H/W combine to form a 'balanced whole' as shown in Fig. 6.7, which corresponds to the abstract pattern of the first atom in Fig. 6.6.[7] It is possible to elaborate on this notion by providing an explicit model of structural balance applicable to any set of kinship relations based on antithetical duality.

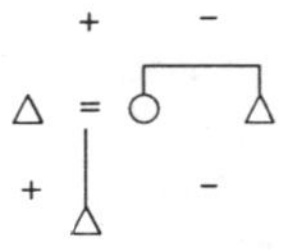

+ = spontaneous gifts of cooked food = "feeding ties"

– = compulsory gifts of raw food (meat) = "blood ties"

FIG. 6.7. The Arapesh atom of kinship

[7] For an example of the second atom in Fig. 6.6, see Needham's (1971a) interpretation of kinship sentiments (ethos) among the (patrilineal) Iatmul, another Sepik society.

The theory of structural balance, as first announced by Heider (1946) and subsequently generalized and formalized by Cartwright and Harary (1956), is well known. A succession of proposed kinship applications was reviewed in Hage and Harary (1983a), and an extended analysis of an African system was given in Hage (1976b). Informally stated, if in an interpersonal system A has a positive relation with B then A will tend to have the same relation—positive or negative—with C that B has. If A has a negative relation with B, then A will tend to have the opposite relation—positive or negative—with C from what B has. Situations that conform to these structures are balanced and presumably stable; those that do not are unbalanced, presumably unstable, and generate their own pressure to change.

A balanced signed graph is the special case for $n = 2$ of an n-colourable signed graph as defined in Chapter 2. In a signed graph S, the *sign of a cycle* is the product of the signs of its lines. Thus every positive cycle has an even number of negative lines, which of course includes the case in which there are no negative lines. The signed graph is *balanced* if every cycle of S is positive. Two criteria for balance from Harary (1953) are contained in the next theorem:

Theorem 6.1. The following conditions are equivalent for a signed graph S.

(1) S is balanced: every cycle is positive.

(2) For each pair of points u, v of S all paths joining u and v have the same sign.

(3) There exists a partition of the points of S into two subsets (one of which may be empty) such that every positive line joins two points of the same subset and every negative line joins points from different subsets.

Thus by condition (3), a balanced signed graph is precisely one that is 2-colourable. In Fig. 6.5 the first and last triangles (3-cycles) are balanced, while the second and third ones are not. The second triangle is, however, 3-colourable. Fig. 6.8 shows a larger signed graph, whose balance can be ascertained either by listing each cycle separately and verifying that it is positive, or, using condition (2), by considering (tediously) each pair of points and verifying that all possible paths joining them have the same sign. For example, all the paths between points A and B are positive, etc. Using condition (3)

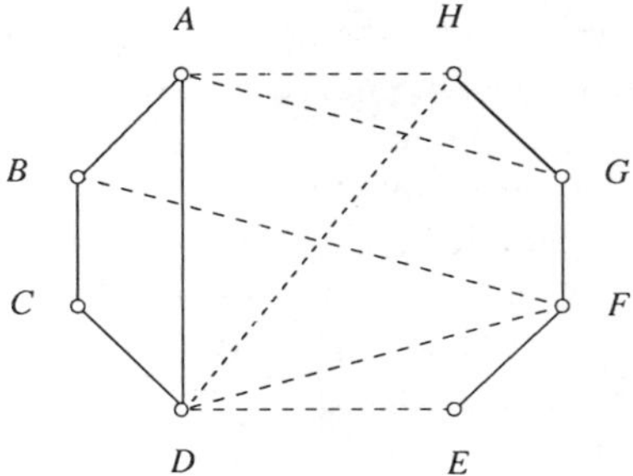

FIG. 6.8. A balanced signed graph

one can readily see that this signed graph is balanced. For *A*, *B*, *C*, *D*, and *E*, *F*, *G*, *H* are clearly the pair of disjoint subsets (of the set of all points) which satisfy condition (3) of Theorem 6.1.

In Fig. 6.9a the Arapesh atom of kinship is represented by a signed graph *S*. The letters F, M, S, U stand for father, mother, son, and mother's brother (U for uncle) respectively. The Class One and Class Two relations, regarded as antithetical duals (positive and negative), are represented by the solid and broken lines. The signed graph includes the two relations not discussed in Lévi-Strauss's analysis: wife's brother/sister's husband (F/U) and mother/son (M/S), which are, as balance theory would predict, negative and positive respectively. One can see that *S* is balanced, because all its cycles (the four 3-cycles and the three 4-cycles) are positive, or, equivalently and more revealingly, because it is bipartitionable as required by condition (3). In Fig. 6.9b, *S* has been condensed into its

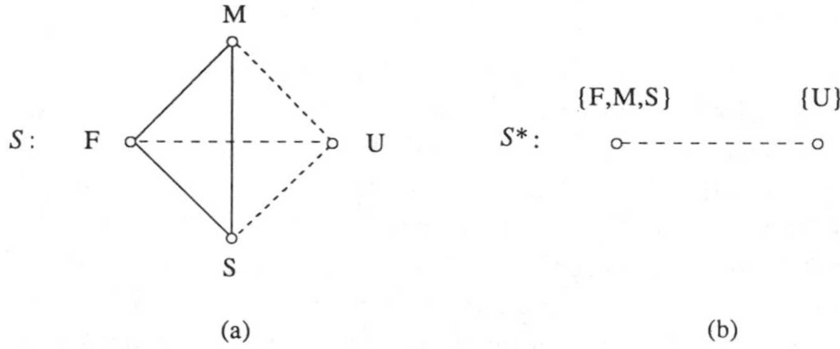

FIG. 6.9. The balanced signed graph of the Arapesh atom of kinship

positive components (two colour sets) to emphasize the culturally defined opposition between a man and the wife and child he has 'grown' and the 'other' who supplied his sister as the man's wife and is owed for the child's blood.

A Generalized Atom of Kinship: Tonga

In an essay on the relation between mythic thinking and social practice, Lévi-Strauss (1985) proposes that myth, unconstrained by infrastructural demands, elaborates a repertoire of logically possible customs, beliefs, and institutions from which particular societies, in effect, make their choice. Reversing functionalist interpretation, mythic speculation precedes social action. A sort of test of this notion could be made by examining the myths and social forms of a group of historically related societies. As part of his demonstration, Lévi-Strauss gives the following example from Polynesia.

In a Pukapukan myth (from Hecht 1977), the island's population originated with sibling incest: an elder brother/sister pair gave birth to a chiefly line, and a younger brother/sister pair gave birth to commoners. As a palliative to the original acts of incest, the chief's sister became a sacred virgin and the commoners were divided into exogamous moieties, thus providing models for 'aristocratic and commoner solutions' to the problem of preventing incest: by avoidance relations and by class relations. A striking implementation of these alternative solutions is found in Fijian societies, where, according to Quain (1948), a strong brother/sister taboo and exogamous moieties are both found, but in complementary distribution.

Such an analysis could be extended in an interesting way by considering the interaction between marriage and rank in Polynesian societies. Recent work, described in Chapter 8 below, has shown how asymmetrical marriage exchange and incestuous unions can each be considered as alternative solutions to the problem of maintaining the rank of chiefly lines. Generalizing the terms introduced by Rogers (1977), the first might be called the 'Tongan solution' and the second the 'Hawaiian solution'. Models for both solutions occur in the Hawaiian version of the Wakea–Papa myth, as presented by Valeri (1985): Wakea (a humanized version of the Polynesian sky father) first marries Papa (a humanized version of the earth mother), who is his matrilateral cross-cousin, and then refuses to give his daughter to another line but instead marries her himself.

Lévi-Strauss's conception of myth as an inventory of differentially implemented social practices suggests a line of research in anthropology in which one would study the relations between mathematical possibility and theoretical as well as ethnographic existence. In this case different anthropologists could be said to choose from the same ideal repertoire of structural models in performing their analyses of social forms. There is in fact an interesting parallel between mathematical and mythical thinking. When mythic thinking is confronted with a cultural relationship it responds by transforming it in various ways:

> Stimulated by a conceptual relationship, mythic thinking engenders other relationships, which are parallel or antagonistic to the first one. If the top is positive and the bottom negative, then the reverse relationship is immediately induced, as though the permutation of multiple axes of terms belonging to the same network were an autonomous activity of the mind, so that any state of a combination would suffice to get the mind moving and, in one surge after another, produce a cascade of all the other states (Lévi-Strauss 1985: 173).

When mathematical thinking encounters an anthropological structure it responds in a similar way. Instead of just scrutinizing the empirical properties of an atom of kinship, mathematical thinking transforms it. In his original formulation, Lévi-Strauss pointed out that the atom of kinship expresses in an economical way the basic relations of consanguinity, filiation, and affinity in a kinship system. It can also be made to express in an economical and more complete way the cross/parallel relation by using the operation of sex duality in graphs.

Fig. 6.10 shows the atom of kinship, (a) as conventionally drawn and (b) modelled as a rooted graph of type 3. The 1s and 0s stand for male and female, the arcs for filiation, the double lines for marriage, and the dashed line for siblingship. The encircled female point is the root of the graph. With three different types of relations (parental, marital, and sibling relations) this is called a rooted graph of *type* 3.

We defined a dual operation on a graph as one which when applied twice results in the original graph. On a graph whose points

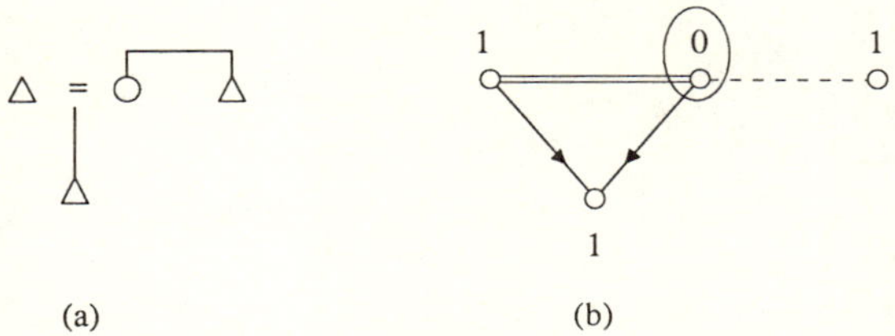

FIG. 6.10. The atom of kinship, (a) as conventionally drawn; (b) modelled as a rooted graph of type 3

are distinguished by sex, the dual operation, called the *sex change operation*, changes the sex of all the points. The graph C^* in Fig. 6.11b is the sex dual of the graph C in Fig. 6.11a: the avunculate for a male is replaced by the amitate for a female.

Taking the root point of the atom of kinship graph as female, and successively changing the sex of the sibling, the offspring, and the sibling and the offspring together, yields four different structures. The sex duals of each of these yield another four, giving a total of eight patterns. Thus the first graph in the first row of Fig. 6.12 shows the atom in Fig. 6.10, the second graph changes the sibling relation (MZ for MB), the third graph changes the offspring relation (D for S) and the fourth graph changes both relations (MZ for MB and D for S). The graphs in the second row are sex duals of those immediately above in the first row.

Such an enumeration is not an idle exercise, for it directs attention to those empirical structures which have already been studied by anthropologists, and those which remain to be studied. We illustrate

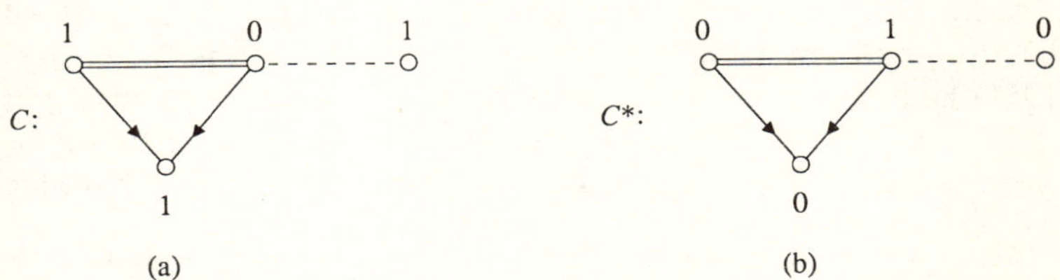

FIG. 6.11. The avunculate (for a male) and amitate (for a female) as sex dual graphs

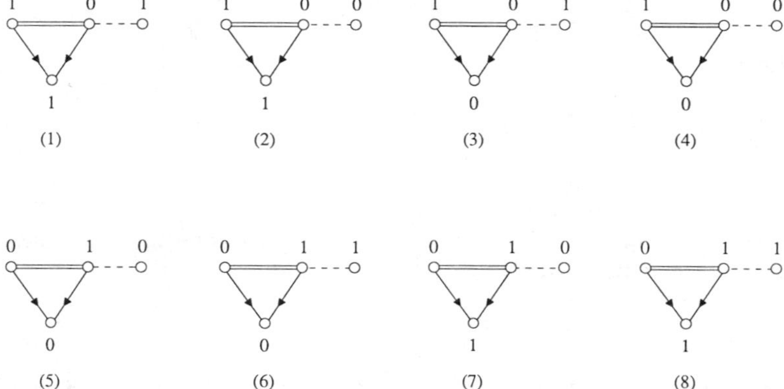

FIG. 6.12. A mathematical generalization of the atom of kinship

the value of this model with reference to a succession of analyses of Tongan kinship, a sort of mythology of Tongan kinship.[8]

Rivers (implicitly) studied the fifth and seventh graphs in Fig. 6.12. In his article, 'The Father's Sister in Oceania' (1910), based on a paper delivered in 1909 and incorporated into his *History of Melanesian Society* in 1914, Rivers reported his discovery, first in Tonga and then in the New Hebrides and Banks Islands, of a 'close relation' between a person and his or her father's sister. Rivers noted that in Tonga a father's sister is honoured even more than a father or father's brother, and that she has considerable power over her nephew and niece, including the right to arrange their marriages and take their possessions. Rivers interpreted this special relation in Tonga as a survival of the transition from matriliny to patriliny. Later commentators have emphasized the religious, economic, and political functions of this relation: Mead (1969 [1930]), Firth (1957), and Rogers (1977) respectively.

A. M. Hocart studied the first graph in Fig. 6.12, the sex dual of the fifth graph. In 'Chieftainship and the Sister's Son in the Pacific' (1915) he focused on the special relation of a man to his mother's brother in Fiji and Tonga. Hocart interpreted the avunculate, as expressed in ritual stealing, not as a survival of matriliny but as a

[8] More accurately, the following analyses of Tongan kinship study certain aspects of the graphs in Fig. 6.12. For a fascinating study of the complementary functions of cross and parallel in-laws in 'neolithic' and 'palaeolithic' origin myths from South America, see Lévi-Strauss (1966, 1970).

logical consequence of the Polynesian theory that chiefs are gods. The nephew's theft, originally of sacrificial offerings, implies the maternal uncle's divine status:

> In Tonga at a formal kava ceremony only the chiefs and their heralds (*matāpule*) are allowed to sit in the ring. All others whether of gentle birth or commoners, sit huddled together behind the kava bowl which is at the bottom of the ring facing the king. Food is laid out before the king; this is afterwards removed to be divided among the people. In the meantime while the kava is being strained a small part of that food is divided into small portions and laid before the chiefs and their heralds. This food is called *fono*. The recipients do not eat it, but kinsfolk come forth from the crowd and carry off the *fono* and eat it up. Not anyone can do so: he must be the chief's grandchild or his sister's child in the classificatory sense; if none of these are present, it must be a stranger, not another class of kinsman.
>
> This custom is the logical consequence of the theory of chieftainship. Not in the Pacific only, but almost universally, food is offered to spirits; a small part is then set aside for them while the worshippers eat the rest. The spirit's share may either be left to rot, or burn, or be carried off by some one. Now the chiefs in the Tongan kava ring are, while the ceremony lasts, gods; the crowd are the worshippers. A portion of the feast is set apart for them but it is carried off and eaten by some human. But why, it may be objected, is not the same done with kava? Because kava is never merely presented to the spirits, then drunk by men: it is poured out at the foot of the post or on the stone where the spirit abides (Hocart 1915: 640).

Another interpretation of the distinction between *kava* and food is given in Biersack (1982), as we note further on.

Radcliffe-Brown studied the first and seventh graphs together, thereby paving the way for the structural analysis of kinship relations. In 'The Mother's Brother in South Africa' (1924) he pointed out that whenever the mother's brother is important, so is the father's sister, but in an opposite way: care and indulgence vs. respect and authority. Radcliffe-Brown accounted for this pattern by invoking a 'tendency' in kinship classification whereby, in patrilineal societies, a mother's brother is regarded as a sort of 'male mother', and a father's sister as a sort of 'female father', as may in fact be implied in kinship terms such as *fae tangata*, 'male mother' for mother's brother in Tonga.[9] The extreme respect shown a

[9] As is well known, Radcliffe-Brown held that 'there are certain fundamental principles or tendencies which appear in all societies or in all those of a certain type. It is these general tendencies that it is the special task of social anthropology to discover and explain' (1952: 18).

father's sister and the outright licence with a mother's brother, as in Tonga, are accounted for by a supplementary tendency in kinship systems: familiarity between persons of the same sex and respect between persons of opposite sex, which thus magnifies the behaviour patterns of these extensions.

A common objection to Radcliffe-Brown's theory, aside from the objection in principle to extensionist theories of kinship, is that it ignores matrilineal societies.[10] We would add that the theory is androcentric in a second way, in that it ignores the third and fifth graphs of Fig. 6.12: what are relations between a woman and her mother's brother, and a woman and her father's sister? A further objection, which is of great significance in the light of modern studies which emphasize the importance of sibling relations in Oceanic social structures—Burridge (1959), Smith (1983), Marshall (1981a)—is that Radcliffe-Brown's theory tends to equate the first with the second and the seventh with the eighth graph, thereby reducing the cross-sibling relation to the parallel sibling relation. This is Schneider's (1981) objection.

Most recently, in an analytical *tour de force*, Aletta Biersack (1982) studied six of the eight graphs in Fig. 6.12 as part of a structural analysis of exchange relations in Tongan kinship. In the avunculate, our first and third graphs, the sister's children have the right to demand goods from the mother's brother, as Hocart emphasized. In the amitate, our fifth and seventh graphs (sex duals of the first and third graphs), the father's sister has the right to demand goods from the brother's children, as Rivers emphasized. The goods consist of *koloa* = 'female goods' such as mats and tapa cloth, and *ngāue* = 'male goods' such as food. These two structures are included in Biersack's 'cross-matrix' of Tongan kinship relations. They jointly define a single exchange structure. Biersack notes that matrilateral cross-cousin marriage, as practiced by Tongan aristocracy, is consistent with this system, so that the cross-matrix is characterized by a unidirectional flow of food, mats, and women from those of inferior to those of superior status, as we show in Fig. 6.13.

Biersack then defines a 'parallel matrix' which includes ego,

[10] Robin Fox has called our attention to the fact that while Radcliffe-Brown ignored matrilineal societies, he encouraged Eggan to extend the analysis to them. Eggan's (1949) study of the Hopi showed that in a matrilineal society, the father's sister is the structural opposite of the mother's brother in a patrilineal society.

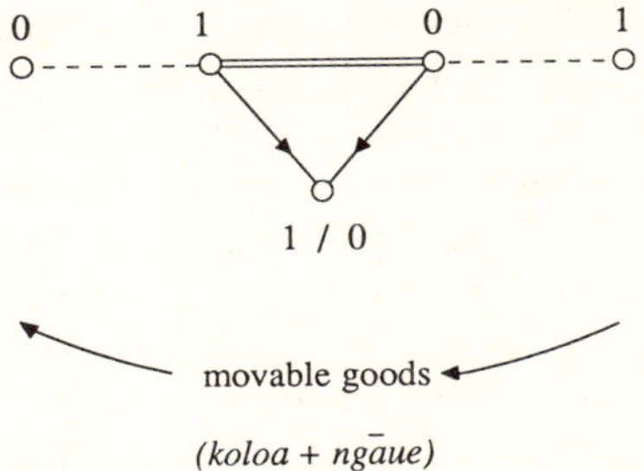

FIG. 6.13. The flow of goods in Tongan cross-relations (based on Biersack 1982)

father, and father's brother, the sixth and eighth graphs of Fig. 6.12. In the cross-matrix, asymmetric relations are defined by superiority in the cross-sibling relation: sister outranks brother. In the parallel matrix, asymmetric relations are based on primogeniture: elder sibling outranks younger sibling. (In traditional Tongan kinship analysis, as Biersack notes, the first asymmetry refers to a difference in rank, the second to a difference in authority or power: Kaeppler 1971, Bott 1981.) In the Tongan political system, titles are ranked by seniority: junior titles are derived from the line of senior titles. Tongans may refer to the relation between chiefs or titles as father/son or elder brother/younger brother. Titles and land associated with them are bestowed by senior chiefs in the *kava* ceremony. Thus in the parallel matrix, there is a unidirectional flow of land titles and *kava* from superior to inferior.

Biersack notes that the circulation of food (*ngāue*) among cross-relatives and drink (*kava*) among parallel relatives is a contrast Tongans themselves make. In the *kava* ceremony, for example, those who attend bring *kava* root and cooked pork (*fono*). The *kava* is distributed among and consumed by chiefs—actual or metaphorical parallel relatives—while the food is taken by cross-relatives. According to Rogers, whom Biersack cites,

> The sovereign is considered too high in rank for any Tongan to carry off and eat his *fono* so a Fijian chief performs the task, but the *fono* of the other title-holders are taken by 'grandchildren' on their senior side chosen because they are of senior kinship status to the title-holders. These *mokopuna* 'grandchildren' are *fahu* to the title-holders; they should be from a line of descent which is related as *mehekitanga* (father's sister) to the title-holder's line (Rogers 1977: 169).

Seen as a problem in structural enumeration, there is an interesting footnote to Biersack's analysis of Tongan kinship. E. Bott, in her study of rank and power in Tonga, considers that Biersack's formulation (as originally presented in 1974) 'understresses the fact that it is only the patrilateral, not the matrilateral, parallel relatives that have authority' (Bott 1981: 78). In reply Biersack notes the restrictions on available ethnographic data and proposes a further line of research:

> My data are drawn from published materials, and to date this material has concentrated on the *male* parallel matrix to the exclusion of the female parallel matrix. The model I develop . . . suggests that transactions among matrilateral parallel relatives systematically contrast with transactions among patrilateral parallel relatives at the same time they are distinctive from transactions among cross-relatives. Perhaps further research will test the hypothesis (Biersack 1982: 209).

In other words, the two remaining graphs in Fig. 6.12—the second and fourth, sororilateral graphs—remain to be studied by Biersack or by someone else for a complete account of this aspect of exchange relations in Tongan kinship.

Structuralistes avant la lettre

The best-known models of marriage exchange in Melanesian anthropology are probably those of A. B. Deacon (1927) for Ambrym. Less well known but certainly of comparable historical and theoretical interest are Reo Fortune's theoretical models of the varieties of cross-cousin marriage. In devising his models Fortune made use of the operation of sex duality in kinship graphs.

To contrast different kinds of relations between descent groups Fortune constructed purely formal models of matrilateral and patrilateral cross-cousin marriage, with sexually dual versions of each. Fortune showed the exchange implications of each type, and like Lévi-Strauss (1969) but unlike Leach (1951) he interpreted matrilateral cross-cousin marriage as an exchange of women *per se*, not as an exchange of women for goods. Fortune published the result of this *Gedanken*-experiment in a short article in *Oceania* in 1933, where it appears to have been completely ignored until after the publication of Lévi-Strauss's work in 1949, when it was noted by

Leach (1951) and Needham (1958, 1968). Fortune's diagrams, evidently prepared by W. E. H. Stanner, are of the same kind as those used in modern kinship analysis (except for the curious retention of the biological Venus symbol for female). Fig. 6.14 reproduces his diagram of matrilateral cross-cousin marriage with patrilineal descent. This figure explicitly shows (*b*) giving wives to (*a*) and (*c*) to (*b*). If augmented to include (*a*) giving wives to (*c*), it could be regarded geometrically as a cylinder.

Fortune described the enduring relation between two descent groups which this marriage rule creates by saying that the men of one line have a 'lien in perpetuity on the sisters of the men of the other' (1933: 2). He observed this system among the Tchamburi (called Tchambuli by Mead 1963, and Chambri by Gewertz 1983) of the Sepik region. He noted that although payments are made for women in marriage, they are not to be regarded as an exchange of goods for women as absolute property. Rather, 'the social drama is conceived as a play between patrilineal lines, women become the object of liens in perpetuity, and this type of entail is validated by payments made against it by those whom it benefits' (1933: 3). Like Hodson eight years earlier, and van Wouden two years later, Fortune clearly recognized the cyclical implications of this marriage rule: 'Thus in [Fig. 6.14] male line (*a*) has a lien in perpetuity on the

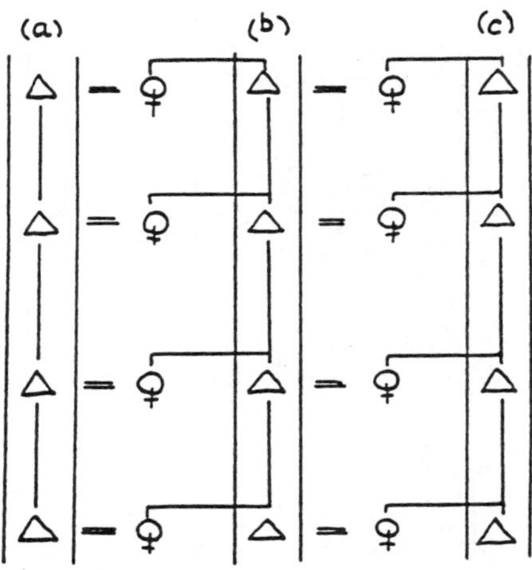

FIG. 6.14. Fortune's (1933) diagram of matrilateral cross-cousin marriage

sisters of male line (*b*). Line (*b*) is to (*c*) as (*a*) is to (*b*), and so on until (*z*) is to (*a*) as line (*a*) is to (*b*)' (1933: 2).

Fortune's second diagram, reproduced in Fig. 6.15, shows patrilateral cross-cousin marriage with patrilineal descent. In this case instead of a lien of one line upon another, the two lines are related by (direct) reciprocity: 'line (*a*) gives a sister to line (*b*) in marriage in alternate generations, while line (*b*) gives a sister to line (*a*) in marriage in the interlocking alternate generations. The writer has not seen this system in operation personally. It occurs, for example, sporadically among the Iatmul tribe of the Sepik River, where Bateson reports the recognition by the natives of the daughter given in marriage in reciprocity for the mother' (1933: 3).

Fortune then constructed the sex duals of these two models, as reproduced in Figs. 6.16 and 6.17, in order to examine the relations between descent groups having these marriage rules. In the case of matrilateral cross-cousin marriage, the theoretical result is that one line of women has a lien in perpetuity on the brothers of the women in another line. Fortune knew of no matrilineal society where the entail in men had been 'perfected'. What happens instead, as in Fiji, is that the men of the matrilineal lines have a lien on the mother's brother's daughter. Men exchange women, whatever the rule of

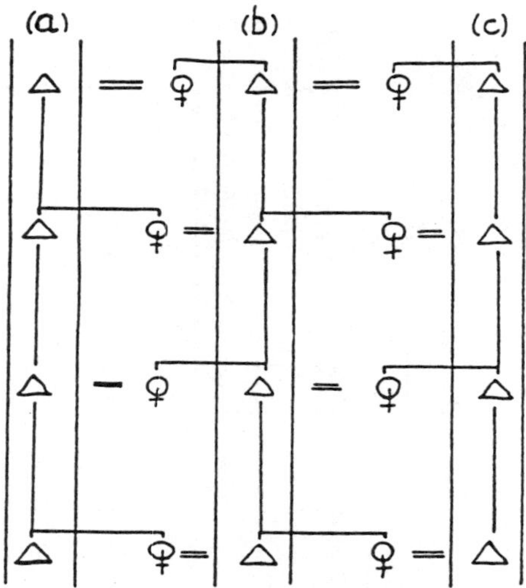

FIG. 6.15. Fortune's (1933) diagram of patrilateral cross-cousin marriage

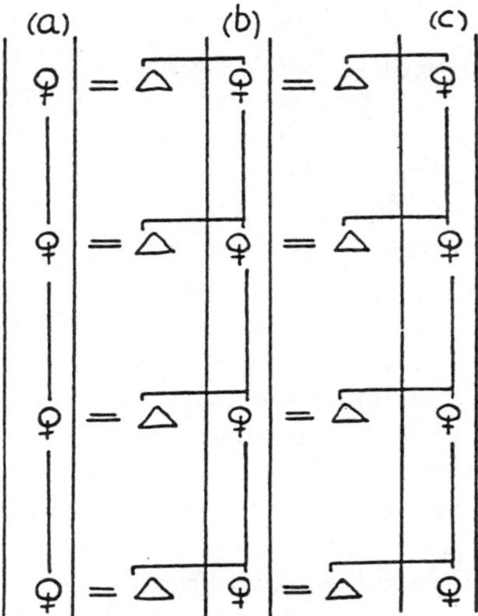

FIG. 6.16. Fortune's (1933) sex dual of Fig. 6.14

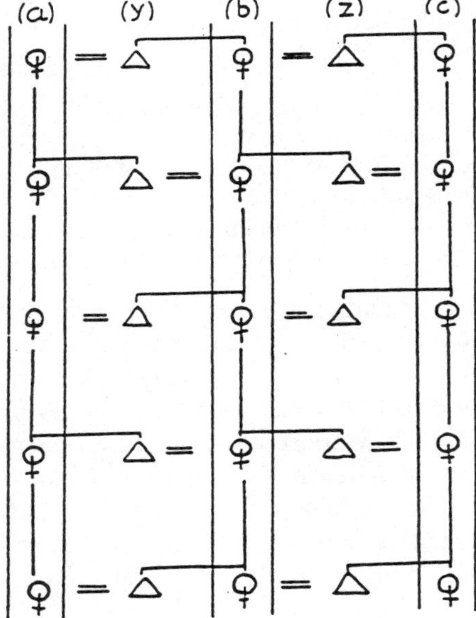

FIG. 6.17. Fortune's (1933) sex dual of Fig. 6.15

descent. In the case of patrilateral cross-cousin marriage, Fortune cited Malinowski on aristocratic marriage in the Trobriands, which has matrilineal lines but which does not consist of women exchanging brothers in alternate generations.

From both a historical and a cognitive view it would be interesting to know how Fortune made his discovery. His article is short, a mere nine pages, and contains no references to any literature—not to W. E. Armstrong (1928), for example, who had tried a kind of do-it-yourself structural enumeration of marriage class systems in an appendix to his monograph on Rossel Island. Fortune's references to Iatmul and Tchamburi societies, however, are significant. It is clear from Mead's (1972) autobiography that when she and Fortune and Bateson were together on the Sepik in 1932–3, they tried to establish significant structural contrasts between Iatmul and Tchamburi. Bateson had been studying the former, which had a rule of FZD marriage, while Mead and Fortune were beginning to study the latter, which turned out to have a rule of MBD marriage. Thus a happy accident of fieldwork undoubtedly facilitated Fortune's grasp of these two marriage rules as aspects of a single conceptual system. Even though nothing was to come of it, Mead and Fortune recognized the significance of his achievement:

> In Tchambuli, while Reo was working out the kinship system, he had one of those flashes of technical insight that make all scientific work worthwhile, as he realized that two types of kinship systems, which hitherto had been described as inherently different, were in actuality mirror images of each other. Gregory came over the afternoon Reo worked this out and marveled at our gaiety. Reo published his finding in *Oceania*, in an article modestly entitled 'A Note on Cross-Cousin Marriage' (Mead 1972: 235).[11]

J. A. Barnes (personal communication) has suggested that Bateson may have contributed to Fortune's discovery by supplying a physical model of marriage exchange. According to Mead, Bateson was fond of using analogies from the physical sciences in their discussions. In his work on marriage class systems in Malekula, Layard (1942) credits Bateson with introducing him to the technique of drawing marriage exchange diagrams on a cylinder in order to display their cyclic structure. Bateson may well have done the same for Fortune.

[11] Fortune's article was actually entitled 'A Note on Some Forms of Kinship Structure'.

The year preceding Fortune's note, F. E. Williams published an article on what he called 'sex affiliation'. Among Koiari-speaking groups of Papua, male children are classed with their father's group and female children with their mother's group. Native interpretations of this classification contain two ideas: a genetic notion, according to which children have the sex characteristics of the same-sex parent; and a marriage rule, according to which a girl, when she marries, must return to her mother's group. The term 'mother's group' does not refer to a matrilineal descent group but to the girl's mother's father's or equivalently the girl's mother's brother's group. Williams defines Koiari groups as (patri-) local not lineal, but Schwimmer (1977) in his commentary on Williams considers that they are rudimentary patrilineal clans. The marriage rule is, in effect, a rule of patrilateral cross-cousin marriage. Williams pointed out that this rule implies an alternating exchange of women between groups:

> By the rule of sex affiliation when a woman born in one group (A) marries into another group (B) her daughter should go back to the original group (A) to marry; their daughters again should go back to (B), and so on. That is to say, (A) and (B) constitute a pair bound permanently together by a succession of unions (Williams 1932: 77).

Williams also pointed out that such a rule could link a number of groups, and he constructed a hypothetical model of a society as a system consisting of a 'multiplicity of reciprocating pairs':

> Group (A), as I have already implied, is not of necessity bound solely to group (B); it may have formed a marriage alliance with group (C) and a number of others as well. (A) and (B), (A) and (C), etc., are then reciprocating pairs. In any generation certain girls of (A) should marry into (B), and certain into (C), because their mothers came from those groups. At the same time, certain girls from (B) and (C) must marry into (A) because their mothers came from that group, and so on. This organization of society into a system or multiplicity of reciprocating pairs is, I think, the logical outcome of sex affiliation combined with marriage by exchange (Williams 1932: 78).

Although the word 'system' is used, there is no necessity for the exchange graph to be connected. Thus it might look like the graph in Fig. 6.18, which has two components.

The rule of sex affiliation combined with sister exchange may occur elsewhere in New Guinea, as suggested by a statement from

FIG. 6.18. A multiplicity of reciprocating pairs in sex affiliation

one of Mead's Arapesh informants: 'It is good that brother and sister should marry brother and sister, that if one clan gives two of its girls to the other, the other clan should reciprocate with two of its daughters' (Mead 1963: 81–2).

Structures *en creux*

Fortune's and Williams's theoretical modelling of exchange structures had no impact on Melanesian studies,[12] not only because of lack of interest in the exchange paradigm and in structural analysis but also because the elementary structures they describe appear to be uncommon in simple form. In New Guinea these rules exist more as structural tendencies (Van der Leeden 1960) or implicit structures (Gell 1975) in complex alliance systems. In Sarmi, for example, there is an

> ideal that female matrilineal descendants of a woman who has married out should marry back again into the original group after having circulated over various local groups for at least three generations. This matrilineal descent reckoning implies a tendency to asymmetrical or circulating connubium, for the inhabitants themselves formulate the ideal as follows, taking the local groups as basic units: a woman of group A marries a man of group B; her daughter marries into group C, etc., until finally the descendant is no longer a close relative and can marry back into group A. It is obvious that the notion of kinship is thus linked to the number of generations, and to the number of local groups which have been 'passed through'. The greater the number of intervening generations, and the further the descendants thereby diverge from 'home', the more traditional the tie of kinship becomes (Van der Leeden 1960: 132).

[12] According to Schwimmer (1977), Williams's entire body of work was so little regarded by British anthropologists that when he died in 1943, he did not merit an obituary in *Man*. The Fortune revival forecast by Lawrence (1980) will undoubtedly extend to Williams.

Such systems do not at all preclude structural analysis: they only require more sophisticated structural models. By way of example, R. C. Kelly (1968) has shown how a determinate marriage structure could be studied on the basis of its negative rules, or as he puts it, *en creux*. His source, ironically enough, is Fortune's (1932) monograph on Dobu.

Dobuan society is organized into matrilineages called *susu* (the term for mother's milk). A number of *susu*, from four or five to ten or twelve, all claiming a common female ancestress, reside in a village. Each such (clan-)village has its own bird totem. A number of villages, of differing, but usually replicated totemic affiliations, define a locality or a district. Villages within a district are related by ties of affinity and are allied in warfare against villages of other districts. Marriage at the district level is virtually endogamous.

According to Fortune, marriage is prohibited (1) between cross-cousins, (2) between members of the same village, and (3) between members of the same totemic affiliation in different villages. In addition, parallel marriages—those which replicate already existing marriage ties between two villages—are discouraged. Thus brother and sister should marry into different groups.

The suggestion of an implicit determinate marriage system is contained in the following Dobuan origin myth:

> Green parrot became pregnant. She laid an egg and brooded over it. It hatched forth—it appeared, not a bird, a human being. The child, a female, grew. The nest collapsed. It fell to the ground.
>
> The husband, child of the White Pigeon, was at sea fishing with nets. The woman, child of the Brown Eagle, his wife, walked the shore looking for shell fish. She heard the child wailing. She took it to her village and suckled it. There it grew up. It became a very beautiful woman. The woman child of Green Parrot married. She bore four children, two male, two female. Her children grew to adult stature. The two brothers married women of the White Pigeon village. The two sisters married men of the Brown Eagle village. These children said 'we are the children of Green Parrot'. They founded our Green Parrot village.
>
> The woman hatched from the egg of Green Parrot was named Bolapas. Bolapas bore the daughters, Negigimoia and Daloyos. Daloyos bore Dosi. Dosi bore my mother (Fortune 1932: 31).

While Fortune found this myth quite illogical, Kelly interprets it not merely as a charter of totemic ancestry but as a model of generalized exchange between totemic groups. The marriages of Bolapas's four children show that White Pigeon gives women to Green Parrot while Green Parrot gives women to Brown Eagle. The totemic affiliations of the adoptive parents imply that Brown Eagle gives women to White Pigeon, thus generating a directed 3-cycle, the simplest possible model of generalized exchange.

Since there are practically no data on actual Dobuan marriages, Kelly proceeds to develop a model of generalized exchange consistent with the totemic one in the origin myth, capable of functioning within the set of negative marriage rules. These rules obviously require more than three groups over time. Kelly's model, which we will put into explicit digraphic form, posits a district with six matrilineal groups (clan-villages), two each of the same totem. In the first generation all six groups can marry in a single cycle of exchange, as shown in Fig. 6.19a. In the next generation, no two adjacent groups could marry (because of the rule prohibiting cross-cousin marriage), nor could any two groups three steps away marry (as they would have the same totem). All the negative rules could be satisfied, however, by means of marriage exchange in two 3-cycles, as shown in Fig. 6.19b. In the third generation, the complete cycle of the first generation could be repeated. The digraph combining these three cycles is shown in Fig. 6.19c (an isograph, as noted in Chapter 2).

It would be very convenient to have a succinct notation for such exchange structures. Our digraphs of Kelly's analysis shown in Fig. 6.19 can be easily characterized in terms of the matrix operations of Chapter 4. Let A be the adjacency matrix of the directed 6-cycle of Fig. 6.19a, labelled 1 to 6 in the usual way, so that

$$A = \begin{bmatrix} 0 & 1 & 0 & 0 & 0 & 0 \\ 0 & 0 & 1 & 0 & 0 & 0 \\ 0 & 0 & 0 & 1 & 0 & 0 \\ 0 & 0 & 0 & 0 & 1 & 0 \\ 0 & 0 & 0 & 0 & 0 & 1 \\ 1 & 0 & 0 & 0 & 0 & 0 \end{bmatrix}$$

For convenience we write $B = A'$, the transpose of A:

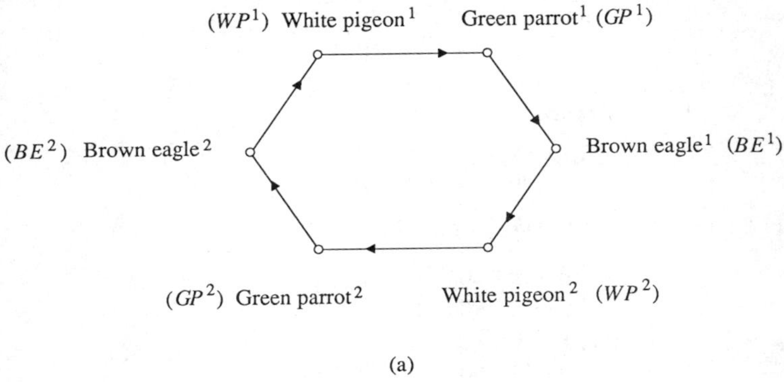

(a)

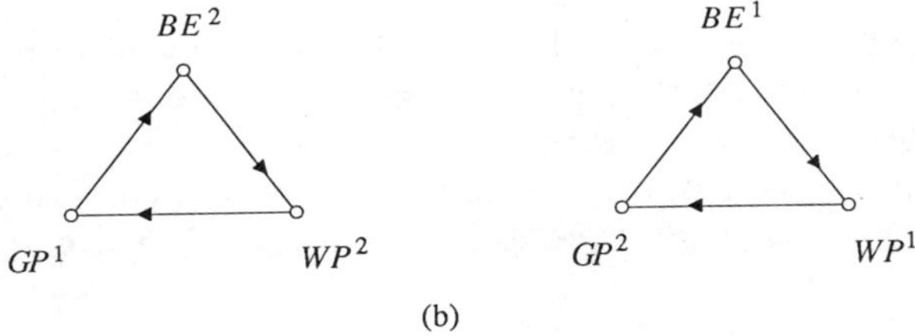

(b)

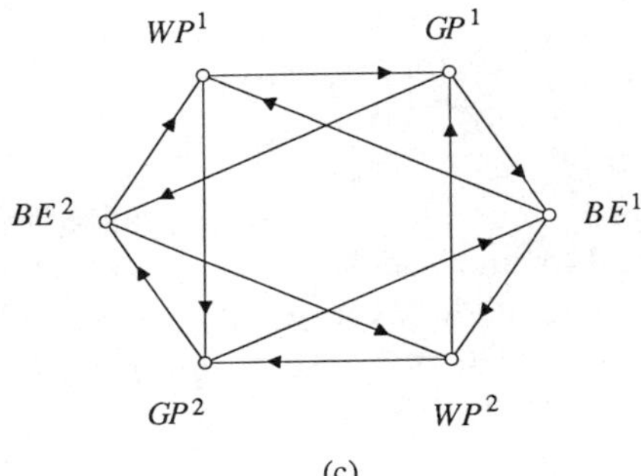

(c)

FIG. 6.19. Digraphs of implicit marriage exchange in Dobu (based on Kelly 1968)

$$B = A' = \begin{bmatrix} 0 & 0 & 0 & 0 & 0 & 1 \\ 1 & 0 & 0 & 0 & 0 & 0 \\ 0 & 1 & 0 & 0 & 0 & 0 \\ 0 & 0 & 1 & 0 & 0 & 0 \\ 0 & 0 & 0 & 1 & 0 & 0 \\ 0 & 0 & 0 & 0 & 1 & 0 \end{bmatrix}$$

Then square this matrix to get B^2,

$$B^2 = \begin{bmatrix} 0 & 0 & 0 & 0 & 1 & 0 \\ 0 & 0 & 0 & 0 & 0 & 1 \\ 1 & 0 & 0 & 0 & 0 & 0 \\ 0 & 1 & 0 & 0 & 0 & 0 \\ 0 & 0 & 1 & 0 & 0 & 0 \\ 0 & 0 & 0 & 1 & 0 & 0 \end{bmatrix}$$

We see at once that B^2 is the adjacency matrix of the digraph $2\vec{C}_3$ of Fig. 6.19b. With this notation, the adjacency matrix of the digraph of Fig. 6.19c is precisely $A + B^2$. The reason is that Fig. 6.19c is the union of the two digraphs above it. As these digraphs are arc-disjoint, $A + B^2$ is the adjacency matrix of their union.[13]

$$A + B^2 = \begin{bmatrix} 0 & 1 & 0 & 0 & 1 & 0 \\ 0 & 0 & 1 & 0 & 0 & 1 \\ 1 & 0 & 0 & 1 & 0 & 0 \\ 0 & 1 & 0 & 0 & 1 & 0 \\ 0 & 0 & 1 & 0 & 0 & 1 \\ 1 & 0 & 0 & 1 & 0 & 0 \end{bmatrix}$$

Kelly's model of a complex form of generalized exchange in Dobu takes up where Fortune and Williams left off. When the model's digraphic basis is made explicit, the way is opened for the mathematical study of many other seemingly intractable marriage systems in Melanesia.

[13] A similar matrix characterization could be given of Eyde's (1983) implicit digraphic model of second cross-cousin marriage in the Admiralty Islands (based on Mead 1934).

On a 'Calculus' of Melanesian Exchange Structures

The need for explicit enumeration techniques is illustrated by certain problems in a recent comparative work on Melanesian exchange structures. In an Appendix to *Gifts and Commodities*, C. A. Gregory (1982) proposes a matrix and combinatorial analysis of kinship and marriage. We cannot take up the entire model, but we can make the mathematical clarification necessary for its future application and, by implication, for similar models which are sure to follow, and in so doing provide a simple introduction to a part of combinatorics.

According to Gregory, five elementary matrices are sufficient to describe the basic exchange structures of many gift reproduction systems in Melanesia. For purposes of exposition, Gregory postulates a system with four marriage groups, practicing restricted or generalized exchange. Translated into the language of graph theory, he defines the structure of the former as a disconnected digraph consisting of two copies of a symmetric pair of arcs, i.e. $2DK_2$. He then shows the three ways to label this digraph in Fig. 6.20, together with their adjacency matrices, which he labels as R_1, R_2, R_3.

Gregory defines generalized exchange as a single directed 4-cycle, and shows what he asserts are the two possibilities in Fig. 6.21. This is not exactly correct, as we shall soon see. The direction of giving is reversed in the two labelled digraphs of Fig. 6.21, and so Gregory calls their adjacency matrices G and G' to indicate that one is the transpose of the other, where he uses G to stand for generalized exchange. But we have been consistently reserving G for graph, so we write A and A' in Fig. 6.21 for these two matrices.

Symbolizing a generic exchange matrix as E, Gregory concludes from his enumeration of exchange structures that restricted exchange exists if $E = E'$, i.e. matrix E is symmetric, and generalized exchange exists if $E \neq E'$.

Gregory's model is unknowingly but inconsistently based on the graph theoretic concept of the *number of labellings of a digraph*. This refers to the number of different (non-isomorphic) ways a given digraph can be labelled. Using this definition, Gregory's assertion that there are three possible restricted exchange structures is correct, but it is known from combinatorics that there are not two but six possible generalized exchange structures, defined as labellings of $\vec{C}_4$, a directed 4-cycle, as shown in Fig. 6.22.

$$R_1 = \begin{bmatrix} 0 & 1 & 0 & 0 \\ 1 & 0 & 0 & 0 \\ 0 & 0 & 0 & 1 \\ 0 & 0 & 1 & 0 \end{bmatrix}$$

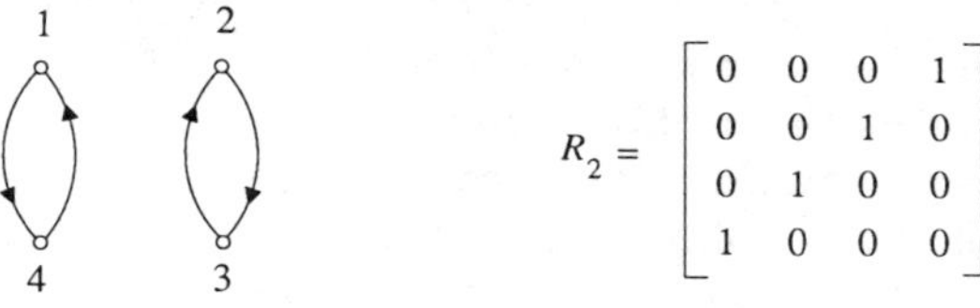

$$R_2 = \begin{bmatrix} 0 & 0 & 0 & 1 \\ 0 & 0 & 1 & 0 \\ 0 & 1 & 0 & 0 \\ 1 & 0 & 0 & 0 \end{bmatrix}$$

$$R_3 = \begin{bmatrix} 0 & 0 & 1 & 0 \\ 0 & 0 & 0 & 1 \\ 1 & 0 & 0 & 0 \\ 0 & 1 & 0 & 0 \end{bmatrix}$$

FIG. 6.20. Gregory's (1982) models of restricted exchange

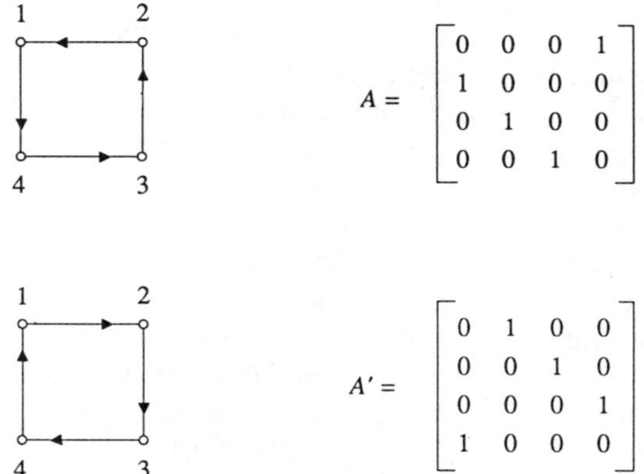

$$A = \begin{bmatrix} 0 & 0 & 0 & 1 \\ 1 & 0 & 0 & 0 \\ 0 & 1 & 0 & 0 \\ 0 & 0 & 1 & 0 \end{bmatrix}$$

$$A' = \begin{bmatrix} 0 & 1 & 0 & 0 \\ 0 & 0 & 1 & 0 \\ 0 & 0 & 0 & 1 \\ 1 & 0 & 0 & 0 \end{bmatrix}$$

FIG. 6.21. Gregory's (1982) models of generalized exchange

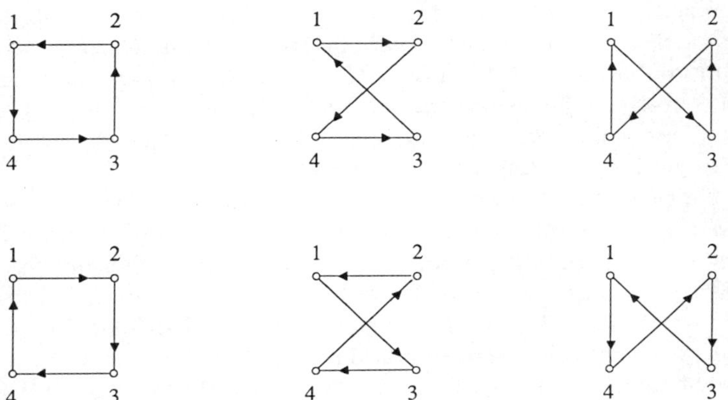

FIG. 6.22. The six labelled directed 4-cycles

Felicitously, the solution to Gregory's particular problem is contained in Harary and Read's (1966) article on the problem of determining the number of ways to label a '1-choice structure'.

A *1-choice structure* is a digraph in which there is exactly one arc from each point. These were called functional digraphs in Chapter 1, and were counted in Harary (1959a). In Fig. 6.23 all the functional digraphs with four points are shown up to isomorphism. The digraphs in the upper left and lower right hand corners are Gregory's models of restricted and generalized exchange.

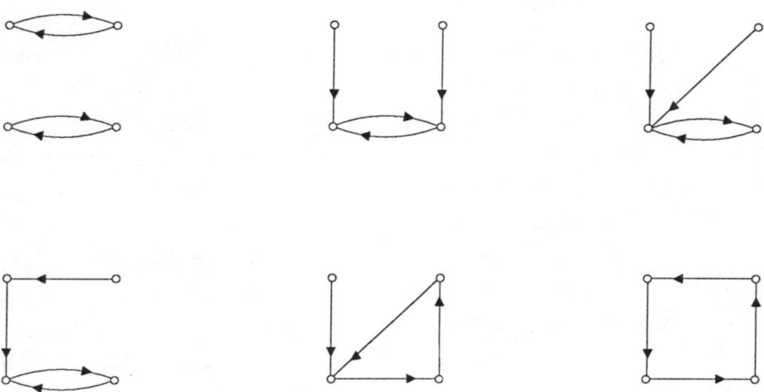

FIG. 6.23. The functional digraphs with four points

When an arc is drawn at random from each point to another point, each of the functional digraphs shown in Fig. 6.23 can occur in more than one way. In Fig. 6.22 we have shown the six different ways in which the last functional digraph in Fig. 6.23 can so occur; these are called *labelled functional digraphs*, since we are now distinguishing the four points from each other.

The proof that there are exactly six ways to label $\vec{C}_4$ is commonly given as an exercise in introductions to combinatorics under the heading of permutations and combinations: Find the number of ways of seating four people at a round table (where it matters who is on one's right and left). The roundness ensures that rotated seating orders are equivalent. The standard reasoning is that person 1 can be seated anywhere and has three possibilities for the first person to his left, followed by just two possibilities for the next person, after which the last person must sit in the one remaining seat.

To give the formula for the number of different labelled digraphs which have the same underlying unlabelled digraph D, we need to define the symmetry number of a digraph. An *automorphism* or *symmetry* of a digraph D is a mapping of the points onto themselves which preserves the arcs. It is well known (see Harary 1969: 161) that the collection of all the symmetries of a digraph forms a (mathematical) group. We will denote the order (number of elements) of this group by $s(D)$ and call it the *symmetry number* of D.

The determination of the number of symmetries of a digraph is not in general an easy matter. However, the structure of the directed 4-cycle (or any other directed cycle) is so regular that it is easy to derive. The equation is $s(\vec{C}_n) = n$, which we now illustrate for $n = 4$ using Fig. 6.24.

In this figure there are four isomorphic labellings of the directed 4-cycle. The first of these is said to be in standard position and is marked (0). The second has all the labels moved one notch clockwise and is designated (1). Continuing, the third has all the number-labels moved two notches clockwise, and is shown as (2). Finally, the last, marked (3), has all labels moved three notches clockwise, which is precisely the same as one notch counter-clockwise. We stop there, as one further notch brings us back to (0). (In mathematical terms this symmetry group is called the cyclic group of order 4.) On the other hand, in terms of the so-called 'new math', these are 'numbers on a clock', i.e. the set $\{0, 1, 2, 3\}$ under addition modulo 4. Thus we know that $s(\vec{C}_4) = 4$.

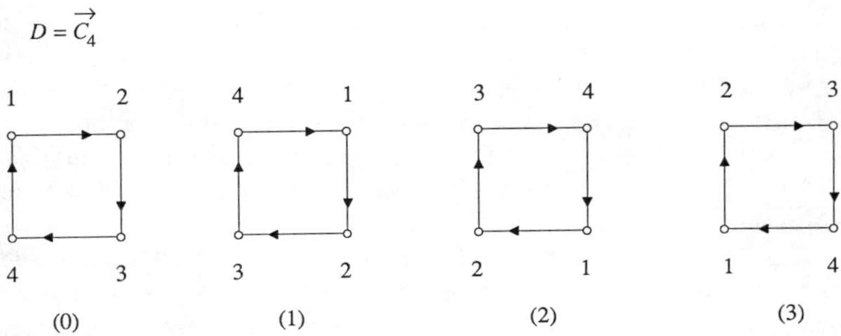

FIG. 6.24. Isomorphic labellings of the 4-cycle to illustrate all its symmetries

Harary and Read found that $l(D)$, the number of labellings of a digraph D with p points, is given by

$$l(D) = \frac{p!}{s(D)}.$$

For example, when $D = \vec{C}_4$, we saw that $p = 4$ and $s(D) = 4$, so $l(D) = 4!/4 = 6$, which is the number of labelled generalized

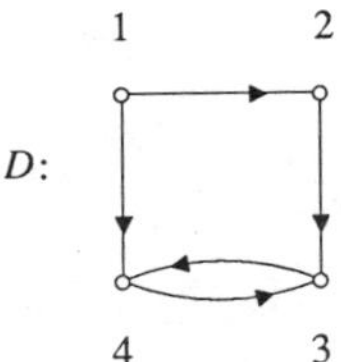

$$A = A(D) = \begin{array}{c} \\ 1 \\ 2 \\ 3 \\ 4 \end{array} \begin{array}{c} \begin{array}{cccc} 1 & 2 & 3 & 4 \end{array} \\ \begin{bmatrix} 0 & 1 & 0 & 1 \\ 0 & 0 & 1 & 0 \\ 0 & 0 & 0 & 1 \\ 0 & 0 & 1 & 0 \end{bmatrix} \end{array} \qquad A' = A(D') = \begin{array}{c} \\ 1 \\ 2 \\ 3 \\ 4 \end{array} \begin{array}{c} \begin{array}{cccc} 1 & 2 & 3 & 4 \end{array} \\ \begin{bmatrix} 0 & 0 & 0 & 0 \\ 1 & 0 & 0 & 0 \\ 0 & 1 & 0 & 1 \\ 1 & 0 & 1 & 0 \end{bmatrix} \end{array}$$

FIG. 6.25. The adjacency matrix and transpose of a non-symmetric digraph

exchange structures in Gregory's model. Similarly, the correct number of labelled restricted exchange digraphs $D = 2DK_2$ in his model is $l(D) = 4!/8 = 3$, which he did obtain. The appendix to Harary and Read (1966) catalogues all the functional digraphs with up to seven points, giving for each digraph its symmetry number $s(D)$ and the number of labellings of D, so it includes all the above information.

With regard to Gregory's proposed matrix criteria for exchange forms, it is obviously not true in general either that $A = A'$ defines restricted exchange or that $A \neq A'$ defines generalized exchange. The condition $A = A'$ only guarantees that the exchange relation is symmetric, while $A \neq A'$ does not guarantee that the relation is asymmetric, as illustrated by the non-symmetric digraph and adjacency matrix in Fig. 6.25. And symmetry and asymmetry are in any case necessary but not sufficient conditions of restricted and generalized exchange, respectively.

7

Binary Operations and Groups

> In my own thinking I drew on the work of Jung, especially his fourfold scheme for grouping human beings as psychological types, each related to the others in a complementary way. Gregory, who tended to use biological analogues, invoked the formal patterns of Mendelian inheritance.
>
> Margaret Mead, *Blackberry Winter*

In developing Mauss's (1935) ideas concerning the cultural symbolism of bodily techniques, Mary Douglas proposes that notions about the body reflect social morphology:

> The social body constrains the way the physical body is perceived. The physical experience of the body, always modified by the social categories through which it is known, sustains a particular view of society. There is a continual exchange of meanings between the two kinds of bodily experience so that each reinforces the categories of the other (Douglas 1973: 93).

As examples she gives correspondences between attitudes towards mental states of dissociation, trance, and the 'relative unarticulateness of social organization' (a number of African societies); restrictions on the ingestion of cooked food and caste purity (India); lateral body symmetry and caste hierarchy (Coorgs); attitudes toward 'bodily margins' and the pressure of political boundaries (Israelites); and pollution rules expressing an 'obsession' with the 'behaviour of liquids' and the 'fluid formlessness . . . of . . . highly competitive social life' (Yurok) (Douglas 1966: 127).

The Yurok case is of particular interest with respect to Papua New Guinea societies, since these are characterized by 'aggressive individualism' (Forge 1972), competitive exchange, and elaborate beliefs concerning the flow of bodily substances. As Gell has observed,

> The highlanders, indeed, seem to have developed an elaborate, almost mathematical, theory of body-contents in equilibrating the relation between intake, output, and internal pressure, perhaps (though the ethnographers have not viewed the matter in this light) as a response to their highly regulated exchange systems; the balanced receipt, storing, increase and redistribution of wealth being reflected in an 'economic' theory of the body (Gell 1975: 252–3).

Caroline Islanders in Micronesia, by contrast, seem to view the body more as a grid than as a container, perhaps as a response to the distinctive properties of their exchange system; the prescribed sequences of travel and exchange and ordered flows of wealth being reflected in a geometrical theory of the body.

We shall analyse two different kinds of ethnophysiological structures: one from Mount Hagen in Highland New Guinea defined by pollution and related beliefs, and one from Lamotrek in the Caroline Islands defined by anatomical concepts. The analysis of these structures requires binary operations on graphs and the application of (mathematical) groups. The binary operations are also useful in the notation of structural forms, as we shall briefly indicate.

A Micronesian Anatomy Graph

On several occasions Alkire (1965, 1970, 1972) has emphasized the 'lineality' of Carolinian thought, as exemplified in the star path compass, the network of sea-lanes (analysed in Chapter 3), knot divination, and even in anatomical concepts.[1] Fig. 7.1 from Alkire (1965) shows the Lamotrekan image of the trunk as a grid of parallel and intersecting lines.

Alkire provides the following brief description:

> In anatomy the trunk is thought of as being divided by six parallel lines, three to each side of the spine, which is itself represented by a central line [Fig. 7.1]. Each of the six lines has nine significant points upon it, while the spinal line has ten. The sixth point from the top on this latter is called *lukh* or center (it falls just above the navel). Thus, there are 64 significant points of reference on this part of the body. When effecting a cure through massage—a common method on Lamotrek—pressure is applied at one of these points and directed along the vertical or diagonal lines which pass from point to point. The basis of the anatomical model, then, is a grid system of parallel and intersecting lines where the significant reference points are found at the points of intersection (Alkire 1965: 126–7).

The resemblance between the anatomy graph and the sea-lane graph in Chapter 3 (Fig. 3.2), between images of the physical body

[1] See also Hage (1978) on the lineality of Carolinian mnemonic and cultural structures.

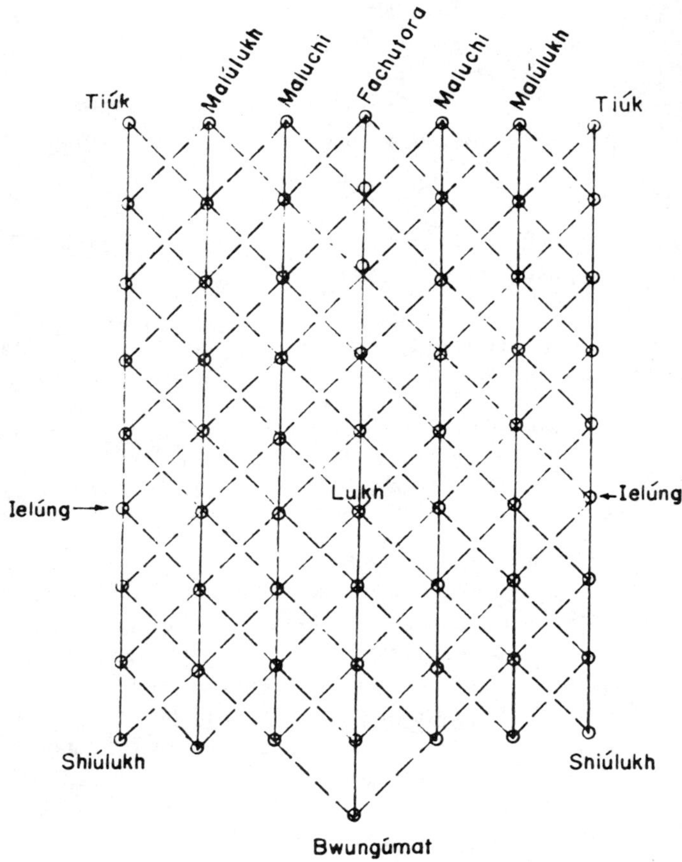

FIG. 7.1. A Lamotrek anatomy graph (from W. H. Alkire, 1965)

and the social body, becomes especially clear when it is recognized that the lines in each are 'subdivided'.

An *elementary subdivision* of a graph G is a graph obtained from G by the removal of some line $e = uv$ and the addition of a new point w and the two new lines uw and vw, i.e. by the insertion of one new point w into the line e. A *subdivision* of G is a graph obtained from G by a succession of elementary subdivisions. If every line of G is subdivided just once, the resulting graph is the *subdivision graph* $S(G)$, as illustrated in Fig. 7.2.

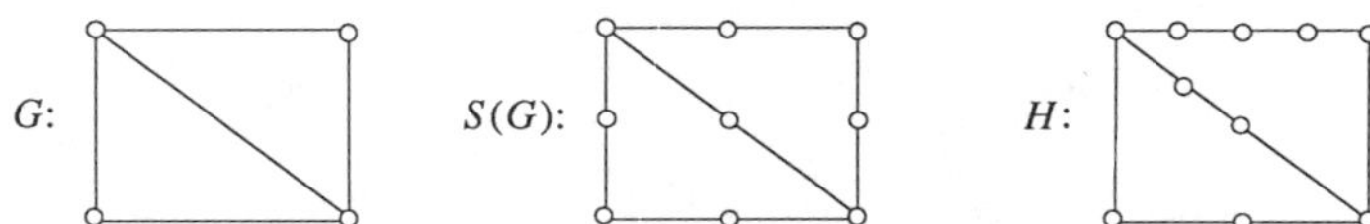

FIG. 7.2. The subdivision graph $S(G)$ and a subdivision H of G

In noting the subdivision of the lines in the anatomy graph, Alkire speculates on possible similarities in the sea-lanes:

During a voyage to Satawal it did not occur to me to question the *pelu* about any subdivision or *lukh* of Uoirekh [the sea-lane between Satawal and Lamotrek]. Nevertheless, as Gladwin (1962: 8–9) has also suggested in his discussion of Trukese cognitive processes, the navigator must be continually aware of his position, so that he knows the times and places where tacking is necessary. It is quite probable that the navigator's image of the seaway, under these circumstances, is subdivided—possibly at named intervals (Alkire 1965: 127).

The sea-lanes are, in fact, subdivided at named intervals in a uniquely Carolinian navigational technique called *etak* (Gladwin 1970, Lewis 1978). In sailing from island u to island v, a third, or *etak* island, w, is used as a reference point. During the course of the voyage the canoe is seen as standing still, while the *etak* island moves back under a fixed sequence of stars. This sequence divides up the sea-lane into a succession of *etak* intervals and allows the navigator to picture the course of his voyage. David Lewis diagrams the system as shown in Fig. 7.3 and relates the famous Puluwatese navigator Hipour's description of its operation:

'You must know about *etak* if you are to understand our navigation,' he said through the interpreter, as he set out his collection in a pattern. 'Here is your canoe.' He pointed to the knife. 'These fishhooks are three islands, the one you are leaving, the one you are bound for and, away out at right angles to your course, the *etak* or reference island. The coins are the star compass points.

'From your home island the *etak* island lies "under" Antares. (I have substituted the Western star names.) By the time your canoe is here'—he shifted the knife to point B—'the *etak* island has "moved" to the star point Aldebaran. When you reach your goal the *etak* island has "moved" further back still. Now it is "under" the Pleiades.

'Of course, the island doesn't really move, but you cannot navigate properly unless you look at it this way. As we picture it, the canoe is

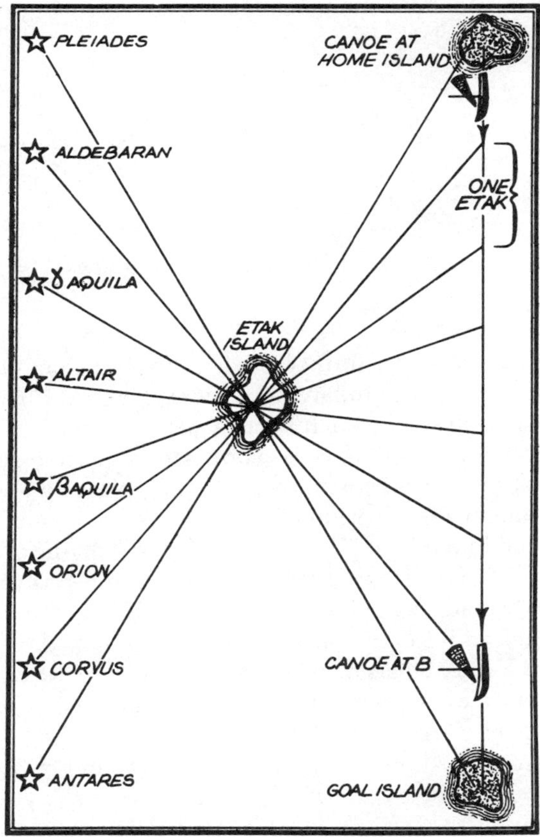

FIG. 7.3. Hipour's model of *etak* (from D. Lewis, 1978)

standing still and so are the stars. The island you have left moves further and further behind and the one you are aiming for comes closer and closer. All the while you will be judging in your mind how far back your *etak* island has moved. Each star point that it moves back means that you have progressed one *etak*' (Lewis 1978: 145–7).

We shall make a brief digression, essential to complete our earlier discussion of planarity. Two graphs G_1, G_2 are *homeomorphic* if both can be obtained from the same graph G by a sequence of subdivisions of lines, or alternatively if there is a graph H obtainable

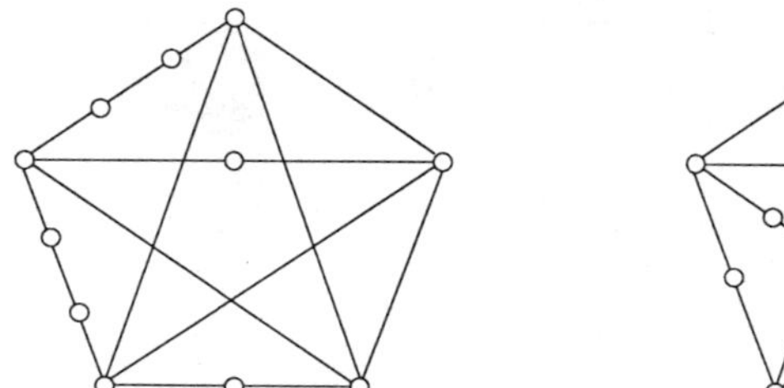
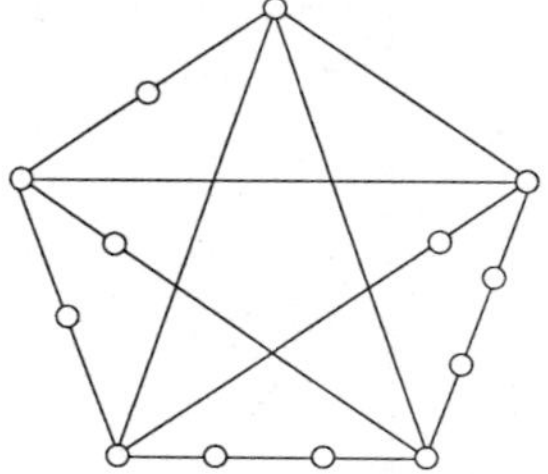

FIG. 7.4. Two homeomorphic graphs

from both G_1 and G_2 by further subdivision. For example any two cycles, regardless of length, are homeomorphic, and Fig. 7.4 shows two homeomorphs of K_5, each being a subdivision of K_5.

In Chapter 4 we cited, in connection with archaeological models of trade networks, the theorem that a graph is planar if and only if it does not contain a subdivision of K_5 or $K_{3,3}$ as a subgraph. Theorem 4.4 has a dual found by K. Wagner (1937) based on the operation of contraction. An *elementary contraction* of a graph G is obtained by identifying two adjacent points u and v, that is, by the removal of u and v and the addition of a new point w adjacent to those points to which u or v was adjacent in G. Alternatively, it is obtained by shrinking the line $e = uv$ to a new point w. A graph G is *contractible* to a graph H if H can be obtained from G by a sequence of elementary contractions. For example, as shown in Fig. 7.5, the first graph (called the Petersen graph) is contractible to K_5 by contracting

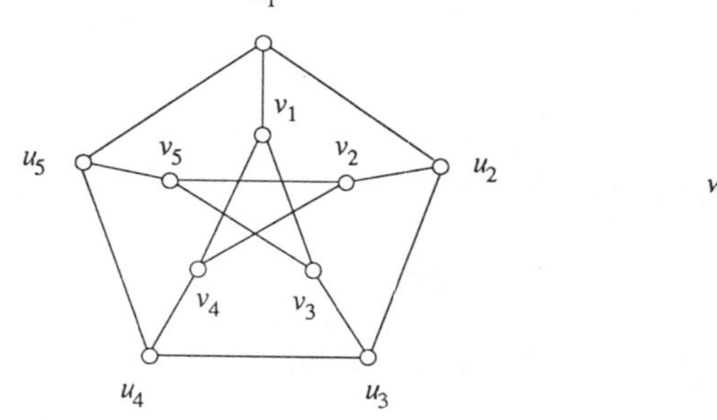

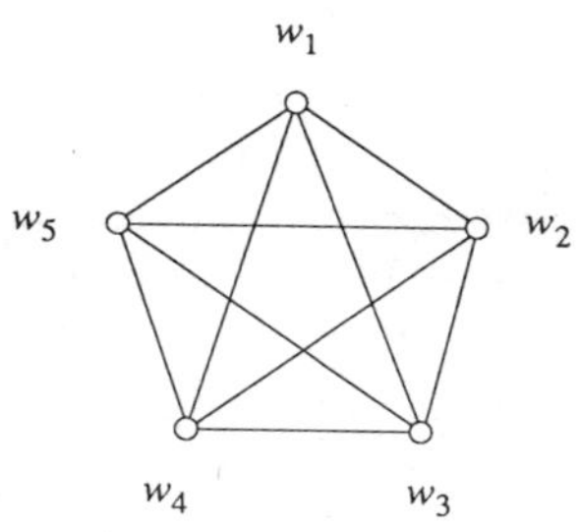

FIG. 7.5. A contraction of a graph

each of the five lines $u_i\, v_i$ joining the pentagon with the pentagram to a new point w_i.

Theorem 7.1. A graph is planar if and only if it does not have a subgraph contractible to K_5 or $K_{3,3}$.

In order to continue our structural characterization of the Lamotrek anatomy concept, we require two binary operations on graphs: the union and the conjunction. For convenience we place these two operations in a taxonomy of five such operations.

Throughout this section, two disjoint (not at all overlapping) graphs G_1 and G_2 have point sets V_1 and V_2 and line sets E_1 and E_2 respectively. Their *union* $G = G_1 \cup G_2$ has, as expected, point set $V = V_1 \cup V_2$ and line set $E = E_1 \cup E_2$. Their *join* is denoted by $G_1 + G_2$ and consists of $G_1 \cup G_2$ and all lines joining V_1 with V_2. These operations are illustrated in Fig. 7.6, with $G_1 = K_2$ and $G_2 = K_{1,2} = P_3$.

To define the (Cartesian) *product* $G_1 \times G_2$, consider any two points $u = (u_1, u_2)$ and $v = (v_1, v_2)$ in $V = V_1 \times V_2$. Then u and v are adjacent in $G_1 \times G_2$ whenever $[u_1 = v_1$ and u_2 adj $v_2]$ or $[u_2 = v_2$ and u_1 adj $v_1]$. The product of $G_1 = P_4$ and $G_2 = P_3$ is shown in Fig. 7.7. Although the definition of the product $G_1 \times G_2$ may appear abstruse at first sight, one can follow its meaning clearly by looking at the labels on the points of Fig. 7.7. In general the distance on the graph which is the product of two paths $P_m \times P_n$ is traditionally called the 'Manhattan metric', in which the two paths refer to the streets and the avenues; this graph is often called a 'grid' or a 'mesh'.

The *composition* $G_1[G_2]$ of two graphs G_1 and G_2 has a very

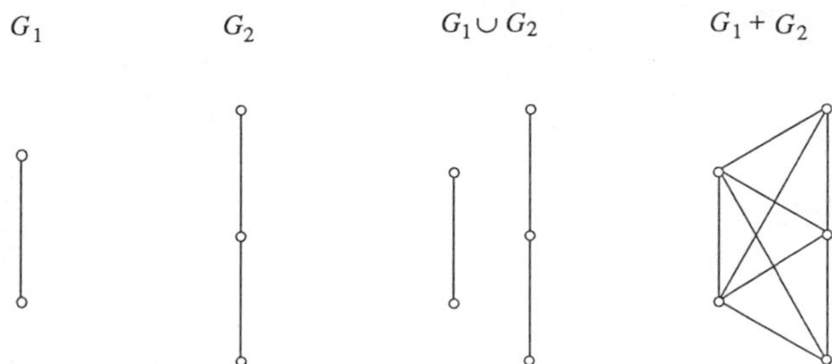

FIG. 7.6. The union and join of two graphs

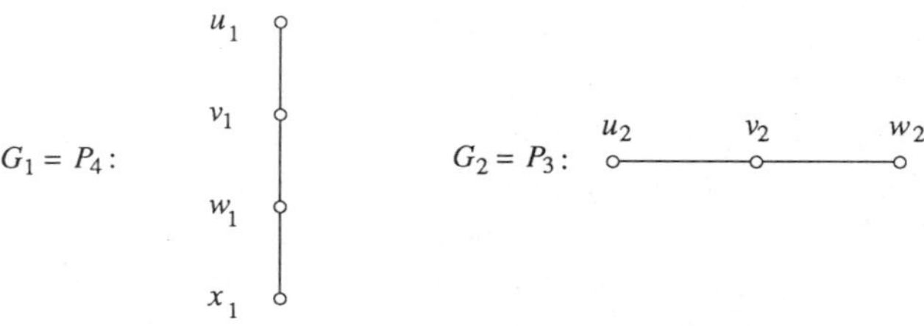

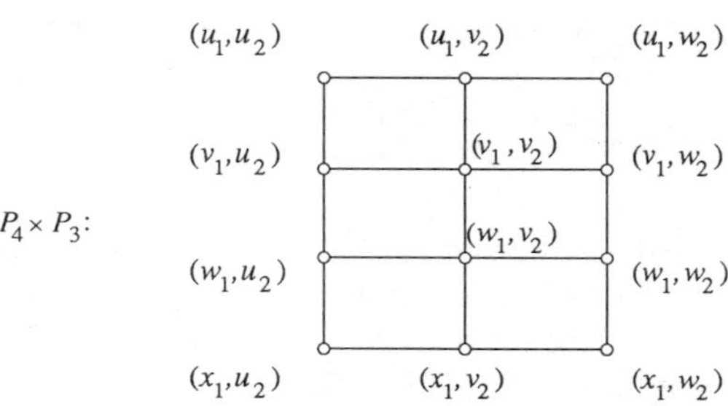

FIG. 7.7. The product of two graphs

interesting construction in which each point of G_1 is replaced by an entire copy of the second graph G_2; then whenever there is a line e in G_1, the two copies of G_2 at its points will be joined with every possible new line between points of the different copies. In other words, we have performed the join of these two copies of G_2. Fig. 7.8 shows the composition $G_1[G_2]$ when these two graphs are paths, $G_1 = P_3$ and $G_2 = P_2 = K_2$. Each point of G_1 has been replaced by a copy of G_2 drawn as a solid line segment, and the lines joining appropriate copies of G_2 are drawn, for illustrative purposes only, as dotted line segments.

The definition of the conjunction G_1 and G_2 of two graphs is a bit more delicate. The notation from logic for the conjunction is $G_1 \wedge G_2$. Its point set is the Cartesian product $V_1 \times V_2$ of the point sets of G_1 and G_2, i.e. the collection of all ordered pairs (u_1, u_2). Then the definition of the conjunction is completed by specifying when two of its points are adjacent, i.e. by characterizing its set of lines. In the *conjunction* $G_1 \wedge G_2$, the points $u = (u_1, u_2)$ and $v = (v_1, v_2)$ are

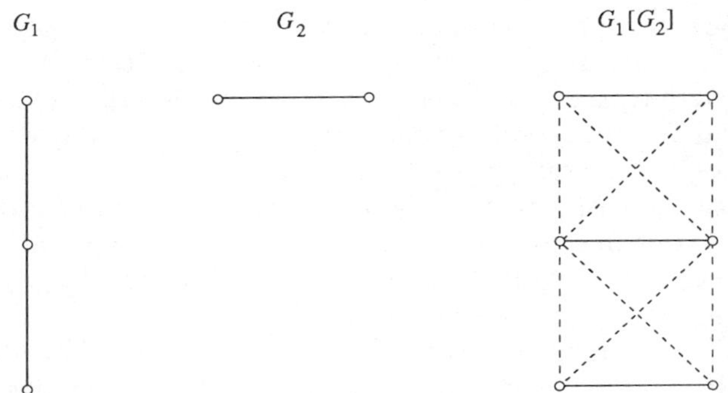

FIG. 7.8. The composition of two graphs

adjacent just when both u_1 and v_1 are adjacent in G_1 and u_2 and v_2 are adjacent in G_2. Before giving an example, we remark that it follows at once from the definition that the conjunction of two complete graphs must also be complete.

We show the conjunction of two graphs which are the paths P_4 and P_5 in Fig. 7.9. The points of the first graph have subscript 1 and those of the second graph have subscript 2. In the conjunction we have labelled only the two points (u_1, u_2) and (v_1, v_2). They are adjacent in $G = P_4 \wedge P_5$ because they satisfy the condition of the definition that u_1, v_1 are adjacent in G_1 and u_2, v_2 are adjacent in G_2.

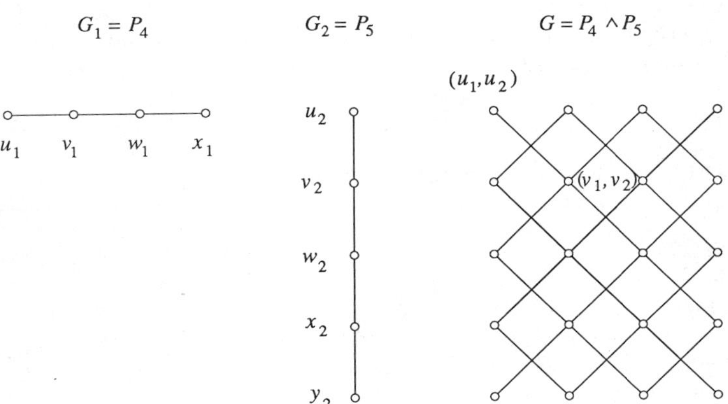

FIG. 7.9. The conjunction of two graphs

This particular conjunction was chosen for illustration because it serves to clarify the graph of Fig. 7.1, as we shall see.

If G_1 and G_2 are (p_1, q_1) and (p_2, q_2) graphs respectively, then for each of the above operations one can calculate the number of points and lines in the resulting graph, as shown in Table 7.1.

We can now describe the Lamotrek anatomy graph using the operations of the conjunction and the union. In the graph of Fig. 7.1, the broken lines are produced by the conjunction $P_7 \wedge P_9$ of the two paths P_7 and P_9. The solid lines are the union $7P_9$ of seven copies of P_9. The addition of a three-point star, $K_{1,3}$, to the middle line (centred at Bwungúmat) gives (as a union of three subgraphs which are not disjoint) the entire graph $G = (P_7 \wedge P_9) \cup 7P_9 \cup K_{1,3}$. Thus G is the superposition (technically, a *factorization* in graph theory) of three line-disjoint subgraphs: the conjunction $P_7 \wedge P_9$, seven copies of path P_9, and a star $K_{1,3}$.

On the Notation of Exchange Structures

It is convenient to be able to express the structure of a given graph in terms of smaller and simpler graphs. It is also of value to have notational abbreviations for graphs which occur frequently. We have already introduced the complete graph K_p, the cycle C_n, the path P_n, and the complete bigraph $K_{m,n}$. We have also noted the following unary operations: the complement $\bar{G}$, the converse D', and the negation S^- of a graph, a digraph, and a signed graph respectively. In Chapter 6 we indicated the potential of matrix characterization of exchange structures. Binary operations on

TABLE 7.1. *Binary operations on graphs*

Operation		Number of points	Number of lines
Union	$G_1 \cup G_2$	$p_1 + p_2$	$q_1 + q_2$
Join	$G_1 + G_2$	$p_1 + p_2$	$q_1 + q_2 + p_1 p_2$
Product	$G_1 \times G_2$	$p_1 p_2$	$p_1 q_2 + p_2 q_1$
Composition	$G_1[G_2]$	$p_1 p_2$	$p_1 q_2 + p_2^2 q_1$
Conjunction	$G_1 \wedge G_2$	$p_1 p_2$	$2 q_1 q_2$

graphs provide an additional means of succinct description. Thus, for example, Fig. 6.18, illustrating F. E. Williams's multiplicity of reciprocating pairs, is the union of the complete graph K_3 with K_4 minus one edge, or $K_3 \cup (K_4 - e)$. It is also possible to give a characterization, if required, based on more than one binary operation, as we now illustrate.

In a seminal article, 'Siblings in Tangu', Kenelm Burridge (1959) called attention to the practical significance of siblingship as opposed to descent in Tangu, a small hunting, gathering, and agricultural society living near the north coast of New Guinea in Madang Province. In Tangu it is siblingship which determines the structure of marriage exchange and co-operation. According to Burridge, descent is even calculated from siblingship rather than conversely. What descent is for Tallensi (Fortes 1949), siblingship is for Tangu. As a direct and indirect result of Burridge's article, many Oceanists have come to focus on the ethnographic primordiality of sibling relations in societies found in Polynesia (Huntsman 1981, on Tokelau) and Micronesia (Smith 1981, on Palau; Marshall 1981b, on Truk) as well as in Melanesia (Goodale 1981, on the Kaulong in New Britain). Burridge's work was a major source of inspiration for Kelly's (1974) models of Etoro social structure. In *SMA* we clarified, by putting into graph theoretic form, the relational structure of Etoro siblingship, and in Chapter 2 of this volume we showed that his model (Fig. 2.14) of marriage exchange and siblingship is a bigraph. In terms of binary operations on graphs, Kelly's lattice graph is quite simply defined as the product $P_4 \times P_5$, which is a grid.

In an article which appeared the year before 'Siblings in Tangu', Burridge (1958) presented a model of sibling and marriage relations. In Tangu, brothers co-operate and brothers-in-law exchange (feasts and dances) with each other. In Tangu conception, if a man married 'just any woman' he would have to 'look two ways', i.e. he would have exchange obligations with two different partners, his sister's husband (ZH) and his wife's brother (WB). But if sisters married brothers of brides, then ZH and WB would be the same person and ego would have only a single exchange obligation. Ideally, a Tangu community consists of two intermarrying sibling sets, as shown in Burridge's diagram reproduced in Fig. 7.10a.

This kinship diagram can be expressed by one which involves complete graphs and two of the binary operations presented above. Each of the two sibling sets is modelled by the complete graph K_8, as

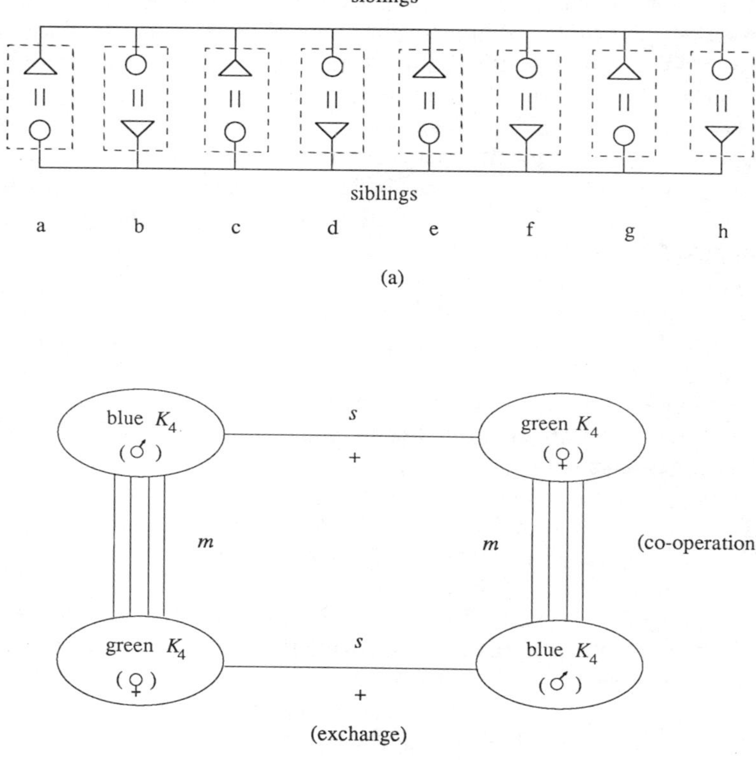

FIG. 7.10. The structure of siblingship and marriage in Tangu, (a) as represented in Burridge (1958), (b) as a combination of operations on complete graphs

it contains eight siblings consisting of four brothers and four sisters. Within each of the two sexes we thus have a copy of K_4. We colour the edges of the K_4 formed by the four brothers blue, and those of the sisters' K_4, green. In the lower half of Fig. 7.10b we place the lower sibling set of Fig. 7.10a, and in the upper part the upper sibling set. Hence the green K_4 in the lower left has as its point set the four sisters and the blue K_4 in the lower right the four brothers. The upper half is labelled similarly. The horizontal line on the bottom joins blue K_4 with green K_4 by the join operation, as every brother–sister pair are siblings, denoted by *s* at that join line. The same is true of the upper horizontal line. The join $K_4 + K_4$ is precisely K_8. Finally, the four vertical lines on each side denote the marriage relation *m*, and the left is the product $K_4 \times K_4$, with the right side similar.

Households *a*, *c*, *e*, *g* (the left side of Fig. 7.10b) co-operate with each other, as do households *b*, *d*, *f*, *h* (the right side), as wives are sisters and husbands are brothers of each other. The exchanges of feasts and dances take place between the two sets of households through the brother and sister ties.

There are undoubtedly more complex interactions between sibling and marriage relations which could be explicated by graph theoretic means. Smith (1983), for example, proposes that Palauan marriage should be seen from the perspective of cross-siblingship, and she maintains that a 'quadratic orientation of HZ/H : W/WB' (1983: 9) clarifies certain aspects of the 'matrilineal puzzle'.

Pollution Beliefs in Highland New Guinea

The following two propositions, one semiological and the other Durkheimian, jointly suggest a line of research on bodily techniques and gender concepts in New Guinea:

(1) The idiom of pollution lends itself to a complex algebra which takes into account the variables in each context.

Mary Douglas, *Purity and Danger* (1966: 9)

(2) L'aventure du sexe, dans ses hantises comme dans sa réussite, traduit en quelque sorte allégoriquement l'aventure de toute confrontation.

Remo Guideri, 'Note sur le rapport mâle/femelle en Mélanésie' (1975: 104)

Beliefs in menstrual pollution are a salient feature of Highland New Guinea societies and figure prominently in analyses of the by now classic topic of sexual disjunction. Aside from ecological interpretations (for example, Lindenbaum 1972), models so far have been mainly typological and taxonomic. Thus Meggitt (1964) defines two areal complexes consisting of beliefs in menstrual pollution, an emphasis on male purification cults, and the relatively low status of women (the Mae syndrome), and the opposites of these (the Kuma syndrome), which are correlated respectively with the negative and positive signs of inter-group relations—marriage with enemies and marriage with friends. Faithorn (1976) treats them as equivalent to other pollution beliefs, such as those concerning semen, while Meigs (1978) treats them as distinct from other danger beliefs, such as those concerning poison. In the first case, there are counter-examples and problems of definition, as in the determination of the relative status of women and the assessment of ambivalence in positive relations (A. Strathern 1969), as well as a tendency to oversimplify the marriage bond by reducing it to a simple asymmetrical relation of dominance, antagonism, or exploitation (Read 1952; Meggitt 1964, 1976; Langness 1967, 1977).[2] In the second case, there are contexts in which beliefs in menstrual pollution should be viewed as elements of a set of contrastive and complementary concepts; and also contexts in which such beliefs, not universally present or equivalently defined in Highland New Guinea, are only one among a number of alternative expressions of a common structure.

We have a double objective. The first is to show how pollution and related beliefs may be treated as part of a system of positive and negative transactions, the underlying structure of which is a transformation group. The second is to show how the empirical properties of such a system may reflect social morphology. The general proposition is illustrated with data from Mount Hagen in the Western Highlands (A. Strathern 1971; M. Strathern 1972). Since we are primarily concerned with presenting a new model for thinking about the structure and interpretation of these beliefs, we work from a highly condensed, but we hope sufficient, summary of social organization. As far as the beliefs are concerned, we should emphasize that their selection is determined not on a priori grounds but on cultural grounds. If the ethnography shows that certain beliefs are related to one another in definite ways, then it would be arbitrary not to treat them as a set and neglectful not to ask what the structure of this set is.

In Mount Hagen society, marriage is with allies. More precisely, Hageners marry into friendly clans rather than enemy clans; since the former are also 'minor enemies', that is, not adversaries in major warfare, the sign is not unequivocally positive. According to A. Strathern (personal communication), allies 'themselves might have marriage links with groups less well-disposed to one's own.

[2] See for example Feil's (1978) criticism.

Moreover, alliances were not entirely stable over time. This is relevant, I think, since it underlines the need, as perceived by Hagen men, to "socialize" in-marrying wives into proper attitudes about their bodies.' In terms of Meggitt's typology, M. Strathern (1972) places Mount Hagen between Mae-Enga and Kuma. There is none the less a marked belief in menstrual pollution, which is intimately related to ideas concerning cooking, poison, and semen.

(1) Ceremonially, men cook (steam) pork and women scrape and cook vegetables. Domestically, it is a wife's duty to provide cooked food for her husband: '. . . providing [cooked] food signifies, for women, their total care for the husband or children. A woman takes offence if the meal is refused ("What! does my food have a bad odour that you spurn it?")' (M. Strathern 1972: 46). A failure to cook, as in the case of a grievance, signifies a repudiation of the marriage bond, and a wife too lazy to cook properly for her husband is said to incur general censure.[3]

(2) Menstruation is described as 'bad cooking', and menstrual blood as 'carelessly given food' (an empirical confirmation of Lévi-Strauss's (1973) intuition of menstruation as a form of 'infracuisine'). It has a bad odour and is dangerous and polluting to men, causing sickness and death.

> Hageners say that men can absorb menstrual blood through the penis, as well as by mouth [together with food]. Blood ingested over a period of time would gradually rise up in the victim's ribcage in two columns which meet at his neck: this is the danger point; some sudden exertion easily snaps the columns and he dies. A nose-bleed at death is symptomatic of pollution (M. Strathern 1972: 167).

Although a woman cannot help her powers, she is responsible for their effects. While she may use them deliberately, for example to get rid of a husband, normally pollution results from carelessness. Menstruation is surrounded by a variety of taboos expressing its antithesis to cuisine ('good cooking'), sexual intercourse ('proper food'),[4] crops and certain religious and social activities, such as the

[3] In Mount Hagen the sharing of cooked food is also used as a basis for the formation of dyadic friendship relations which 'project a notion of kinship amity'. More generally, food notions constitute a powerful symbolism for the expression of common substance and 'form an integral part of concepts of descent, kinship and locality ties' (A. Strathern 1977: 510).

[4] Adultery is 'stolen food', sexual intercourse on improper occasions is the 'wrong food', and the meal consumed during the Female Spirit cult (A. Strathern 1970, 1979) is 'powerful food' (food bespelled by ritual experts).

'shining, healthy splendour' of ceremonial dancing and being able to attract wealth in exchange relations. According to A. Strathern (personal communication),

> A men's house spoilt by menstrual blood or by the presence in it of poison is said to become cold, and its occupants will not be able to draw in wealth goods to it. Menstrual blood, inappropriately given, thus reverses the appropriate flow between groups via the affinal bond. Menstrual blood is also inimical to fresh green crops as they grow; in other words it is not only bad food in itself but actively destroys both persons ordinarily nourished by good food and the crops which should be converted into good nourishment for people.

(3) Poison, the 'destructive impoverishing counterpart of nourishment and of wealth' (M. Strathern 1972: 184), is inherently toxic, causing sickness and death. It is made from substances which are rotten, directly or by association (which presumably by similarity corrodes victims): snake skin (because of the deadliness of the snake's attack), rotting corpses (because of the association with death), powder scraped from stones near which dying plants have been observed, and so on. It is similar to menstrual blood in being administered with food[5] and 'in the likelihood that men, primarily husbands, will be the victims. Poison differs in that it always implies deliberate intent and is not attributed to carelessness as menstrual pollution can be' (M. Strathern 1972: 174). In-marrying wives administer poison on behalf of their male kinsmen, who tend to be its provisioners. Thus poisoning accusations may reflect the current state of micro-political relations, which fluctuate from amity to enmity. Although a woman may sometimes (but not always) be 'out of her mind' when she administers it (as a result of being hypnotized by her kinsmen), she is responsible for its effects.

(4) Semen or 'grease', together with good womb blood, forms and nourishes the foetus, a process which requires repeated, dutiful acts of intercourse.[6] The result of frequent intercourse is physical depletion.

[5] The notion of woman as in some respects an *anticuisinière* is not unique to New Guinea, cf. the French drinking song cited by Sperber (1968): 'Elle ne mettra plus / de l'eau dans mon verre. / La poison, la guenon / elle est morte.'

[6] Semen can be polluting not to the woman but to her child in the latter part of pregnancy, after it has been properly formed and acquired a spirit, and during breast-feeding, when milk 'contaminated with semen [would] destroy rather than nourish' (M. Strathern 1972: 168). Purely as a conjecture, semen may, in the language of Héritier's (1982) 'symbolics of identity and difference', contaminate milk as an

Men in particular are weakened by frequent indulgence in sexual intercourse, a dogma which perhaps encourages the observance of the long *post partum* taboo . . . But in addition to making men grow old before their time, frequent intercourse has the specific effect of causing the flesh to fall loosely—a tight, bulging skin being an indication of good health (M. Strathern 1972: 164–5).

Sexual intercourse is regarded as pleasurable, however, and there is a 'positively stated desire for children. To have several children, women say, means that they will receive plenty of meat at feasts, for themselves and their offspring together' (M. Strathern 1972: 134).

Thus a husband ought to provide semen and a wife cooked food, and a wife ought not to give blood or transmit poison. All four transactions are metaphorically and transformationally related, and their underlying structure reflects that of the social system. All are expressed in a common code, just as in the case of certain American Indian myths (Lévi-Strauss 1973) cooking and its 'periphery' define both the 'physiology' and the 'pathology' of the marriage bond: its complementarity and its internal and external threats.[7] This code operates in language and also in ritual: in dining (eating proper) and in sex (proper eating) and in menstrual taboos and poisoning procedures.

In the case of menstruation, the fact that women pollute perhaps accords with their position in the interstices of structure (Douglas 1966), with their loyalties divided between two groups of men (as 'women in between'). On the other hand, there is a positive side to pollution beliefs, since women express their concern for their husbands' well-being by the elaborate care they take not to endanger them—not to give food carelessly, part of a general expression of *de bonnes manières* (Lévi-Strauss 1978):

Women regard the taboos which they observe not as an oppression but as the means by which they can protect their menfolk (sons, husband), as they put it: so that the 'good work of the men' is not spoiled. It is a bad, careless, lazy woman, they say, who does not pay proper attention to the rules (M. Strathern 1972: 172).

In the case of poisoning, its anticulinary status is clearly revealed not only by its methods, but also by its divination procedures which

'accumulation of the identical'. See also J. A. Barnes (1974) on postnatal physical transmissions.

[7] See Hage and Harary (1983b) for another example from New Guinea.

are based on an inability to cook or to make a cooking fire (omens whose occurrence is caused by ancestral ghosts), and its punishments 'which operate in the form of culinary mediation':

> Divination was conducted either by the taro or the fire-thong test. The first was employed if a member of the dead person's settlement was directly suspected. Each person was required to choose a taro corm, and these were then cooked in small separate ovens. The one whose taro remained raw when all others were cooked was convicted, for the ancestors of the dead man were supposed to reveal his guilt in this way . . . The fire-thong test was used when the dead man's kin had captured someone from another group whom they suspected. In one case a clan inveigled one of their minor enemy groups up to their ceremonial ground with promises of cooked pigs as a peace offering, then closed in and forced them to try the test: if they could not make fire, they were to be pronounced guilty (A. Strathern 1971: 83).

The punishment for women, which included roasting above a fire, could perhaps be said to remove them from the 'world of rottenness' or to transform them by moving them towards the cooked (Lévi-Strauss 1970).

Seen from the standpoint of male ethnophysiology, an examination of these transactions discloses two levels of structural interpretation. Internally, the structure of this system is algebraic. A male is both repleted and depleted (r/d), positively and negatively (+/−), as shown in the graph of Fig. 7.11, the product $K_2 \times K_2$, which relates the transactions to one another and shows their relative distances. Thus blood has, as M. Strathern emphasizes, 'close associations' with both poison and food, and the former is the 'impoverishing counterpart' of the latter.

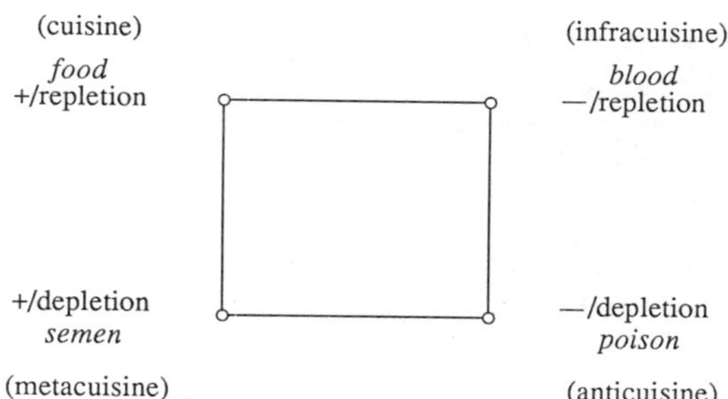

FIG. 7.11. Mount Hagen ethnophysiological structure

The set of transactions is defined by four transformations:

α = interchange the effects (r/d).

β = interchange the signs (+/−).

γ = interchange both the effects and the signs.

i = interchange neither the effects nor the signs.

These transformations form a mathematical group. Considering each transformation as equivalent to an automorphism of a graph, the group of this system is isomorphic to that of the graph shown in Fig. 7.12.

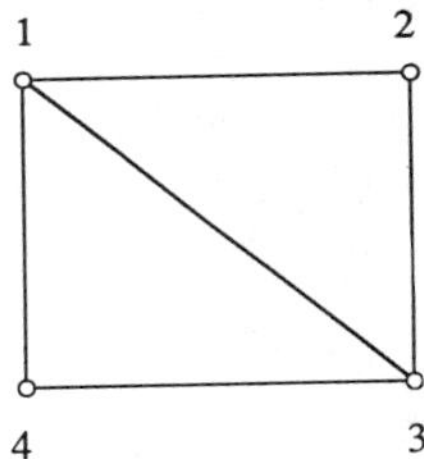

FIG. 7.12. A graph whose automorphism group has order 4 but is not cyclic

Let us explain what is meant by a group. Martin Gardner has given an excellent intuitive definition: 'Roughly speaking, it is a set of operations performed on something, with the property that if any operation in the set is followed by any operation in the set, the outcome can also be reached by a single operation in the set. The operations are called the elements of the group, and their number is called the order of the group' (1980: 20). The elements could include such diverse actions as arithmetic operations on numbers, permutations of the attributes of objects, or Gardner's example of four commands: 'right face', 'left face', 'about face', 'do nothing'. In the last example, 'right face' followed by 'about face' can be reached by 'left face', etc. We have already encountered this group in the last section of the preceding chapter, as it is a disguised form of the automorphism group of $\vec{C}_4$. Still another presentation of the cyclic group of order 4 was given there as the numbers {0, 1, 2, 3} under addition modulo 4, i.e. regarded as 'numbers on a clock'.

The set of elements of a group must satisfy four axioms, which Gardner informally describes as follows:

1. Closure: The product of any pair of operations is equivalent to a single operation in the set.
2. Associativity: If the product of any two operations is followed by any operation, the result is the same as following the first operation with the product of the second and the third.
3. Identity: There is just one operation that has no effect, in this case doing nothing.
4. Inverse: For every operation there is an inverse operation such that executing an operation and then its inverse is equivalent to executing the identity operation. In this example left face and right face are inverses of each other, whereas do nothing (the identity) and about face are each their own inverse (Gardner 1980: 20).

Formally stated, the non-empty set A together with a binary operation, denoted by the juxtaposition $\alpha_1 \alpha_2$ for α_1, α_2 in A, constitutes a *group* whenever the following four axioms are satisfied:

Axiom 1 (*closure*) For all α_1, α_2 in A, their 'product' $\alpha_1 \alpha_2$ is also an element of A.

Axiom 2 (*associativity*) For all α_1, α_2, α_3 in A, $\alpha_1(\alpha_2 \alpha_3) = (\alpha_1 \alpha_2)\alpha_3$.

Axiom 3 (*identity*) There is an element i in A such that $i\alpha = \alpha i = \alpha$ for all α in A.

Axiom 4 (*inversion*) If Axiom 3 holds, then for each α in A, there is an element denoted by α^{-1} such that $\alpha\alpha^{-1} = \alpha^{-1}\alpha = i$.

We shall see that the automorphisms α of any graph G constitute the elements of a group. For convenience of exposition, we will repeat a number of definitions given in various preceding chapters.

In order to show clearly that the collection of all automorphisms of a graph constitute a group, we review the definitions of isomorphism given in Chapter 2 and automorphism given in Chapter 6. For this purpose, consider two graphs $G = (V, E)$ and $G' = (V', E')$, both with p points. As before, we label the points in V by the first p positive integers 1, 2, . . ., p. With this labelling of G fixed, we say that G and G' are *isomorphic graphs* provided it is possible to label the points in V' as $1'$, $2'$, . . ., p' in such a way that

(7.1) Points i and j are adjacent in G if and only if points i' and j' are adjacent in G'.

This means that whenever i, j are adjacent in G we must have i', j' adjacent in G', and also if i and j are not adjacent in G then i' and j' are not adjacent in G'. This condition, which is spelled out here, was put more succinctly in Chapter 2 in the phrase 'there is a one-to-one correspondence between V and V' which preserves adjacency'.

An example of isomorphism in graphs was given previously. Fig. 2.7 shows two isomorphic graphs with rather different appearance, each of which is the union of two (undirected) 7-cycles on the same set of seven points.

Fig. 7.13 shows two isomorphic presentations of the complete bigraph $K_{3,3}$. In G we have i, j adjacent if and only if one of these points has an odd label and the other an even label. Obviously the same statement holds in G' for two points i' and j'. Hence we have a one-to-one correspondence between V and V' which preserves adjacency, i.e. an explicit isomorphism between G and G'.

In these terms, we clarify in more detail the definition of an automorphism of a graph given in Chapter 6. Now we have just one graph $G = (V, E)$ at hand. We say that α is an *automorphism* of G if α maps $V = \{1, 2, \ldots, p\}$ onto itself in such a way that isomorphism is preserved. In other words α is a relabelling of the same set V of points either with the same labels or with a different labelling $1', 2', \ldots, p'$. To illustrate, we show in Fig. 7.14 the four labellings of the graph G of Fig. 7.12 which give all the automorphisms of G.

The first of these, the identity automorphism i, leaves all the labels of Fig. 7.14 unchanged. In conventional notation for permutations, i is written:

$$i = (1)\,(2)\,(3)\,(4).$$

The second automorphism α maps points 1 and 3 onto themselves but interchanges points 2 and 4:

$$\alpha = (1)\,(3)\,(2\,4).$$

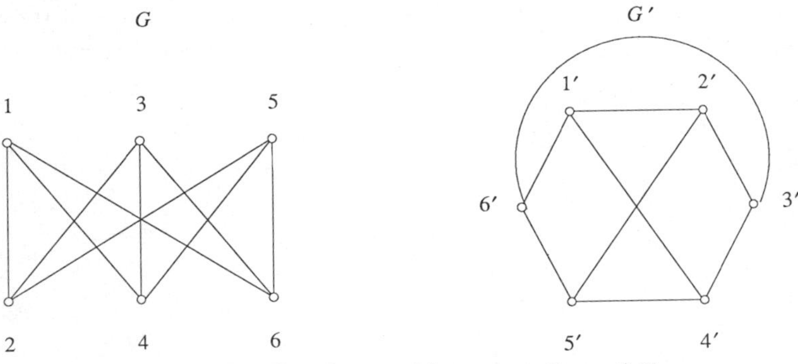

FIG. 7.13. Two isomorphic presentations of $K_{3,3}$

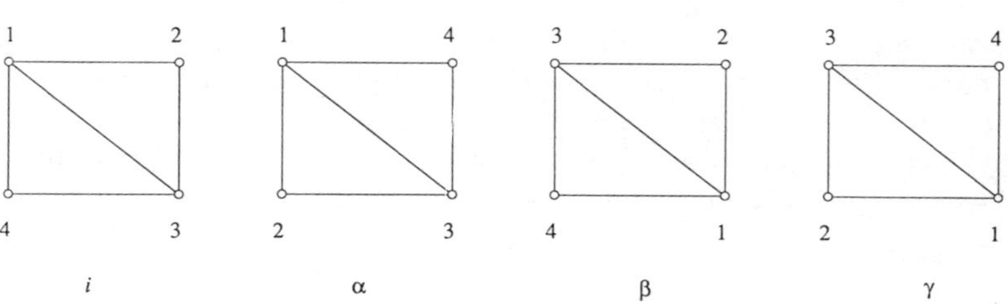

FIG. 7.14. The automorphisms of the graph in Fig. 7.12

Continuing, automorphism β fixes points 2 and 4 but interchanges 1 and 3:

$$\beta = (2)(4)(1\ 3).$$

Finally, automorphism γ interchanges both the point pair 1, 3 and the pair 2, 4:

$$\gamma = (1\ 3)(2\ 4).$$

Clearly, each of these 1–1 mappings of $V = \{1, 2, 3, 4\}$ onto itself preserves adjacency, and these are all the possible automorphisms of G. Now we require a realization for these mappings, which are called permutations, and of the abstract binary relation in the definition of a group. To do this we say that the product αβ is the mapping obtained by first permuting by α and then by β. It is immediately seen in Fig. 7.14 that $\alpha\beta = \gamma$. A summary of all products of these four automorphisms is given in the 'group multiplication table':

Γ(G):

	i	α	β	γ
i	i	α	β	γ
α	α	i	γ	β
β	β	γ	i	α
γ	γ	β	α	i

This table gives the result of performing a first operation followed by ('multiplied by') a second operation, and shows that the axioms of a group are satisfied.

We note that the group of the graph of Fig. 7.12 is the 4-group, which Lévi-Strauss (1978, 1981) refers to in his analyses of the roles of menstrual blood, liver, scalps, and dandruff in the outcomes of complementary male activities (hunting and warfare), and of menstrual blood, sperm, scalps, and honey in affinal exchanges, as expressed in certain American Indian myths. To evoke the 4-group Lévi-Strauss uses the symbols x, $-x$, $1/x, -1/x$ which stand for the additive and multiplicative inverses of numbers. This arithmetic group and the automorphism group of $K_4 - e$ are isomorphic as may be verified when

$$\begin{aligned} \alpha &= x \to -x, \\ \beta &= x \to 1/x \\ \gamma &= x \to -1/x \\ i &= x \to x \text{ (no change).} \end{aligned}$$

This group, sometimes known as the Klein group, is also fundamental in the work of Piaget (1971), as noted in *SMA*.

Theorem 7.2. Let $G = (V, E)$ be a graph. Let $\Gamma(G) = \{i, \alpha, \beta, \gamma, \ldots\}$ be the collection of all the automorphisms of G. Then $\Gamma(G)$ is a group under the binary operation 'is followed by'.

At the second, external, level of interpretation, the male physiological system as conceived by Mount Hageners is, metaphorically speaking, isomorphic to their social system. Although pollution beliefs are not consistently correlated with the *sign* of marriage exchange as Meggitt (1964) proposes, the structure of the *entire system* in which they figure as terms may well be correlated with the form of economic exchange. Recalling Gell's (1975) statement concerning the Highlanders' economic theory of the body, the elaborate system of physiological repletion and depletion defined above can be interpreted as a reflection of the 'alternating disequilibrium' of wealth flows in *moka* (A. Strathern 1971), the highly developed system of competitive exchange described in Chapter 2.

The model we propose is a general one, not restricted to any particular set of beliefs or concepts. Thus beliefs in menstrual pollution and more particularly in poisoning are not universal in

Highland New Guinea. However, similar types of structures can be expressed by other beliefs, such as those relating to sorcery or witchcraft. They can also be expressed not by the transaction of complementary substances in the same social relation but by the transaction of the same substance in different relations. For example, the Etoro on the Papuan Plateau, who lack beliefs in menstrual pollution and poisoning (Kelly 1976: 41), conceive that the life force (*hame*) which is concentrated in semen is transacted in four relations,

> interrelated through a general equation whereby receiving semen : life, growth, and vitality :: losing semen : weakness, senescence, and death. This formulation also constitutes the specific ideational content of a more general and fundamental conceptual orientation of Etoro cosmology *viz.* the concept that the total system is closed and bounded such that accretion at one node necessarily entails a corresponding depletion at another (and vice versa) (Kelly 1976: 47–8).

In Etoro, a male provides semen for both a foetus and a male youth (the latter by oral insemination) to augment their growth, thereby depleting himself. He loses semen in heterosexual intercourse (which, in contrast to homosexual intercourse, is 'anticultural' and regarded with ambivalence) and in witchcraft. Since a male, ideally, is inseminated by a sister's husband and since a witch uses the victim's life force to augment the growth of its own offspring, a male is parentally and affinally repleted in the first part of the life cycle (as child and protégé) and affinally and indirectly parentally depleted in the second part (as inseminator/husband and parent/victim), which again results in a 4-group.

An examination of Papua New Guinea ethnographies may reveal a variety of empirical manifestations of the same structure as well as different types of abstract structures (groups of different order and type). There may also be different concrete logics correlated with morphological differences in exchange systems. For example in some societies there is symmetrical and reflexive or autopollution—'male menstruation', as in Arapesh (Mead 1940) and Wogeo (Hogbin 1970)—which could be interpreted as in Hage (1981), along the lines proposed by Douglas (1975), as a reflection of dual organization.

We have proposed a semiological as opposed to a typological or taxonomic model for the analysis of beliefs in menstrual pollution in Papua New Guinea, and, by implication, for such beliefs generally.

The model has three advantages. Semantically, a particular belief is not arbitrarily isolated from others in terms of which it is defined, nor is the set of which it is a member determined in advance by what it includes or excludes. There is an internally logical basis for a seemingly bizarre belief that a man might be filled with menstrual blood until his neck snaps off and he dies, and menstrual blood itself may have unexpected relations with apparently quite heterogeneous substances. Structurally, the model permits the comparison of related belief systems, not term for term, but on the basis of an underlying system of relations, which is also the level to seek valid correlations between symbolic and social structures. From this point of view it may be quite irrelevant whether a given society has or does not have a belief in menstrual pollution, since the same structure can be expressed in a variety of ways. Sociologically, a particular social relation such as the marriage bond is more comprehensively characterized either by the multiple transactions, conjunctive as well as disjunctive, which define it, or by its place in a set of contrasting relations.

8

Logic of Relations

In Fiji all things go in pairs, or the sharks will bite.

Lauan proverb from A. M. Hocart, *The Northern States of Fiji*

In *Culture and Practical Reason*, Sahlins uses a relational model to suggest an isomorphism between social and material classification in Fiji. According to Sahlins, the commutability within, and incommensurability between, certain classes of goods can be seen as a transposition of the scheme of social distinctions, which unite and oppose classes of men, onto the plane of objects. But in considering this world of objects one does not know exactly what is intended when it is said that a relation of 'transitive equivalence' obtains among whales' teeth, pigs, and turtles, because of their common interchangeability with men:

Whale's teeth are 'the head of all things'. Their only social measure is a human being: they give a claim on the services, in war or in work, of those who accept them as an offering; they secure the wife to the husband, compensate the warrior for his tribute of a cannibal victim, the father-in-law for the death of the wife, the mother's kin for the birth of her child. Whale's teeth today are sometimes displayed on the upper posts of the house, but customarily they are concealed in the rear section, cohabiting with other objects of potency (*mana*) [in] the sleeping place of senior men. Valuable as they are, whale's teeth cannot be exchanged for any ordinary useful thing or food—anymore, one might say, and for similar reasons, than a younger brother could eat the food of the firstborn. Only two things move regularly against whale's teeth—turtle and pig—the 'head' of all sea foods and the 'head' of all land foods. But then, a pig may be substituted for a man in sacrifice; and turtle is the 'fish that lives' (*ika bula*) or the 'man-fish' (*ika tamata*), for whom, should the chief's fisherman fail in the hunt, he shall be obliged to substitute himself. There is a transitive equivalence among whale's teeth, pig, and turtle, based on their common interchangeability with men (Sahlins 1976: 36–7).

If an equivalence relation is meant, then the phrase 'transitive equivalence' is redundant, for an equivalence relation is transitive by definition. To establish that whales' teeth, turtle, and pig are in an equivalence relation one would have to establish that the relations of reflexivity, symmetry, and transitivity all hold. But in fact these objects may not be so related, for Sahlins only says (1) that a pig can be substituted for a man and (2) that a man can be substituted for a turtle. We do not know if the converses of these statements are true, nor do we know if pigs and turtles can be substituted for each other, nor do we know how whales' teeth fit in.

In order to put the study of relational properties of exchange structures on a firm footing, we propose the use of graph theoretic models. These have the advantage of providing the most natural and intuitive representation possible. Our aim is not only to give precise definitions of individual relational properties but also to show how these properties combine to create more complex structures constituting a potential repertory of analytical models.

Relations and Graphs

A rigorous definition of a relation requires some basic concepts from set theory. A *set S* is just a collection of elements. For example, the alphabet A is the set of letters $a, b, \ldots, z$. It is customary mathematical notation to write $A = \{a, b, \ldots, z\}$ which is read 'A is the set consisting of the elements $a, b, \ldots, z$.' An *ordered pair*, written (x, y), is a set of two elements, with x called the first and y the second. In other words it is a 2-term sequence. An *ordered triple* (x, y, z) is, as expected, a 3-term sequence. A *relation R on a set V* is precisely a set of ordered pairs of elements of V. The *domain* of a relation R is the set of all the first elements of the ordered pairs in R; its *range* is the set of all second elements.

As noted in *SMA*, the historical origins of this formulation go back to Peirce (1933), who illustrated these concepts using kinship relations. Thus the relation 'mother' is the set of all ordered pairs (x, y) such that x is the mother of y; its domain is the set of all women who have children, and its range is the set of all children. 'Sister' is the relation whose domain is the set of all females who have a sibling, and its range is the set of all people who have a female sibling. 'Uncle' is a composite relation, being the union of the two relations, 'brother of a parent of' and 'brother-in-law of a parent of'. Each of these two is called a *relative product* or a composition of two relations, meaning one relation 'parent of' followed by another relation 'brother or brother-in-law of'.

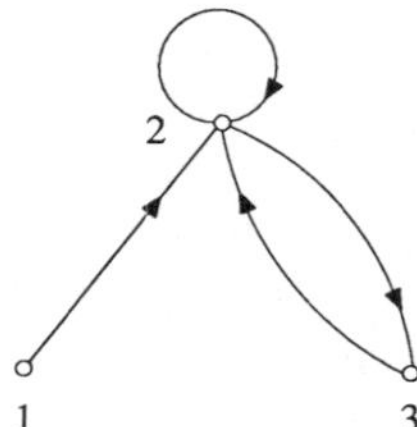

FIG. 8.1. A relation

A relation R is represented graphically by taking the elements of V as points, and drawing an arc from u to v whenever the ordered pair (u, v) is in R, and also an arc from point u to itself, called a *loop* if (u, u) is in R. Fig. 8.1 shows a relation R on $V = \{1, 2, 3\}$ having four arcs one of which is a loop; here $R = \{(1, 2), (2, 3), (3, 2), (2, 2)\}$.

Because a relation is a set, and as no set lists any of its elements more than once, there can never be more than one arc from one point to another (although there may be two arcs, one in each direction, as in Fig. 8.1). It is most convenient and intuitively clear to exploit the graphical representation of a relation when developing the rich variety of properties of relations. Fig. 8.2 shows some of the relations on three points.

In order to make this section mathematically self-contained, we must repeat several definitions given earlier, while introducing new ones. Relation R is *reflexive* if there is a loop at every point; it is *irreflexive* if no point has a loop. In Fig. 8.2, (c) is reflexive and (b) is irreflexive; (a) and (d) are neither. As an empirical example, if we represent two or more social groups as points and the relation 'intermarries' as arcs, endogamy determines a reflexive relation and exogamy an irreflexive one.

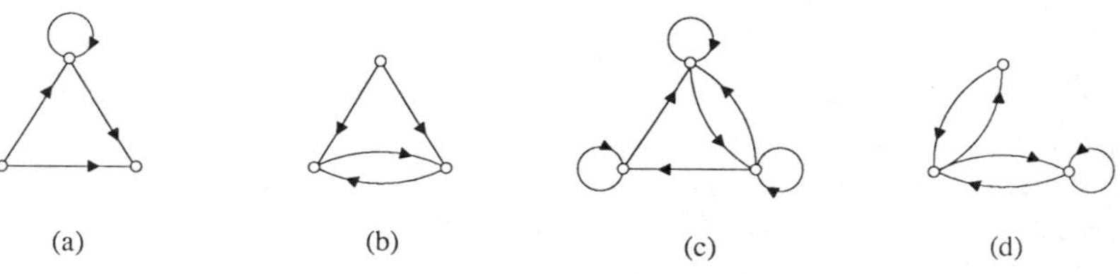

FIG. 8.2. Four relations on three points

Whenever there are two arcs of the form (u, v), (v, u) they form a *symmetric pair* of arcs. Then R is *symmetric* if every arc (other than a loop) is part of a symmetric pair; R is *asymmetric* if no arc is part of a symmetric pair. In Fig. 8.2, (d) is symmetric and (a) is asymmetric; (b) and (c) are neither. We saw in Chapter 2 that in Lévi-Strauss's (1969) classification of marriage systems, restricted exchange is a symmetric relation and generalized exchange is asymmetric.

It is customary to draw a symmetric relation by replacing each symmetric pair of arcs by an undirected line, thus obtaining a graph, as in Fig. 8.3 and as in preceding chapters.

The next property of a relation depends on the concept of a *2-path* from u to v. This consists of two arcs of the form (u, w) and (w, v) with the three points u, v, w distinct.[1] Then relation R is *transitive* if, whenever there is a 2-path from u to v, the 1-path or arc (u, v) is also in R. By an *intransitive* relation is meant one that is never transitive, that is, whenever there is a 2-path from u to v the arc uv cannot be present. In Fig. 8.2, (a) and (b) are transitive; (d) is intransitive, and (c) is neither. (See also Fig. 1.13 for further illustrations.) To continue with our marriage example, if, for all ordered triples, u gives women to v and v gives women to w but u does not give women to w, the relation is intransitive. Such a triple is intransitive whether or not it is also cyclic.

FIG. 8.3. A symmetric pair of arcs (a) and a line (b)

[1] There are two schools of thought concerning the definition of transitivity (König 1936; Harary 1961a). One school defines a relation R on a set S as transitive if for all x, y, z in S, not necessarily distinct, xRy and yRz imply xRz. The other school, to which we subscribe, easily avoids the paradox implicit in the preceding definition by stipulating that this implication holds whenever x, y, z are distinct. By paradox we mean that there are binary relations which are symmetric and transitive but not reflexive, as is well known. However, if one takes the first definition of a transitive relation, in which the three elements x, y, and z are not necessarily distinct, then one can prove that symmetry and transitivity imply reflexivity, as follows. If R is a relation which is symmetric and transitive, then R is reflexive. *Proof.* Let x, y be elements of the set S on which R is a binary relation. If xRy, then yRx since R is symmetric. Now xRy and yRx imply by the first silly definition of a transitive relation that xRx, i.e. that R is reflexive. That is the presumed proof of this false assertion, and that is the entire reason why we reject the definition of transitivity in which the three elements are permitted to overlap and why we insist that they must be distinct.

In a transitive relation the existence of a 2-path from u to v implies the existence of a 1-path as well. If there are no 2-paths in R, the relation is said to be *vacuously transitive*.

Relation R is *complete* if for each pair of points u, v of R at least one of the two arcs (u, v) and (v, u) must occur. The first three relations in Fig. 8.2 are complete. A marriage exchange relation is complete if for every pair u and v of groups, u gives women to v or v gives women to u or both.

Intransitivity and completeness interact, as may be illustrated in connection with an asserted identity between the pecking order and the vassalage (*nggali*) relation in Fiji. According to Lévi-Strauss,

> There are societies characterized by hierarchical and intransitive cycles and quite comparable to the pecking order—as, for instance, in the Fiji Islands, where the population was organized until the beginning of the twentieth century into fiefs which were interconnected by relations of fealty, so that fief A might be a vassal of fief B, B of C, C of D, and D of A. Hocart described and explained this structure, which at first sight seems unintelligible, by pointing out that in Fiji two forms of vassalage exist—vassalage by right and vassalage by conquest. Fief A might thus be traditionally a vassal of B, B of C, and C of D, whereas fief D might have recently become, as a result of an ill-fated war, a vassal of A. Not only is this structure the same as that of the pecking order, but—and this passed unnoticed—anthropological theory out-distanced mathematical interpretation by several years, since the latter is based on the distinction between two variables which operate with a time-lag between them—and this corresponds exactly to Hocart's [1952] (posthumous) description (Lévi-Strauss 1963a: 323).

This particular relation of vassalage is intransitive, but the pecking order is (non-vacuously) intransitive only in the special case that it contains three hens which peck each other cyclically rather than transitively. As soon as the flock contains four or more hens its digraph, which is always complete, cannot logically be intransitive. This was noted in *SMA* (chapter 5), and it may be verified by inspection of all the asymmetric complete 4-point digraphs in the bottom row of Fig. 8.13.

It is convenient to use the following abbreviations for the seven properties of relations we have defined:

r	reflexive	$\bar{r}$	irreflexive	c	complete
s	symmetric	$\bar{s}$	asymmetric		
t	transitive	$\bar{t}$	intransitive		

The *Cartesian product* $V \times V$ of a set V with itself is the collection of all ordered pairs (u, v) where both u and v are in V. We say that R is a *relation on the set* V if R is a subset of $V \times V$. The *universal relation* on V is the entire set $V \times V$. It is called universal because it is universally true that (u, v) is in it for every choice of u, v in V, including $v = u$. When (u, v) is in relation R, it is often convenient to write uRv.

In an axiom system there must be undefined terms called *primitives* in order to avoid circular definitions, and *axioms* (or *postulates*, a synonym) in order to avoid circular reasoning. The seven basic properties of relations just introduced, r, $\bar{r}$, s, $\bar{s}$, t, $\bar{t}$, c, can now be regarded as individual axioms. Combinations of these seven symbols will serve as axiom systems for eleven different types of structures. In all of these it is understood that the primitives are a set V of elements and a subset R of $V \times V$, and that the axioms include the assertions:

1. V is finite,
2. V is not empty (but R is permitted to be the empty set of ordered pairs).

When there are no other axioms, this constitutes a simple axiomatization of a relation.

We can now list axiom systems for eleven kinds of relational structures.

Structure	*Axioms*
1. Digraph	$\bar{r}$
2. Graph	$\bar{r}$, s
3. Oriented graph	$\bar{r}$, $\bar{s}$
4. Similarity relation	r, s
5. Equivalence relation	r, s, t
6. Partial order	$\bar{r}$, $\bar{s}$, t
7. Complete order	$\bar{r}$, $\bar{s}$, t, c
8. Tournament	$\bar{r}$, $\bar{s}$, c
9. Parity relation	$\bar{r}$, s, t
10. Antiequivalence relation	$\bar{r}$, $\bar{s}$, $\bar{t}$
11. Antiparity relation	r, $\bar{s}$, $\bar{t}$

The twelve diagrams of Fig. 8.4 illustrate first a relation R with none of these seven properties, and then one of each of these eleven structures, some of which we have already seen. We will now show

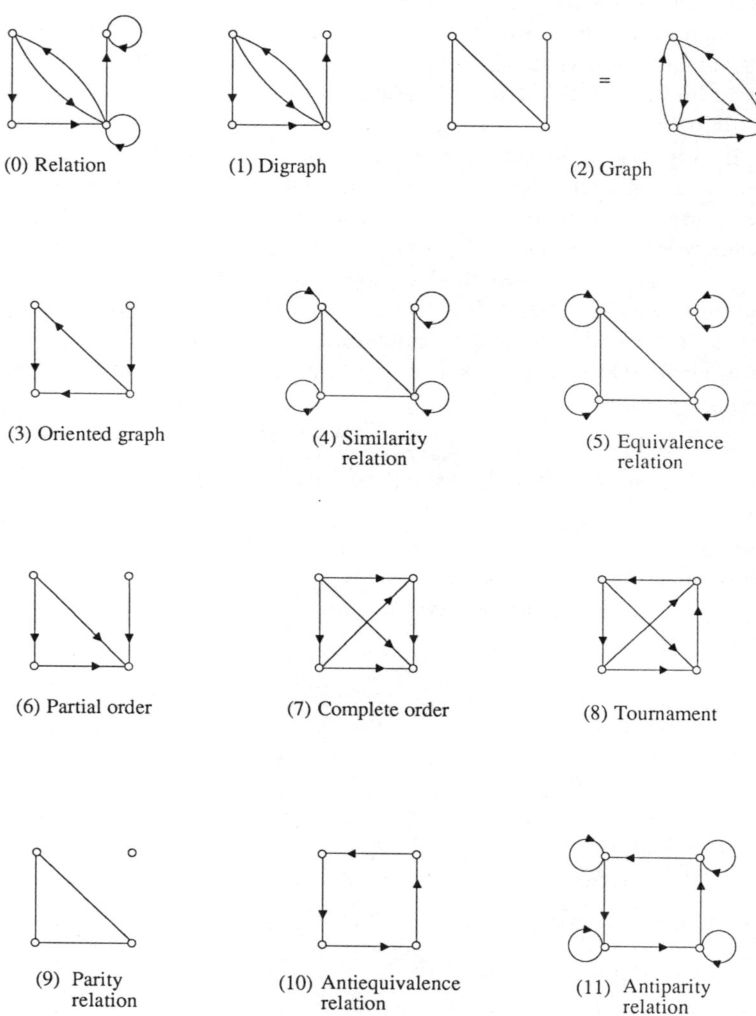

FIG. 8.4. Twelve types of exchange structures

explicitly that all of them occur as exchange structures in Oceania. First of all, however, we need to distinguish, for purposes of anthropological application, between a network N and a relation R.

Networks vs. Relations

R. F. Salisbury (1956: 639) defines an asymmetrical marriage system as one in which 'certain groups give out significantly more women in marriage than they receive, while other groups receive more women than they give'. By this definition a marriage system as opposed to a marriage rule (as in Lévi-Strauss's *The Elementary Structures of Kinship*) refers to actual statistical patterns of marriage exchange between groups. Informants need not be conscious of such differences and they may even deny that they exist. An asymmetrical system can occur when there are differences in status or in the distribution of resources between wife-exchanging groups.

Salisbury constructs a model of such an asymmetrical system consisting of a chain (path) of societies linked by reciprocal exchange of women for valuables, with the source of valuables placed at one end of the chain. Assuming a constant supply of valuables at the source and a decreasing supply with increasing distance from the source, societies at one end of the chain will tend to accumulate women, and societies at the other end will tend to give more women than they receive, while societies in the middle will tend to break even. The expected differences in the relative abundance and scarcity of women result in predictions about features of social organization in societies at either end of the chain, including early vs. late age of marriage, low vs. high bride-price, and low vs. high rates of polygyny. In an application of the model to New Guinea societies, Salisbury defines such a marriage chain running from the coast near Bogadjim on Astrolabe Bay near the source of shells, pigs, axes, and cloth, through the Siane area up to the Fore area near Mt. Michael in the Eastern Highlands. Available evidence (C. H. Berndt 1953, R. M. Berndt 1954, Lauterbach 1898) offers some support for the predicted differences in social organization.

The type of model Salisbury proposes is an important one because it takes an entire region as a unit of analysis and relates aspects of social organization to network position. It should be recognized, however, that it is quite literally a network model, in which different amounts of different classes of goods (women and valuables) are exchanged. It is not a digraph model defined by a single relation which either has or does not have the property of asymmetry. Clearly, as Leach (1957) was quick to point out, Lévi-Strauss and Salisbury are not talking about the same type of structure.[2]

A *network* satisfies the following axiom:

Network axiom. For each arc $x = (u, v)$ in R, there is an associated real number $f(x)$.

The letter f comes from the word 'function'. The number $f(x)$ is called the *value* of arc x in the network, which is denoted by $N = (R, f)$. The values which $f(x)$ may attain serve to specify the type of network at hand. In Chapter 5, for example, we used a network model of a Markov chain. Here the underlying structure is an arbitrary relation R, and the following two stipulations hold.

Markov axiom 1. For each arc x, whether or not it is a loop, the value of $f(x)$ satisfies the inequality $0 < f(x) \leqslant 1$.

The number $f(x)$ stands for the probability of arc x. Thus this axiom says that each arc in R is *possible*, as its probability is positive.

Markov axiom 2. For each point v of N, with outdegree d, let the arcs from v be $x_1, x_2, \ldots, x_d$ and let $p_i = f(x_i)$. Then

$$p_1 + p_2 + \ldots + p_d = 1.$$

The seven relational properties pertain only to the presence vs. the absence of arcs. For that reason, these mathematical models are generally oversimplifications of real anthropological phenomena, because they do not show the intensity or strength of a relation. Such features can be shown by converting a graph into a network. Conversely, the values of the arcs of a network can be ignored by converting it into a graph. One might, for example, use quantitative measures to establish a threshold for determining whether or not a relation can be said to exist. The first three examples we give—of a relation, a digraph, and a graph—are graphical interpretations of networks, as the quantities are not taken into account. Clearly, the choice of model depends on the aim of analysis (and very often on

[2] According to Leach (1957: 343), 'the notion of asymmetry with respect to marriage systems has consistently been held to refer to a customary rule whereby marriage of a man with one of his female cross-cousins is prohibited, while marriage with the other is allowed or even preferred. The discussion has been concerned solely with the significance of such legal rules, and has nothing to do with the statistical frequency of any particular type of marriage between kin or non-kin.'

the availability of data). If one wishes to study differences in the flow of goods, then a network model is appropriate; if one wishes to study purely relational properties of exchange structures then a graph model is appropriate. Sometimes both types of models are applicable to the same exchange structure. The *kula* ring, for example, has been modelled as a graph to study centrality (Brookfield and Hart 1971; Hage 1977; Irwin 1983) and as a Markov chain to study the flow of valuables (Chapter 5).

Relation: Siassi Marriage Exchange

Marriage exchange in Siassi exemplifies a relation. The Siassi Islands are a group of coral islets in the Vitiaz Strait between New Guinea and New Britain. The inhabitants are famous as seafaring middlemen in the trade between New Guinea, New Britain, and neighbouring islands (Harding 1967). According to M. P. Freedman (1970), the most significant social units in a Siassi village are men's houses. These are named groups whose membership does not depend on any specific demonstrated or stipulated genealogical connections. Efforts to maintain numerical parity among them suggest a highly cognatic composition. Men's houses have two primary functions: the distribution of ceremonial food and the regulation of marriage.

Freedman describes the marriage system as a 'random connubium', by which he means non-systematic exchange of women between men's houses, each of which has about the same number of exchange partners. Such a system is said to prevent the formation of village blocs. The marriage exchange relation for one Siassi village, Mandok, at the time of Freedman's investigation, is shown in the following matrix, adapted from the one in Freedman. Each men's house is represented by a row and a column, and the entries are $a_{ij} = 1$ if i gives women to j and $a_{ij} = 0$ if not.

As Peirce demonstrated, the properties of a relation are reflected in its matrix.[3] The Siassi marriage relation has none of the seven properties defined earlier as can be determined by inspecting its

[3] Peirce (1933) introduced matrix analysis to the study of relations: 'Imagine all the dyads (or ordered pairs) of individuals in the universe to be arranged in a *matrix* (Cayley's term, though the application of the conception to the logic of relations was first made by the author)' (1933: 377).

	Mandogsala	Tabobpugu	Pandanpugu	Tapugu	Simban	Bedbedang	Panuboga
Mandogsala	1	1	1	1	1	1	1
Tabobpugu	1	0	1	1	1	1	1
Pandanpugu	1	1	1	1	1	1	1
Tapugu	1	1	1	0	1	1	0
Simban	1	1	0	1	0	1	1
Bedbedang	1	1	1	1	1	0	1
Panuboga	1	1	1	0	0	1	0

matrix entries. It is neither r nor $\bar{r}$, as there are both 1s and 0s on the diagonal. Marriage between men's houses is thus not perfectly exogamous. It is not s, because $a_{3,5} = 1$ but $a_{5,3} = 0$; and it is not $\bar{s}$, because $a_{1,2} = a_{2,1} = 1$. The relation is not t, because $a_{5,1} = a_{1,3} = 1$ but $a_{5,3} = 0$; and it is not $\bar{t}$, because $a_{1,2} = a_{2,3} = a_{1,3} = 1$. Finally, R is not complete, because $a_{4,7} = a_{7,4} = 0$ (the only unjoined point pair). Because the matrix has so few 0s, marriage in Siassi would seem to tend more toward a complete than a random connubium.

Digraph: Rossel Island Marriage Exchange

Marriage exchange on Rossel Island (Yela) in the Louisiade Archipelago as described by W. E. Armstrong (1928) exemplifies a digraph, an irreflexive relation. Rossel Islanders, famous as producers of the red spondylus shell necklaces exchanged as valuables in the *kula* and as money internally, are practically endogamous except for a few marriages with their main trading partner, Sudest Island (Tagula). At the time of his visit in 1921, Armstrong estimated a population of about 1,500 divided into 15 or so exogamous, matrilineal clans each with three linked totems: a plant, a bird, and a fish. Armstrong could not discover any significant rules or restrictions governing marriages between clans. On the basis of 40 recorded marriages he produced the following list of clans *A*–*M*, showing the marriages of men into other clans:

A: *B*, *C*, *D*, *E*, *F*, *G*
B: *C*, *H*
C: *D*, *I*
D: *A*, *B*, *F*, *G*, *H*
E: *G*
F: *H*, *M*
G: *A*, *B*, *C*, *D*, *E*
H: *F*
I: *C*
J: *A*
K: *G*, *I*
L: *A*
M:

We note that a list is a third alternative presentation of a digraph, in addition to the more usual pictorial and matrix forms. As Needham (1983) observed in his essay on Wittgenstein's arrows, one can grasp such a system more easily by turning this relation around to show groups of men giving women to rather than taking women from other groups. The converse listing, which makes the system more directly comparable to the Siassi matrix, is

A: *D*, *G*, *J*, *L*
B: *A*, *D*, *G*
C: *A*, *B*, *G*, *I*
D: *A*, *C*, *G*
E: *A*, *G*
F: *A*, *D*, *H*
G: *A*, *D*, *E*, *K*
H: *B*, *D*, *F*
I: *C*, *K*
J:
K:
L:
M: *F*

Inspection of the list will show that the marriage exchange relation is a digraph, as it has only the property $\bar{r}$. It is thus almost as structureless as the Siassi Island relation.

Because of Rivers's work (1914), Armstrong was concerned to discover possible evidence of dual organization. He concluded, however, that his list was 'inconsistent with a dual organization, because, for example, *A* must belong to a different moiety from *B* owing to the *A* male marriages, and also to the same moiety owing to the *D* male marriages' (Armstrong 1928: 44). Graph theoretically, this marriage structure contains a 3-semicycle, which contradicts the criterion for dual organization, a bipartite graph, as discussed in Chapter 2.

Graph: Lesu Marriage Exchange

Marriage exchange in Lesu, a small village on the east coast of New Ireland in the Bismarck Archipelago, is regulated by two exogamous matrilineal moieties, *Telenga* = fish hawk (*Pandion leucocephalus*) and *Kongkong* = eagle (*Haliaetus leucogaster*). The fish hawk and eagle moieties, which are widespread in New Ireland, are subdivided into clans. According to Powdermaker (1933) the rule of moiety exogamy is strictly observed, but extensive genealogical data disclose no pattern of marriages between clans or sets of clans. At the clan level the marriage exchange relation is a digraph ($\bar{r}$) like Rossel Island, but at the moiety level it is a graph ($\bar{r}$ and s), as shown in Fig. 8.5. The odd and even numbers represent clans in the eagle and fish hawk moieties respectively.

Oriented Graph: Carolinian Tribute Relations

The Yapese Empire is defined by a tribute relation that joins fourteen low coral islands in the Carolines to the high island of Yap. As emphasized in Chapter 3, hierarchical relations also obtain

FIG. 8.5. A graph of marriage exchange in Lesu

among certain of the low islands. Thus, for example, Lamotrek ranks above the adjacent islands of Elato and Satawal, from which it receives seasonal tribute payments. The Satawalese send bread-fruit and coconuts and the Elatoans send live turtles. According to Alkire (1965), Puluwat and Pulusuk were formerly a part of this system. And according to Damm and Sarfert (1935), Pulusuk also paid tribute to Puluwat, as did Pulap and Namonuito, in the form of taro, bread-fruit, coconut, and fish. This relation is obviously $\bar{r}$, since no island pays tribute to itself, and s since no two islands pay tribute to each other, and it is therefore an *oriented graph* as shown in Fig. 8.6. Note that it is not transitive.

A particularly interesting expression of Lamotrek's superiority is the gift of turtles:

> A symbolic act directly related to the recognition of Lamotrek's suzerainty over the outlying islands . . . is carried out by individuals of Elato, Satawal, Puluwat and Pulusuk whenever they meet Lamotrekans on one of the uninhabited islands. These islanders must give the Lamotrekans at least one head from any turtles taken on these islands. One will remember that the heads of turtles taken on Lamotrek are offered to the paramount chief during distribution (Alkire 1965: 148).

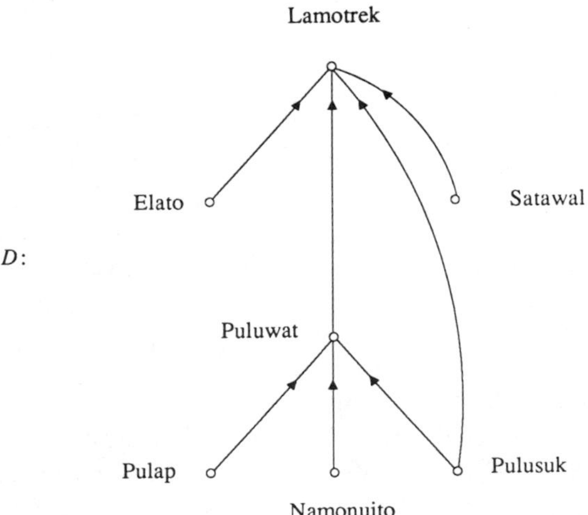

FIG. 8.6. An oriented graph of a tribute relation in the Caroline Islands

As occurs elsewhere in Oceania, from Fiji (Hocart 1952) to the Marquesas (Rolett 1986), turtle symbolism figures prominently in ideas about rank.

Similarity Relation: The Trukic Dialect Chain

In the social sciences the concept of a 'similarity relation' is best known in psychology (see e.g. Restle 1959). Sometimes it is called a 'proximity relation', as in Coombs (1964). The underlying idea is always that there is a collection of objects, each of which has a real number value associated with it. There is also a fixed positive number t, for 'threshold'. Whenever the magnitude of the difference between the values of two objects is less than t, they are said to be similarly related to each other, or, more briefly, *similar*. Thus every object is necessarily similar to itself. Also it is obvious that if one object is similar to another object, then the second object must be similar to the first. But it is not always true that if, among three distinct objects, the first is similar to the second and the second to the third, then the first must be similar to the third. For the difference in values between the first and third objects may exceed the threshold even though each of their values is within the threshold of the value of the second object. For these reasons, *similarity relations* have been described axiomatically as binary relations which are reflexive and symmetric, but 'not necessarily transitive'. Logically, this third stipulation could not be more superfluous. For if a condition is not necessarily true, this is captured simply by not asserting that it is true. Similarity relations are naturally expressible as graphs, and their concepts are subsumed within the framework of graph theory, as pointed out in Harary (1964).

To illustrate a psychological application, experimental tests are performed on the similarity of pairs of colours. In the corresponding abstract system, there is a collection of stimulus events a, b, c, . . . and a binary relation denoted by M called 'matching'. The stimuli stand for colours, and relation M for indistinguishability between two stimuli. Galanter (1956) postulates that the relation M is symmetric and reflexive but 'not necessarily transitive'. The central assumption is that two stimuli are taken as similar whenever the set of stimuli which match with the first is identical with the matching set of the second stimulus.

The concept of a similarity relation has a natural application to the study of dialect chains in anthropological linguistics.[4] Here the objects are individual dialects joined by a relation of mutual intelligibility. Such chains are obviously *r* and *s* but 'not necessarily *t*', because it is not always true that if, for three dialects, the first is mutually intelligible with the second, and the second with the third, then the first is mutually intelligible with the third.

In Chapter 3 we described the vast overseas exchange system in the Caroline Islands stretching from Palau, Yap, and Ulithi in the west to Truk and the Lower Mortlocks some 1,600 kilometres to the east. While the languages of Yap and Palau (and also of the Marianas, which was a part of this network) belong to distinct language groups, those spoken on Truk and on all the low islands belong to the Trukic subgroup of nuclear Micronesian.[5] In the preface to his Trukese grammar, Dyen (1965) posed the following question (cited by Quackenbush 1968) concerning the relation among these languages:

> A brief contact with a native of Ulithi, whom I met on Guam, led me to conclude that the language spoken there is closely related with Trukese, but it was also clear that Ulithians and Trukese could not possibly understand each other's language. This leads to the interesting question: What is the linguistic relationship of the languages or dialects lying between Truk and Ulithi? Is there only a gradual increase of differentiating features in the languages or dialects as one progresses in one direction through the islands lying between those two? Is the gradual increase of differentiating features only such that the languages or dialects which are somewhat remote from each other are mutually unintelligible while those which are geographically neighbors are always mutually comprehensible or nearly so? In either case, how many different languages are there? (Dyen 1965: p. x).

E. M. Quackenbush (1968) answered Dyen's question by showing that the Trukic languages or dialects are part of a single chain or continuum. Quackenbush identified 17 dialects (or 'dialect areas'). His list includes five pairs judged as equivalent, and one (Carolinian) dialect from Saipan in the Marianas. This leaves the 12 dialects, 11 excluding Saipan, shown in Table 8.1. For convenience we have preserved Quackenbush's numbering: the slashed numbers represent

[4] For an application to phonemics see Harary and Peterson (1961).

[5] Nuclear Micronesian includes Gilbertese, Nauruan, Marshallese, Kusaiean, Ponapean, and the Trukic languages. It excludes the languages of Yap, Palau, the Marianas (Chamorro), and the two Polynesian outliers, Nukuoro and Kapingamarangi.

TABLE 8.1. *Dialects in the Trukic language group (from Quackenbush 1968)*

Number	Name of dialect (area)	Islands included in dialect area
1	Sonsorol	Sonsorol, Pulo Anna, Merir
2	Tobi	Tobi
3/4	Ulithi	Ulithi, Fais, Ngulu, Sorol
5/6	Woleai	Woleai, Eauripik, Lamotrek, Faraulep, Elato, Ifaluk
7	Satawal	Satawal
8	Saipan	Saipan
9/10	Puluwat	Puluwat, Pulusuk
11	Pulap	Pulap
12	Namonuito	Namonuito
13	Murilo	Hall Islands
14/15	Nama, Satawan	Upper, Lower Mortlocks
16/17	Truk (Fanapanges, Moen)	Truk

dialects he regarded as equivalent for purposes of his study. The series runs generally from west to east.

Interviews with an informant from each of these dialect areas generated the graphs shown in Fig. 8.7. Both graphs are generally strung out in an east–west orientation. (At the ends of the first graph, 2 (Tobi) is slightly farther west than 1 (Sonsorol), and 14/15 (Nama, Satawan) slightly farther east than 16/17 (Truk).) In the first graph, G_1, the Trukic L-complex, the lines join pairs of mutually intelligible dialects. In G_2, constructed from Quackenbush's overlapping box diagram, which of course constitutes still another independent discovery of Venn diagrams, the criterion of mutual intelligibility is relaxed to include borderline or partial as well as complete mutual intelligibility. A third graph, not shown, also constructed from a box diagram (Venn diagram), is based on strictly lexical evidence using Swadesh's (1954) criterion that dialects sharing 81 per cent or more of their basic vocabulary belong to the same language. The only difference between it and G_2 is the absence of the line joining point 2 and point 3/4 in G_3, thus providing some independent confirmation of Quackenbush's research. (The shared basic vocabulary of 2 and 3/4 is 78 per cent.)

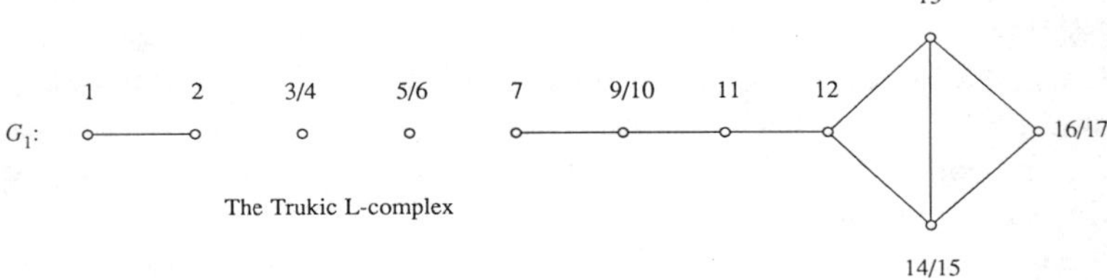

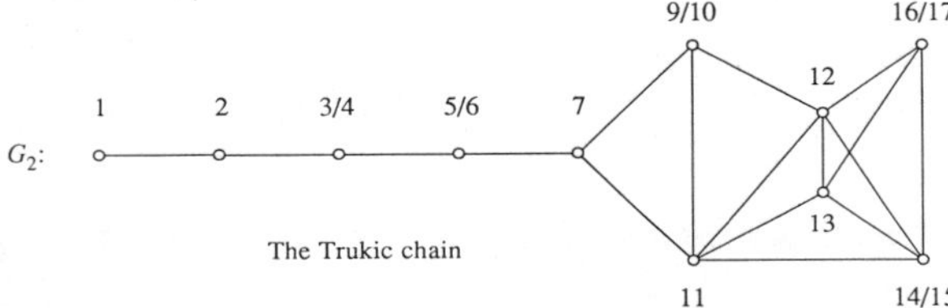

FIG. 8.7. Similarity relation (with loops omitted) of Trukic dialects (adapted from Quackenbush 1968)

Both graphs are similarity relations (for convenience the loops on each point are omitted), and they all confirm Dyen's surmise that Ulithians and Trukese could not possibly understand each other. Indeed, even Puluwatese and Pulapese, whose islands are adjacent with Truk in the sea-lane graph of Fig. 3.2, could not understand Trukese. Thus when these and many other pairs of adjacent islanders traded with each other communication rested on bilingualism or semi-bilingualism. Since some islanders' languages are more prestigious than others this relation was not always symmetric. People from islands adjacent to Truk, for example, learned Trukese but not conversely, and communication between Ulithians and Satawalese depended on the bilingualism of the latter.

If a language is defined as a maximal set of mutually intelligible dialects, then, in the graph of a dialect chain, the maximal complete subgraphs (*cliques* of a graph) define the individual languages. In answer to Dyen's question asking how many languages there are in the Trukic chain, we use the graph G_2, and see that there are eight languages, as there are four K_2 *cliques*, and two each of K_3 and K_4. One interesting consequence of this graphical definition of a language is, as Quackenbush recognizes, that a dialect may belong to more than one language. Thus all but two dialects in G_2, Sonsorol, point 1 at the far western end of the chain, a language which Quackenbush describes as the most archaic of the Trukic languages, and Truk, point 16/17, belong to at least two different languages. Pulap, point 11, and Namonuito, point 12, in the east, belong to three different languages.

Such a definition has practical implications for comparative studies. Blust (1984), for example, in a critique of Marshall's (1984) analysis of sibling classification in Oceania, objects to treating the Trukic chain as 17 distinct languages. Quackenbush had already reduced this number to 12, as shown in Table 8.1, and graph analysis reduces it to eight.

Equivalence Relation: The Neighbourhood of a Dialect

An *equivalence relation* is *r*, *s*, *t*. Consider for example the set of integers, and let R be the relation such that uRv if and only if the difference $u-v$ is even. We see that R is *r*, since every difference $u-u = 0$. It is *s*, for if $u-v$ is even, then $v-u$ is even. And it is *t*, for if $u-v$ and $v-w$ are both even, then $u-w$ is even. By definition, then, R is an equivalence relation. We observe that R is not *c*, because no odd number is in relation R to any even number. Abstractly, equivalence relations play the role of an equal sign. Thus every equivalence relation is a similarity relation, but not conversely.

Quackenbush's study of the Trukic dialect chain is based on concepts from Hockett (1958). We can take this opportunity to simplify Hockett's implicit graph theoretic model and to provide a linguistic application of a social structure model introduced in Chapter 3. Hockett distinguishes between an 'L-simplex' and an 'L-complex'. The first refers to a maximal set of mutually intelligible dialects (or idiolects) and is thus a maximal complete subgraph of a dialect chain. The second refers to a maximal set of linked dialects and is thus a connected component of a graph. In the graph G_1 in Fig. 8.7, there are eight cliques which are L-simplexes, while obviously there are four components, i.e. L-complexes.

A potentially useful simplification of a dialect chain would consist of grouping together all dialects that are structurally equivalent. This can be done by defining the 'neighbourhood of a dialect'. In Chapter 3 in the Micronesian sea-lane graph (Fig. 3.2) we defined the neighbourhood of an island u as the set of all islands v adjacent to u. Analogously, we can define the neighbourhood of a dialect u in the similarity graph.

Let us say that two points u and v of a graph are *N-equivalent*, denoted by uNv, if their closed neighbourhoods $N[u]$ and $N[v]$ consist of exactly the same points, $N[u] = N[v]$. This is an equivalence relation, for it is r, s, t, and expresses the condition for two dialects to be regarded as structurally equivalent.

An *r-graph* is not really a graph but is obtained from one by the addition of a loop at each point, making it reflexive, i.e. it is precisely a similarity relation, as it is r, s. We use capital letters A, B, C, . . . for the closed neighbourhoods of points a, b, c, . . ., respectively. In psychophysics a 'matching graph' means an r-graph whose points represent physical stimuli (such as colours) and whose lines indicate subjective identity. In the present context, the points stand for dialects and the lines for mutual intelligibility. We illustrate some of these concepts with the graph M of Fig. 8.8.

In this figure it is understood that there is a loop at each point even though these are not drawn. The neighbourhoods of all the points of M are denoted by corresponding capital letters:

$$\begin{aligned} A &= \{a, b, c\}, \\ B = C &= \{a, b, c, d\}, \\ D &= \{b, c, d, e\}, \\ E &= \{d, e, f, g\}, \\ F = G &= \{e, f, g\}. \end{aligned}$$

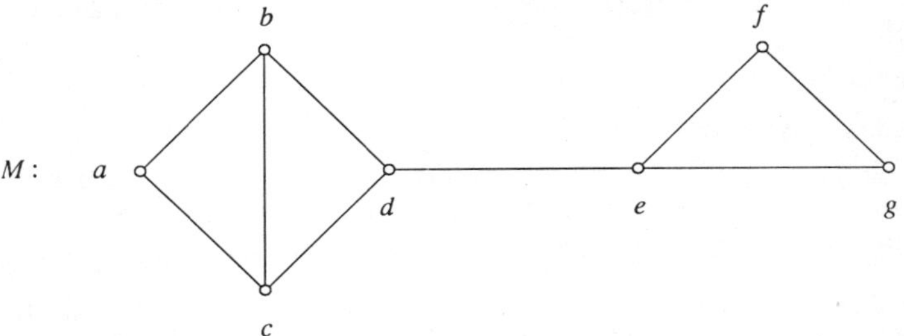

FIG. 8.8. A matching graph M, with loops not shown

Since $B = C$ and $F = G$, it follows that points b and c are N-equivalent, as are points f and g.[6] If we denote the combined points by bc and fg, then we get the *reduced graph* M^* obtained from Fig. 8.8 on coalescing (identifying) N-equivalent points, as shown in Fig. 8.9.

The graphs G_1^* of the reduced Trukic L-complex G_1 and G_2^* of the reduced Trukic chain G_2 from Fig. 8.7 are shown in Fig. 8.10. Thus in both G_1 and G_2, the dialects to the north and south of Truk, 13 (Murilo in the Hall Islands) and 14/15 (Nama, Satawan in the Upper and Lower Mortlocks), are structurally equivalent.

FIG. 8.9. The graph M^* obtained from Fig. 8.8

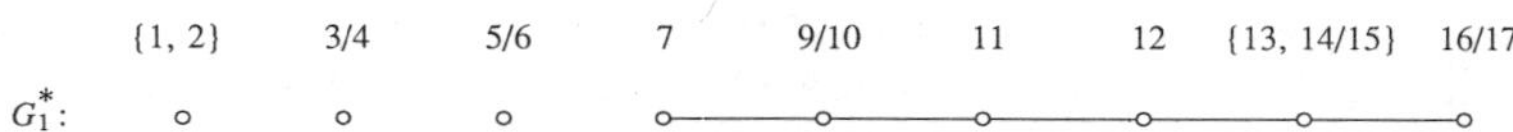

The reduced Trukic L-complex

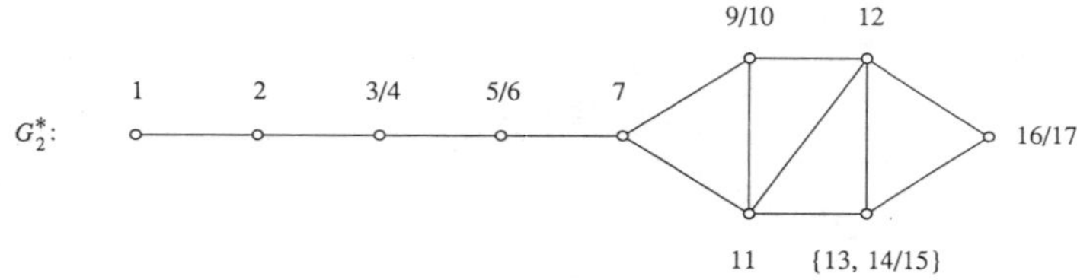

The reduced Trukic chain

FIG. 8.10. Structural simplifications of the Trukic L-complex and Trukic chain in Fig. 8.7.

Orders: Big Man, Chief

In most Melanesian societies political statuses are differentiated but only partially. The commonest form of leadership is 'big man' polity

[6] There is a classic sin of notation here, as the letter G has often been used for a graph but here it stands for the closed neighbourhood of point g. There should be no confusion, as the graph at hand is M.

(Sahlins 1963). Melanesian big men, in contrast to Polynesian chiefs, are informal, non-hereditary leaders who achieve renown and gain a following through their own initiative, especially through success in competitive feasting and exchange as in the *moka* and *te* systems of Highland New Guinea described in Chapter 2. In Mount Hagen society in Highland New Guinea, people distinguish between broad classes of major and minor big men as opposed to ordinary men and 'rubbish men'. There is 'rough agreement as to who can and who currently cannot lay claim to [big man] status' (A. Strathern 1971: 187). In Siuai society in the Solomon Islands (Oliver 1955), a ladder of competitive feasting arranges big men in a partial order.

A *partial order*, often called a partially ordered set or poset, is $\bar{r}$, $\bar{s}$, t. The prototypical partial order in mathematics is that of a *proper subset*, which we denote here for two sets A and B by $A < B$, that is, A is contained in B but is not identical with B. We verify that this relation is a partial order:

$\bar{r}$ $A < A$ never holds, as $A = A$.
$\bar{s}$ if $A < B$, then $B < A$ is impossible.
t if $A < B$ and $B < C$, then of course $A < C$.

This is called a partial order because it satisfies all the axioms of a complete order except for c, completeness. Thus there may well exist two sets A and B for which neither $A < B$ nor $B < A$ holds, in which case A and B are *incomparable*. (We note that a partial order is sometimes defined as r, $\bar{s}$, t and exemplified by subset $A \subset B$, not necessarily a proper subset. It would be better and clearer if partial order were always used to mean an $\bar{r}$, $\bar{s}$, t relation. Then the r, $\bar{s}$, t case should be called a *reflexive partial order*.)

Competitive big man feasting appears to have reached its apogee in Siuai. According to Oliver (1955), a Siuai big man, called a *mumi*, acquires renown by feasting a rival big man with pigs (given together with gifts of shell money). The recipient must repay with a feast of at least equal value if he is not to lose renown and become ranked lower 'in the minds of fellow Siuai' (1955: 393). If the debtor-guest can only repay with a feast of approximately equal value then further competition between them is called off. If one *mumi* vanquishes another, perhaps after a succession of feasts which exhaust and impoverish the loser and his followers, then he may seek out a worthier opponent. Theoretically there is no limit to such a progression, but it does not appear to go beyond Siuai tribal boundaries. At the time of Oliver's stay four *mumi* were at the top in their own areas and two of them were starting to compete for highest rank in Siuai.

The exchange of competitive feasts generates a rank hierarchy given the following assumptions: (1) No *mumi X* outranks himself. (2) No two *mumis X* and *Y* outrank each other. (3) A *mumi X* who challenges an opponent *Y* (who already outranks *Z*) and wins, then outranks *Z* as well as *Y*. The result is a partial order by definition, as (1) = $\bar{r}$, (2) = $\bar{s}$, and (3) = t.[7] This is illustrated in Fig. 8.11.

A *complete order* is $\bar{r}$, $\bar{s}$, t, c. An arithmetic example is the relation 'greater than' on the set of all positive integers. Clearly this relation is $\bar{r}$: no integer is greater than itself. It is $\bar{s}$: for any two integers u and v, if $u > v$, then $v \not> u$. It is t: if $u > v$ and $v > w$ it follows that $u > w$. It is also c: for any two different integers u and v, either $u > v$ or $v > u$. (A *reflexive complete order* means a relation that is r in addition to $\bar{s}$, t, c.)

In most Polynesian societies rank is theoretically determined by genealogical seniority.[8] In Western Polynesia the relative rank of chiefly titles is given public and dramatic expression in the *kava* ceremony. Writing of Futuna, Burrows says:

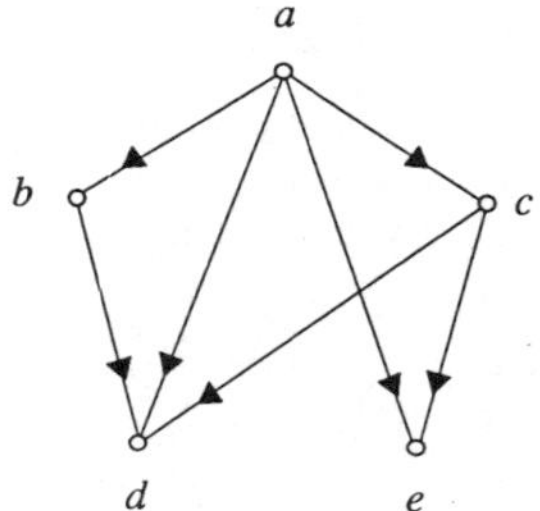

FIG. 8.11. A partial order of *mumi*-hood in Siuai

[7] D. Oliver (personal communication) notes that individual opinions on the relative status of past competitors diverge: in the case that two *mumis X* and *Y* quit competing after equal feasts (or after a series of feasts whose culmination was equal), it appears that the partisans of each would assert the superiority of their respective leader. Even third parties would, in Oliver's experience, assert the superiority of one over the other, depending on factors such as kinship or residential proximity.

[8] In practice, there are of course divergent interpretations, disputes, and periodic reshufflings of the rank of chiefly titles. See for example Valeri (1985, chapter 5) on the relation between genealogical rank and the realities of wealth and power in Hawaii.

> Formal recognition of rank appears most conspicuously in the order of serving *kava* on public occasions. Chiefly titles are metaphorically called '*kavas*' from the right they confer to certain positions in the order of serving *kava*. Thus Tu'i Angaifo once said, 'There are only two old *kavas* in Singave.' '*Kava*' is also used to mean royalty, in a figure of speech equivalent to the European 'throne' or 'crown'.
>
> At formal *kava* parties, the chiefs sit along the front of the house with their backs to the wall logs. They take their places without regard to the position of the wall posts, which are more widely spaced than those of Samoan houses, and have no association with rank. Thus seated, the chiefs form the *alofi* or *kava* 'circle'. The senior chief—the king, if he is present—sits in the center. At his right sits the chief next in rank, or one of the king's adjutants (*mua*) who is regarded on these occasions as sharing, or rather representing, the king's supreme rank. At the king's left sits the chief whose rank is third in order in the district, or the other *mua*. The rest of the chiefs are similarly seated in order of rank, successively to right and to left of the king or ranking chief present (Burrows 1936: 91–3).

Fig. 8.12 from Burrows shows the arrangement of a district *kava* ceremony on Futuna. The numbers indicate the serving order, with 1 = the king, 2 = his *mua*, 3 = the Tu'i Saakafu, etc. Clearly, this is a complete order, as it satisfies $\bar{s}$, t, c and also $\bar{r}$ since no one can be served before himself.

A complete order, which is always a transitive tournament, is further illustrated in the next section in Fig. 8.13.

Tournament: The Tongan Solution

A *tournament* is a relation which satisfies $\bar{r}$, $\bar{s}$, c. The name of this relation derives from the structure of round-robin tournaments in

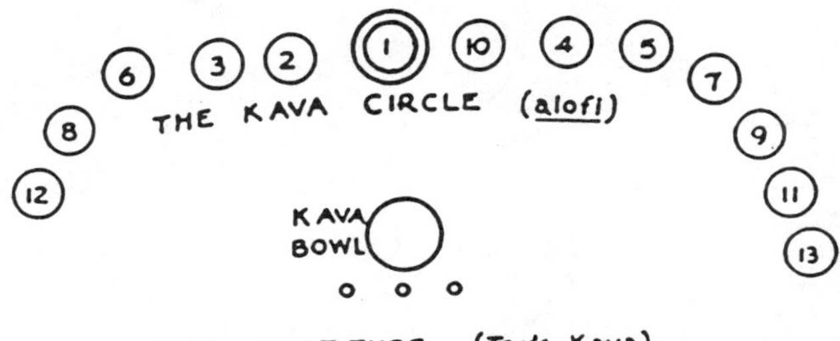

FIG. 8.12. A district *kava* ceremony in Futuna (from E. G. Burrows, 1936)

which players engage in a game that cannot end in a tie and in which every player plays every other exactly once. Tournaments are realized by a great many empirical phenomena in addition to round-robin competitions. Tournaments appear in the method of scaling known as 'paired comparisons' in the study of consumer preferences, in scheduling problems, in voting theory, and in dominance relations as in a flock of hens. Fig. 8.13 shows all the non-trivial tournaments with at most four points.

Two types of tournaments are of particular anthropological interest. A *transitive tournament* is t in addition to $\bar{r}$, $\bar{s}$, c and is therefore a complete order. The first 3-point digraph in Fig. 8.13, also called a *transitive triple*, and the first 4-point digraph are transitive tournaments. If the digraph of a rank system is a transitive tournament and thus a complete order, this means that every position is uniquely ranked in a linear sequence: $1 < 2 < 3 < \ldots < n$. A *strong tournament* is strongly connected. The second 3-point digraph in Fig. 8.13, also called a *cyclic triple*, and the last 4-point digraph are strong tournaments. If the digraph of a marriage system is a strong tournament with three or more points, then it contains a spanning cycle of exchange (as well as a cycle of each length $k = 3, 4, \ldots, p$), as noted in Chapter 2. Both types of tournaments are fundamental in Tongan social structure.

In Tonga 'the ranking of individuals within the . . . family, termed *lahi* (abundance, plenty, greatness) is the key to the organization of

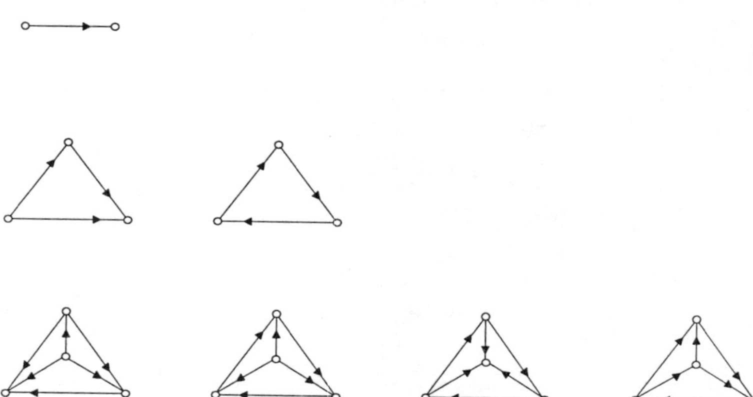

FIG. 8.13. The tournaments with two, three, and four points

... society in every stratum' (Gifford 1929: 19). A sister and her children outrank her brother and his children. A father outranks his children, and the children (being on the chiefly (*'eiki*) side of the family) outrank their mothers. Fig. 8.14 presents a digraph of this system, showing ego (a child) outranked by his father (F), his father's sister (FZ), and his father's sister's children (FZC), while ego's father (F) is in turn outranked by his own sister (FZ) and her children (FZC). Ego's father's sister (FZ) is outranked by her own children (FZC). The digraph satisfies $\bar{r}$, $\bar{s}$, c, and t and is thus a transitive tournament.[9]

This system of kinship ranking created a marriage problem in Tonga and in other Polynesian societies, for it meant that a high chief would be outranked by his sister's children.[10] G. Rogers (1977) gives three solutions, which we represent digraphically in Fig. 8.15 by showing in all three instances two chiefly lines *A* and *B* where *A* acts in such a way as not to be outranked by his sister's children: the Hawaiian solution of ritual sibling marriage (Fig. 8.15a); the Pukapukan solution of permanent celibacy for the sister—the 'sacred maid' (Fig. 8.15b); and the Tongan solution of ethnic exogamy—marriage to a foreign group which did not count in the internal system of rank (Fig. 8.15c). It appears that in some societies celibacy and ethnic exogamy were alternative solutions. According to Kaeppler,

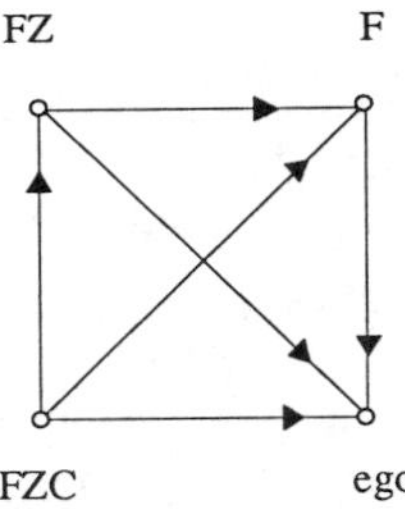

FIG. 8.14. A Tongan kinship tournament

[9] In Polynesian culture, rank is pervasive, embracing individuals, groups, and even the gods. The rank structure of the four major gods in the Hawaiian pantheon as described by Valeri (1985) is the transitive tournament shown in Fig. 8.14, with the points FZC, FZ, F, and ego relabelled as Ku, Lono, Kane, and Kanaloa respectively.

[10] See T. Mabuchi (1960, 1964) on the 'spiritual predominance of the sister' in Oceania.

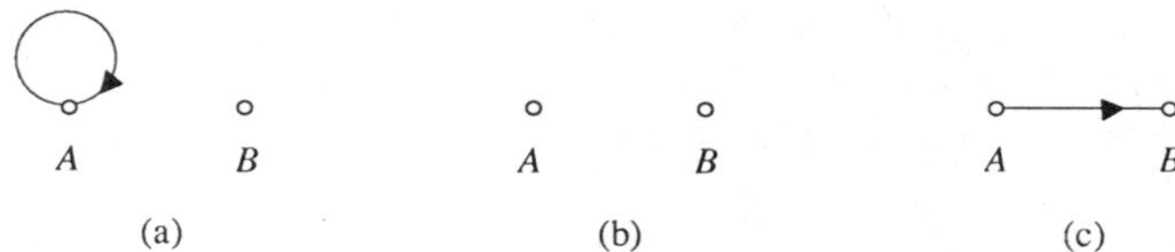

FIG. 8.15. Three solutions to the problem of rank and marriage of the sister in Polynesia

. . . Samoa, too, occasionally needed an appropriately-related social order into which a high-ranking woman could be married. If a Samoan chief managed to acquire all four so-called Tu'i titles, and thus become the Tafaifa, his eldest female descendant would be too high to marry a Samoan and, in fact, to do so would have been considered incest. Her only possible marriage alliance was with a Tongan. Salamasina, the sister of the present Samoan Head of State, although not the daughter of a Tafaifa, remains unmarried because she is too closely related to the other three royal families (Shore 1976: 295). In order to retain the dignity and honour of her brother and the lines from which she descends, she must either not marry, or marry out (Kaeppler 1978: 249).

The Hawaiian solution was elegant in theory but compromised in practice by the requirements of political alliance. As Valeri has pointed out,

'un chef peut rarement se permettre de refuser l'alliance avec ses collatéraux ou avec d'autres lignées puissantes et concurrentes: une alliance qui replie constamment une lignée sur elle-même, qui conserve à chaque fois pour elle-même le potentiel de parenté contenu dans ses sœurs, élève peut-être son rang, mais s'isole et s'expose à la révolte des collatéraux frustrés' (Valeri 1972: 41).

The Hawaiian solution also entailed fratricide as a corollary:

Strategically, a king attempts to establish a line that is superior in rank to all others; hence, rank being transmitted bilaterally, he must monopolize all the high-ranking women of the kingdom. This implies royal incest (which is the ideal case) or if the king is an upstart, forced hypogamy. To pave the way for this latter marriage or to exclude his rivals from sharing the high-ranking women (his sisters) with him, the king must destroy them—unless, of course, they resign themselves to being excluded. Hence the sacrifice of the rivals (the brothers) is the logical correlate, at least in extreme cases, of marriage with high-ranking sisters (Valeri 1985: 165).

Actually the Tongan solution was more refined than indicated by Rogers, for, as Leach (1972), Biersack (1974, 1982), and Bott (1981) have shown, it managed to integrate relations of power and rank in a single 'marriage connubium'. It was an imaginative structural variant of Austronesian political duality.

Dual power structures consisting of opposed lines or classes of titles occur in a number of Polynesian[11] and Micronesian societies. Usually one line is senior, sacred or semisacred, and the ultimate source of formal authority, while the other is junior, more secular, and charged with practical, executive responsibilities. In Ponape in Micronesia, the two (matrilineal) lines headed by the Nahnmwarki and the Nahnken are related by symmetrical marriage exchange—even though they are metaphorically designated as royal 'fathers' and royal 'children' (Riesenberg 1968). In Samoa the two classes of chiefs, the *ali'i* and the *tulāfale*, are related by the exchange of 'female' and 'male' goods, *toga* and *'oloa*, analogous to the dowry and bride-price given at marriage (Mead 1969 [1930]). In Tonga the two lines are related by asymmetrical marriage exchange in the following way.

The sacred king of Tonga, the Tu'i Tonga, married the eldest daughter of the secular ruler, the Tu'i Kanokupolu, thereby reaffirming his superiority to the latter as wife-taker to wife-giver. Along with his daughter the Tu'i Kanokupolu gave, in accordance with the obligations of wife-givers to wife-receivers, political support and food. The Tu'i Tonga in turn gave his eldest sister, the Tu'i Tonga Fefine, to a chief in a Fijian group, the Fale Fisi. Although outside the Tongan system of rank, the Fale Fisi gained in prestige as well as receiving political protection. The eldest daughter of this marriage, the Tamahā, married the Tu'i Kanokupolu, thereby elevating his personal rank and that of his children in accordance with the Tongan concept that personal rank can be inherited from both parents. In this case the rule by which wife-givers provide political support and food to wife-takers was waived; the Tamahā brought the gift of high rank. Bott describes the first two marriages as prescribed and the third as a result of 'pressure to "marry in a circle" ' (1981: 56).[12] The connubium or strong tournament is shown in Fig. 8.16; it is a cyclic triple.

[11] See Goldman (1970) and Shore (1982).

[12] The strong tournament in Fig. 8.16 is implicit in Leach's (1972) Fig. 1, and it is a subgraph of the 4-point asymmetric connubium (which is not a tournament) shown in Biersack (1974). Bott (1981) also mentions non-prescribed, non-institutionalized marriages of Fale Fisi women to Tu'i Tonga men and Tu'i Kanokupolu women to Fale Fisi men, and gives a network model of this system.

Tu'i Tonga

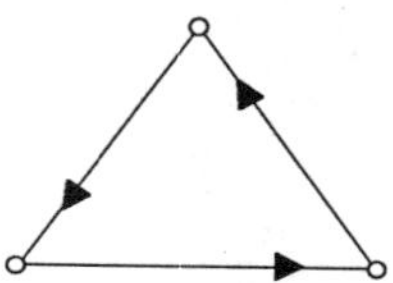

Fale Fisi Tu'i Kanokupolu

FIG. 8.16. A Tongan marriage tournament

Parity Relation: Inner Structure

Exchange defines relations within as well as between groups, and in both cases graph theoretic considerations can lead to the discovery of structural necessity. In Chapter 2 we described Kelly's (1974) lattice graph model of Etoro marriage exchange as bipartite. According to Theorem 2.2 a graph G is bipartite if and only if it contains no cycles of odd length. This rules out triangles (3-cycles), from which we conclude that G is $\bar{t}$ (as well as $\bar{r}$ and s by the definition of a graph). In Etoro conception, any two lineages that exchange women with a common third lineage are 'brothers' and may therefore not exchange women with each other. We hypothesized that the brother relation in Etoro is t, which means that any two lineages joined by a sequence of brother links are brothers to each other even if they do not exchange women with a common third lineage. For example, in Fig. 8.17a, lineages 1 and 3 are brothers (as both are adjacent to point 2), as are 3 and 8, so, by transitivity, 1 and 8 are brothers. This relation is $\bar{r}$, s, t and thus satisfies the axioms of a *parity relation* (Harary 1961b). A parity relation is a graph in which every connected component is complete.

We have already seen that a grid, the product of two paths, is bipartite in the illustration of Fig. 7.7. This also follows from the fact that it is a planar graph in which every interior face is a square, i.e. every region other than the exterior is a square. In Fig. 8.17a the two colour sets $V_1 = \{1, 3, 6, 8\}$ and $V_2 = \{2, 4, 5, 7\}$, giving two disjoint sets of brothers, are the exchange units in Etoro society, the implicit moieties. Their respective brother relations, the inner structure of the two moieties, are the two complete graphs G_1 and G_2 of Fig. 8.17b.

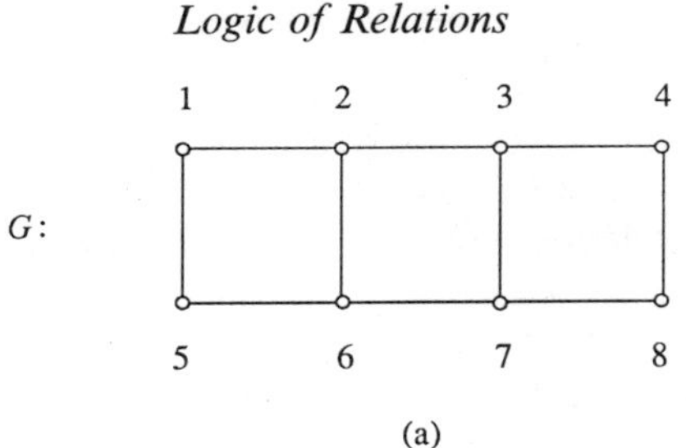

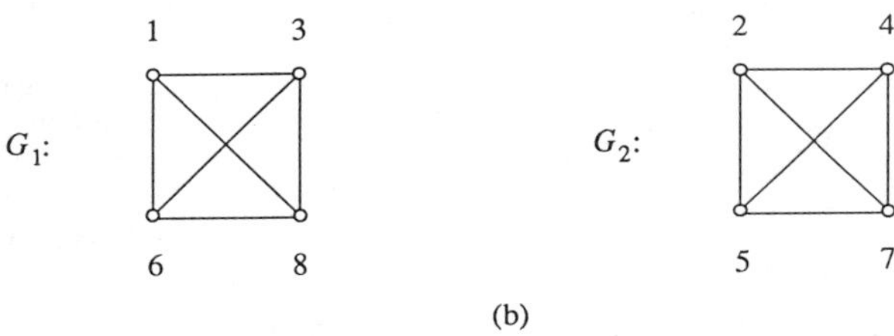

FIG. 8.17. Etoro marriage structure: (a) marriage exchange between lineages; (b) the parity relation between lineages

The operations on graphs which model marriage between Tangu sibling sets (Fig. 7.10) bear some structural similarity to Fig. 8.17. In both instances the Cartesian products represent the marriage relation and the complete graphs K_4, siblingship.

We now state the basic structural property of a parity relation.

Theorem 8.1. A graph is a parity relation if and only if every component is complete.

Antiequivalence Relation: Ritual Exchange

Exchange between moieties, or between any two groups, can take the form of a symmetric pair of arcs, or the union of symmetric pairs of arcs, as when subgroups within each moiety or pair of clans exchange with each other. In this case the structure satisfies $\bar{r}$, s, and is a graph. But exchange between moieties can also take the form of a directed cycle of even length, or a directed path. All directed cycles and directed paths satisfy $\bar{r}$, $\bar{s}$, $\bar{t}$ and are therefore *antiequivalence relations*. We have already given an example of a directed 4-cycle of marriage exchange between pairs of clans—Wagner's (1967) model of Daribi described in Chapter 1. An example of a directed path in ritual exchange is described by Bateson (1958) in *Naven*. The directed path is a good example of the asymmetry in dual organization noted by Lévi-Strauss (1944).

In Iatmul society in the Sepik region of New Guinea, male initiation is regulated by two pairs of moieties A, B and x, y which intersect to give the four groups A_x, A_y, B_x, B_y. Each of these groups is divided into named initiatory grades, 1, 3, 5 in the A groups and 2, 4, 6 in the B groups, let us say. The entire series 1–6 is ranked by seniority. Men of A_{x1} initiate youths of B_{y2} who initiate youths of A_{x3}, continuing down to A_{x5} who initiate youths of B_{y6} when they come of age. The sequence is parallel for Ay and Bx, with two corresponding groups in each set, say A_{x3} and A_{y3}, initiated at the same time. When A_{x1} die out, a new grade A_{x7} appears. Moiety B then becomes senior to A, as recognized by a ceremonial brawl. The digraph of this relation is shown in Fig. 8.18.

The Iatmul use two different kinship metaphors to describe the relation between the initiatory grades. Each grade is spoken of as the 'elder brother' of the grade immediately below it, and at a certain point in the ceremonies it is also spoken of as the 'mother' of the junior adjacent grade. These two metaphors are logically consistent:

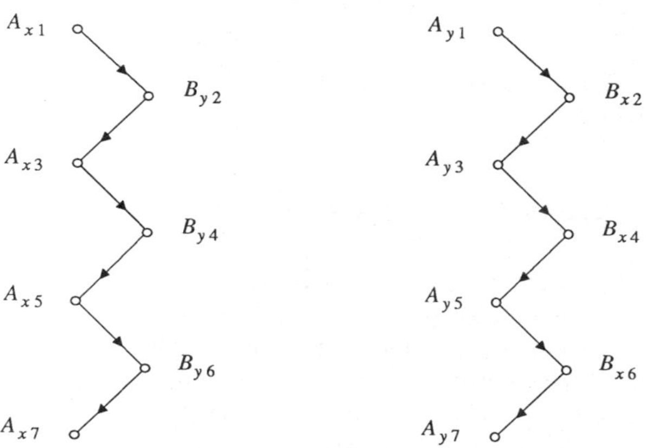

FIG. 8.18. An antiequivalence relation in Iatmul initiation

'mother' satisfies $\bar{r}$, $\bar{s}$, $\bar{t}$, as does elder brother when defined as 'immediately elder brother'. Both are therefore antiequivalence relations.

Antiparity Relation: Whales' Teeth, Turtles, Pigs, and Men

It is clear from Sahlins's (1976) analysis of Fijian exchange that whales' teeth, turtles, pigs, and men are all related in some way or, better, related in multiple ways. In addition to exchange relations obtaining between various of these objects, as when a Fijian chief makes a gift of whales' teeth to the fishermen who catch turtles for him, there are relations of substitutability as well, defined as 'x can take the place of y in an offering'. Sahlins mentions that a pig can be substituted for a man in sacrifice, and that a man can be substituted for a turtle when the fishermen are unsuccessful in the hunt. But this is not all, for according to Hocart (1929) whales' teeth can be substituted for men in offerings (*soro*) to a chief. Suppose that the converses of these statements are not true and that neither whales' teeth nor pigs can be substituted for turtles (nor for each other), and grant, as a purely formal condition, that each object can be substituted for itself. Then the substitutability relation over whales' teeth, pigs, men, and turtles satisfies the axioms r, $\bar{s}$, $\bar{t}$ and is an *antiparity relation*, as shown in Fig. 8.19.[13]

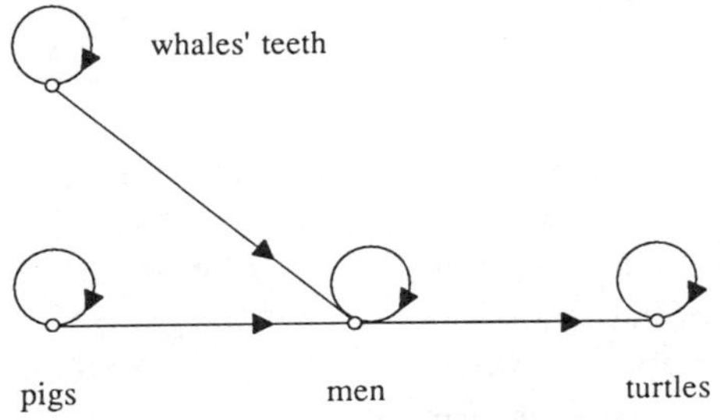

FIG. 8.19. The substitutability relation of Fijian objects as an antiparity relation

[13] For the sake of completeness: the digraph also shows that turtles cannot be substituted for whales' teeth or pigs.

If this relation actually holds then it implies something interesting about the supreme value of turtles. Hocart notes an identity between turtling celebrations and man-slayer's consecration, and he tentatively suggests that the turtle is the god-victim: 'It may be ... that the turtle is the god-victim; of this there is no real evidence, but only suggestion, such as the name of Temple of the Turtle borne by a clan of Navatu, and the fact that the chief fisherman sleeps with the turtle, which is suggestive of a sacred marriage' (Hocart 1952: 55).

Our interpretation of these Fijian data is provisional, offered only as a hypothesis intended to stimulate further research, and given to make a general point. In his discussion of 'transitive equivalence' Sahlins makes reference to the concept of 'spheres of exchange', or 'separate circuit[s] of transactable items' (1976: 37). Firth introduced this term in *Primitive Polynesian Economy* to designate 'separate series of exchanges ... the goods in which are not completely convertible into those of the other series' (Firth 1939: 340). Firth distinguishes three such series in Tikopia: the 'food series', the 'bark-cloth sennit series', and the 'bonito-hook, turmeric cylinders, and canoes series'. From Firth's discussion, for example his remarks on the non-equivalence of bark-cloth and sennit in certain contexts, and his characterization of items in each series as 'not completely convertible', it is clear that the relational properties of such spheres or circuits may be more complex than is commonly realized. The elucidation of these properties does not depend just on more detailed ethnography but on using a structural model right from the beginning of the investigation.

Conclusion

In concluding our analysis it is fair to ask how graph theory has enlarged our understanding of exchange in Oceania. Let us review our results in the light of the reasons given in Chapter 1 for using graph theory in this book.

First, graph theory provides a rich and precise language for describing and classifying the great variety of exchange structures found in Oceania. We have used graph theoretic concepts to describe trade networks (Chapters 3, 4, 5), social networks (Chapter 4), and kinship structures (Chapters 2, 6); to classify types of marriage and ceremonial exchange (Chapter 2); to give generic characterizations of social structures (Chapter 2); and to clarify seemingly intractable marriage arrangements (Chapter 6).

Surely all anthropologists would agree on the need for accurate and revealing descriptions and classifications of exchange structures. But, one may ask, is graph theory really necessary to accomplish this task? Is it necessary, for example, to define such an elementary concept as a cycle in a digraph? As shown in Chapter 2, it is if we want to eliminate the confused interpretations of the forms of marriage exchange and their misleading extension to forms of ceremonial exchange. But do we need such a large repertory of graph theoretic concepts? We do if we want to take account of the empirical diversity of exchange structures. By using graph theoretic concepts we can eliminate the current hodgepodge of general purpose concepts, *ad hoc* definitions, and mathematical metaphors which blunt ethnography and restrict theoretical analysis. We have emphasized the use of theorems. But do we really need, say, a theorem on bigraphs to define dual organization? We do if we want to recognize its variable and sometimes implicit or hidden forms, such as 'chequerboards', 'lattices', 'trees', etc., and if we want to prove, not merely assert, its compatibility with exchange cycles of different lengths. Theorems enable us to think deductively about structural problems.

Secondly, graph theory provides techniques for calculating quantitative features of exchange networks. Our network analysis of economic dominance and political stratification in the Caroline Islands trade network in Chapter 3 was based on quantifiable concepts of centrality and betweenness in graphs. Without such concepts there is no way to establish that one community is more advantageously located than another, and no way to test network theories of economic performance and social organization.

In Chapters 3, 4, and 5 we noted the common interests of anthropologists, archaeologists, and geographers in studying quantifiable aspects of social, trade, and transport networks. Our contribution to network analysis consists of expanding the range of models and applications and clarifying concepts such as centrality, connectivity, and planarity, thereby giving the enterprise a broader and more solid foundation.

Thirdly, the matrix methods associated with graph theory permit rapid and accurate analyses of large exchange systems. In Chapter 4 we used the adjacency, distance, and reachability matrices of a graph to calculate centrality, betweenness, and various types of path structures in trade and social networks. These methods are convenient for analysing smaller networks, and absolutely indispensable for analysing larger networks such as the Caroline Islands and Late Period Mailu networks described in Chapters 3 and 4.

Fourthly, graph theory provides models which, in the absence of historical data, can be used to simulate the behaviour of certain exchange systems. In Chapter 5 we used one such model, that of a regular Markov chain, to estimate the distribution of valuables in the *kula* ring in order to offer a general theory of the relation between social stratification and access to wealth. This analysis expanded the network alternative to economic and demographic theories of stratification proposed in Chapter 3.

Fifthly, graph theory provides a means for enumerating structural forms. In Chapter 6 we used the operation of sex duality in graphs to generate the cross/parallel relations implicit in the atom of kinship. We then showed how, in the case of Tongan kinship, a long line of theorists, beginning with Rivers and Hocart, had succeeded only in studying various subsets of these relations instead of taking them as a comprehensive set defining a complete system of exchange.

As pointed out in Chapter 6, a number of anthropologists from Lowie to Lévi-Strauss have recognized the analytical utility of structural enumeration. When the number of logical possibilities is more than a few, mathematical techniques based on graph theory or on interactions between graph theory and group theory become useful or even essential for specifying the structure of a set.

Sixthly, through its binary operations, graph theory provides a means for exact descriptions and concise notations of many complex exchange structures. We can say, for example, '$P_4 \times P_5$' instead of 'a lattice structure with 20 units, four down the side and five across the top, with lines joining the units horizontally and vertically but not diagonally'. Through its interaction with group theory, graph theory provides an explicit and generally applicable semiotic model. By using the model of the 4-group in Chapter 7, we showed that pollution beliefs in Highland New Guinea may best be analysed not in isolation as is usually done, but as part of a system of metaphorically related, mutually defining bodily concepts. These concepts define the marriage bond in terms of its significant transactions rather than its singular antagonisms. Collectively they reflect the exchange system of a society. It turns out that the symbols x, $-x$, $1/x$, $-1/x$ in *Mythologiques* do have a mathematical meaning with useful empirical applications.

Finally, graph theory provides clear and visually appealing models of the logical properties of exchange structures. As pointed out in Chapters 1, 2, and 8, anthropologists often use terms like symmetry and transitivity in purely idiosyncratic ways. Such terms do have legitimate uses, such as making deductions about permissible and impermissible marriage exchanges and showing the equivalence of empirically different kinship relations (Chapters 2 and 8). We therefore provided, in Chapter 8, a set of precise definitions and showed how relational properties combine to form larger structures, such as the orders, parity, and similarity relations which define rank hierarchies, exchange structures, and dialect chains.

In *The History of Melanesian Society*, published in 1914, Rivers foresaw that mathematics would be helpful in the analysis of kinship relations. Much later, in his 'Introduction à l'œuvre de Marcel Mauss', Lévi-Strauss (1950) emphasized that mathematics would be essential for decisive progress in the analysis of social relations in general. In completing this book in 1986 on the 250th anniversary of Euler's discovery of graph theory in 1736, we concluded that the time had indeed come for the comprehensive mathematical analysis of social life conceived as a system of relations. In this study we have shown that graph theoretic models are naturally suited to the study of exchange structures and are capable of handling considerable variety and complexity. The results in this book provide directions for future research on exchange by social networkers, structuralists, ecologists, semioticians, and archaeologists, as well as anthropologists with more traditional interests. For each area of study in anthropology that involves structure, there is a branch of graph theory that can serve as the appropriate mathematical model.

APPENDIX 1

TABLE OF NAMED SEA-LANES IN THE CAROLINES AND MARIANAS

Sea-lanes between Island Pairs	Name of sea-lane on: Lamotrek	Woleai	Satawal (S) Faraulep (F) Yap (Y)	Puluwat
1. Guam–Gaferut	Mûtau-uol		Medoal (S)	
2. Yap–Ulithi	Mûtaurupal	Matewairuwepal	Medoairubal (S) Mateuari ruepane (F) Madori e pal (Y)	
3. Ulithi–Sorol	Mûtaupengakh		Lamuafan (S) Madau limar (Y)	
4. Ulithi–Fais	Mûtaumarfach	Matewalimwarefash	Medamarefas (S) Metaue mare vatje (F) Madau vorovoch (Y)	
5. Yap–Ngulu	Chumahos (?)		Djumagos (S) Madau marafach (Y)	
6. Ngulu–Palau	Mûtaumal		Madau mal (Y)	

7. Faraulep–Woleai	Faligůmatar	Faaliigemwatur	Faligomodur (S)	
			Fanike matouru (F)	
			Gomatur (Y)	
8. Woleai–Eauripik	Můtaulībwul	Matewalibul	Medoalibul (S)	
			Metau ni polu (F)	
			Madau luvul (Y)	
9. Woleai–Ifaluk	Faliorůma	Faaliyaroma	Falearoma (S)	
			Fani aroma (F)	
			Fal yoroma (Y)	
10. Faraulep–Ifaluk	Můtaupugakh	Metawepengag	Medabanar (S)	
			Metaue panaki (F)	
11. Ifaluk–Elato	Hapilerap	Gapiileeirub	Abilorub (S)	
12. Elato–Olimarao	Haisa			Metawenėkaayssa
13. Olimarao–Faraulep	Hapilůmůtau		Audja (S)	
14. Lamotrek–Satawal	Uoirekh	Wooireg	Woira (S)	Oowirek
			Uoi reki (F)	
			Oreg (Y)	
15. Lamotrek–West Fayu	Leharakh		Abiliwos (S)	(no name)
			Garedol (Y)	
16. Satawal–West Fayu	Můtaupunakh		Pogedul (Y)	Metawėpengas
17. Satawal–Pikelot	Můtausu		Gauudul (Y)	Metawůwů
18. Fais–Faraulep	Falůmůhol		Falepi (Y)	
19. Fais–Eauripik	Hapilimahal	Faalimogal		
20. Sorol–Eauripik	Gilůmar	Matewaligilimwar	Medoairubal (S)	
21. Elato–Lamotrek	Mutol	Matol	Madol (S)	
22. Satawal–Puluwat			Abilalei (S)	Åpinåålley
			Kapini nei (F)	
			Gopil e lai (Y)	

Sea-lanes between Island Pairs	Name of sea-lane on: Lamotrek	Woleai	Satawal (S) Faraulep (F) Yap (Y)	Puluwat
23. Satawal–Pulusuk			Failubei (S) Faisog (Y)	Nůkasė
24. Puluwat–Pulap			Faisob (S)	Faasopw
25. Pulusuk–Truk			Luguliwal (S) Voluvolol (Y)	Nůkůniiwan
26. Woleai–Olimarao			? (no name) (S)	
27. Ifaluk–Eauripik			Merol (S)	
28. Woleai–Sorol			Falemogol (S) Metau ni kini mare (F) Madau airen (Y)	
29. Truk–Nama			Failemo (S)	Faaynėmė
30. Fais–Sorol			? (no name) (S)	
31. Palau–Yap		Metawemwal	Medaumal (S) Metau u male (F) Madau man (Y)	
32. Pulap–Ulul*			Faluruai (S)	Faynunuway
33. Truk–Ruo†			Lugidol (S)	Åperu
34. Pikelot–West Fayu			Medaoi (S)	Metawaychim
35. West Fayu–Gaferut			Sibaalol (S)	(no name)
36. Woleai–Fais		Gapilimogol	Kapini ma kano (F) Gopinemogon (Y)	

37. Puluwat–Truk		Eruani (F)	Neeyaruaan
		Yorual (Y)	
38. Puluwat–Pulusuk		Metaue panaki (F)	Mettow pėngėk
		Madau aigir (Y)	
39. Lamotrek–Olimarao		Audja (S)	Leekara
40. Rota–Guam		Magusedemak (S)	
41. Rota–Tinian		Magus (S)	
42. Saipan–Tinian		Magusigidigit (S)	
43. Pulap–Truk		Arual (S)	Faaset
44. Puluwat–Pikelot		Medoalimel (S)	Metawenimwan
45. Pulap–Pikelot		Medoalimel (S)	(no name)
46. Faraulep–Gaferut		Medoalimel (S)	
		Madau limol (Y)	
47. Pikelot–Guam		Medoal (S)	Metaweniwȯn
48. Ifaluk–Olimarao	Mů̇taualifa		
49. Olimarao–West Fayu			Nůkůniikara
50. East Fayu–Murilo		Isalifai (S)	
51. East Fayu–Nomwin			Esenifay
52. Puluwat–East Fayu			Nnwůreniyeng
53. Puluwat–Pisaras*		Madau ailam (Y)	Fåånůmooch
54. Puluwat–Ulul*		Madau aiz (Y)	Faanunuway
55. Truk–Pisaras*			Wonupwenimwayr
56. Truk–Ulul*			Wonupwėlȯl
57. Truk–East Fayu			Wonupwėnekėėcha
58. Truk–Nomwin			Nůůkwiiton
59. Truk–Murilo			Lugidol; Åperu
60. Nama–Losap			Faaynėmė
61. Pulusuk–Pikelot			Faaynůkasė
62. Pulusuk–Pulap			Metawaneengi

Sea-lanes between Island Pairs	Name of sea-lane on:			
	Lamotrek	Woleai	Satawal (S) Faraulep (F) Yap (Y)	Puluwat
63. Pulusuk–East Fayu				(no name)
64. East Fayu–Ulul*				Faanifaw
65. East Fayu–Piseras*				Fåånifaw
66. East Fayu–Magur*				Metawenuur
67. East Fayu–Ono*				Metawenipwėn
68. Namonuito–Pikelot				(no name)
69. Namoluk–Losap				Aapȯng
70. Namoluk–Etal				Metawachikėchik
71. Etal–Lukunor				Metawachikėchik
72. Etal–Satawan				Metawachikėchik
73. Lukunor–Satawan				Metawachikėchik
74. Pulap–East Fayu				(no name)
75. Pulap–Satawal				Metawenȯȯch
76. Lamotrek–Pikelot				(no name)
77. Pikelot–Saipan				Metawėpengas
78. Ulul*–Pisaras*				Sopunȯȯch
79. Nomwin–Murilo			Aroalu (S)	Aroalu‘
80. Faraulep–Guam		Matewaliwoal		

* Namonuito
† Murilo

Sources: Lamotrek (Alkire 1965), Woleai (Sohn and Tawerilmang 1976), Satawal (Damm and Sarfert 1935), Faraulep (Krämer 1937), Yap (Müller-Wismar 1918), Puluwat (Gladwin n.d.).

The following points should be noted concerning this table: (1) All sea-lanes from Alkire are in his monograph (1965) except for no. 48 which is a personal communication. (2) Damm and Sarfert (1935) give a sea-lane Pulusuk–Lukunor, but Gladwin (1970) emphasizes that direct trips from Puluwat, and presumably Pulusuk, are never made. Instead the trip is via Truk, Losap, and Namoluk. (3) Additional names, but not sea-lanes, are given in Burrows and Spiro (1957). Their map is difficult to decipher, but there appear to be definite correspondences with sea-lanes 3, 8, 9, 10, 14, and 36 in this table. (4) For the lists in Müller-Wismar, Damm and Sarfert, and Krämer, no attempt is made to reproduce the baroque phonetic symbols, nor is any attempt made to standardize the orthography of the other sources. (5) When names differ for the same sea-lane, this may be due to dialect or linguistic variation or to differences between the two schools of navigation in the Carolines. An entry of 'no name' or (?) means that the sea-lane exists but a name for it was not elicited. (6) In the German lists, sequences of three islands are interpreted as two adjoining sea-lanes, e.g. Lamotrek–Olimarao–Faraulep = Lamotrek–Olimarao and Olimarao–Faraulep. (7) Four sea-lanes in *G* (Fig. 3.2) not identified by name and not given in the table are Sorol–Yap, inferred from sailing directions in Burrows and Spiro (1957) and from statements by the early Woleaian informant Kadu cited in Damm and Sarfert (1935), West Fayu–Saipan from McCoy (1973), and West Fayu–Guam and Namonuito–Saipan (Damm and Sarfert 1935 and Lewis 1972).

APPENDIX 2
SIMULATIONS OF ALTERNATIVE *KULA* RINGS

TABLE A2.1. *The predicted distribution of valuables in Malinowski's (1922) second version of the* kula *ring.*

Kula community	Proportion of armshells	Rank	Proportion of necklaces	Rank
1 Kitava	0.1065	1=	0.1020	1=
2 Kiriwina	0.0266	11=	0.0340	11=
3 Sinaketa	0.0666	5	0.0680	5
4 Kayleula	0.0133	12=	0.0113	13
5 Vakuta	0.0488	8	0.0453	8
6 Amphletts	0.0599	6	0.0340	11=
7 NW Dobu	0.1065	1=	0.1020	1=
8 Dobu	0.0355	10	0.0510	7=
9 SE Dobu	0.0533	7	0.0688	4
11 East End Islands	0.0266	11=	0.0523	6
12 Wari	0.0133	12=	0.0357	10
13 Tubetube	0.0999	2	0.0828	2
14 Misima	0.0400	9	0.0382	9
16 Woodlark	0.0699	4	0.0510	7=
17 Alcesters	0.0733	3	0.0765	3
18 Marshall Bennetts	0.1065	1=	0.1020	1=
19 Wawela	0.0266	11=	0.0227	12=
20 Okayaulo	0.0266	11=	0.0227	12=

TABLE A2.2. *The predicted distribution of valuables in Belshaw's (1955) version of the* kula *ring.*

Kula community	Proportion of armshells	Rank	Proportion of necklaces	Rank
1 Kitava	0.1129	1=	0.1132	1=
2 Kiriwina	0.0282	8=	0.0377	6=
3 Sinaketa	0.0706	3	0.0755	3
4 Kayleula	0.0141	9	0.0126	9
5 Vakuta	0.0518	6	0.0503	5
6 Amphletts	0.0635	4	0.0377	6=
7 NW Dobu	0.1129	1=	0.1132	1=
8 Dobu	0.0376	7	0.0566	4=
9 SE Dobu	0.0565	5=	0.0566	4=
10 East Cape	0.0282	8=	0.0283	7
13 Tubetube	0.1129	1=	0.1132	1=
16 Woodlark	0.0847	2	0.0566	4=
17 Alcesters	0.0565	5=	0.0849	2
18 Marshall Bennetts	0.1129	1=	0.1132	1=
19 Wawela	0.0282	8=	0.0252	8=
20 Okayaulo	0.0282	8=	0.0252	8=

TABLE A2.3. *The predicted distribution of valuables in Brunton's (1975) version of the* kula *ring.*

Kula community	Proportion of armshells	Rank	Proportion of necklaces	Rank
1 Kitava	0.0966	1=	0.0992	1=
2 Kiriwina	0.0241	8=	0.0331	6=
3 Sinaketa	0.0604	3	0.0661	3
4 Kayleula	0.0121	9	0.0110	9
5 Vakuta	0.0443	6	0.0441	5
6 Amphletts	0.0543	4	0.0331	6=
7 NW Dobu	0.0966	1=	0.0992	1=
8 Dobu	0.0322	7	0.0496	4=
9 SE Dobu	0.0483	5=	0.0496	4=
10 East Cape	0.0241	8=	0.0248	7=
13 Tubetube	0.0966	1=	0.0992	1=
16 Woodlark	0.0724	2	0.0496	4=
17 Alcesters	0.0483	5=	0.0744	2
18 Marshall Bennetts	0.0966	1=	0.0992	1=
19 Wawela	0.0241	8=	0.0220	8=
20 Okayaulo	0.0241	8=	0.0220	8=
21 Bonvouloir	0.0966	1=	0.0992	1=
22 Panamoti	0.0483	5=	0.0248	7=

APPENDIX 3

THE GEODESIC COUNTING MATRIX OF A GRAPH

GIVEN a connected graph G with p points v_i, let g_{ij} be the number of geodesics connecting v_i and v_j. (Of course $g_{ji} = g_{ij}$ as G is symmetric.) Then the *geodesic counting matrix* $gcm(G) = [g_{ij}]$ tells the number of geodesics between every pair of points. Note that each diagonal entry $g_{ii} = 1$ as the distance from a point to itself is 0, so that the trivial path consisting of v_i alone is the unique v_i, v_i geodesic. Also for two adjacent points v_i and v_j, $d(v_i, v_j) = 1$; hence there is again just one geodesic and we have $g_{ij} = 1$. We have already established by these remarks that for the complete graph, the matrix $gcm(K_p) = J_p$, the universal matrix in which every entry is 1.

There are many other graphs G for which $gcm(G) = J$; we then call G a geodetic graph, following Buckley (1984), as there is always a unique geodesic joining each pair of points. Both connected graphs with $p = 3$ are geodetic, and there are just two graphs with four points which are not: the cycle C_4 and the random graph $K_4 - e$.

It is well known and very easy to verify from the definition that in any tree T there is a unique path between any two points, *a fortiori* a unique geodesic. In fact a tree can be equivalently defined as a graph having a unique path between every pair of points. Thus every tree is a geodetic graph, as is every complete graph.

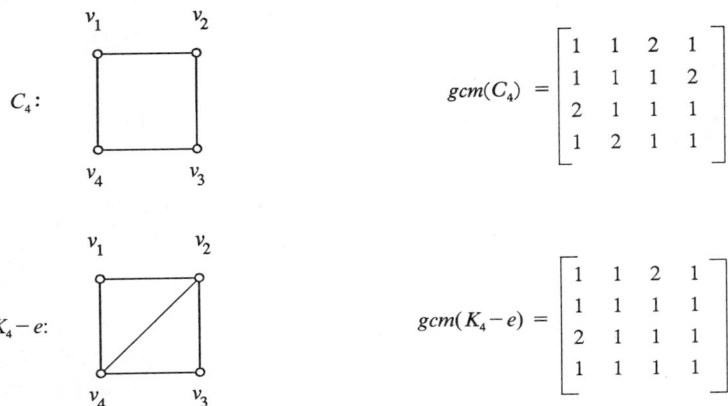

FIG. A3.1. The geodesic counting matrices for two small graphs

The matrix $gcm(M)$ provides a new concept in graph theory which has not previously been studied. Consequently theoretical unsolved problems abound. Two of these open questions are as follows:

(1) Call M a geodesic counting matrix, or briefly a *g-c* matrix, if there is some graph G for which $gcm(G) = M$. Which square matrices of positive integers are *g-c* matrices? We already know that J_p is a *g-c* matrix and have seen two other examples in Fig. A3.1.

(2) Among all graphs G with p points, which matrices $gcm(G)$ have the largest average value of their entries? We illustrate in Fig. A3.2 with the complete bipartite graph $K_{2,n}$ using $n = 3$.

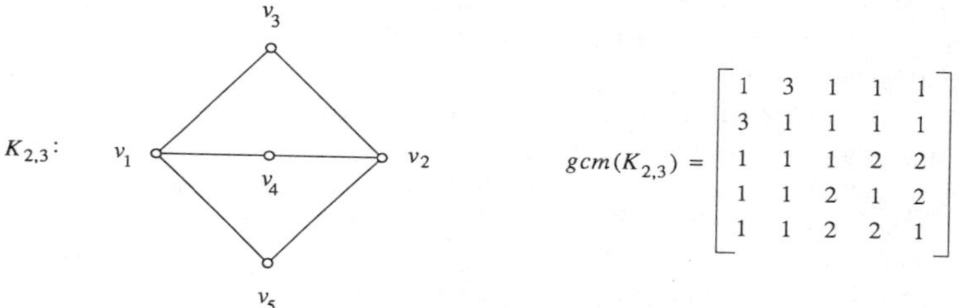

FIG. A3.2. A complete bipartite graph and its *g-c* matrix

APPENDIX 4
REFERENCE LIST OF THEOREMS AND PROOFS

AN essential part of our analysis of exchange structures involves the use of mathematical theorems. We now illustrate the nature of a theorem and its proof. (We will not give the proofs for all theorems but they can be found in the references.)

The only field which has theorems is mathematics. A theorem is a statement with a hypothesis and a conclusion which has a logical proof based on the axiom system under study. A proof is a sequence of steps, beginning with the hypothesis as given and ending with the desired or stated conclusion. Each step in the proof must be justified by a reason for its correctness. An acceptable reason for a step can be any of the following: the hypothesis, an axiom, a definition or notation, a previous theorem, a previous step, or a law of logic.

We now present in step-by-step fashion a simple proof of what may well be the easiest theorem of graph theory, namely, $\Sigma d_i = 2q$.

Theorem 2.1. The sum of the degrees of the points of any graph G is twice the number of lines of G.

Theorem 2.1 (restated). Hypothesis: G is a graph.
Conclusion: $\Sigma d_i = 2q$.

Proof. Step 1. If G has no lines at all, $q = 0$ and each degree $d_i = 0$, so (1) holds.

2. If G has exactly one line, $q = 1$, then two points have degree 1 and all other points have degree 0, so $\Sigma d_i = 2$. Thus (1) holds in this case.

3. In the same way, each line of G contributes exactly 2 to the sum of the degrees.

4. Writing Step 3 in algebraic form,

$\Sigma d_i = 2 + 2 + \ldots + 2$ (q summands) $= 2q$.□

Note: The symbol □ is now in current usage in the mathematical literature as a replacement for Q.E.D. It is an expression of joy, relief, and accomplishment to celebrate the end of a proof.

Theorem 2.2. A graph G is bipartite if and only if it contains no cycles of odd length.

We give a proof of this theorem which expands on our discussion of the basic logical structure of all theorems and proofs.

Proof. There are two parts to the proof, the 'if' part and the 'only if' part. The if part states that if no cycle of G has odd length then G must be bipartite. The only if part is the converse statement; as it is easier (the so-called easy half of the proof), we prove it first.

Proof of the only if part. Here the hypothesis is that G is bipartite and the conclusion to be demonstrated is that every cycle, if any, has even length. In particular, if there are no cycles there is nothing to prove. Therefore consider a bipartite graph G having one or more cycles and let Z be any one of its cycles. We now show that Z has even length. Since G is bipartite, it is possible to regard its point set V as the union, $U \cup W$, of two disjoint non-empty subsets U and W, where U can be regarded as the set of umber points of G and W as the set of white points. Without loss of generality, consider cycle Z as beginning at an umber point u. By the definition of bipartite, the first line of Z must join u to a white point, w, and the next line of Z joins w to a different umber point u'. Thus on returning from the original point u to the set U for the first time, the length of this beginning path in Z is so far 2. Do it again: go from u' to w' to u'' and thus far we have length 4. At every stage, when we have arrived at an umber point the length is even. In particular, when we have finished traversing the entire cycle we are in the set U, so the length of Z must be even.

Proof of the if part. This proof is a bit more difficult. Here the hypothesis is that G has no odd cycles and we are to prove that G is bipartite. The method of proof will be to colour each point of G either white or umber in such a way that every line joins two points with different colours. If G is not connected, then each component is a connected graph to which the following procedure applies. Hence we may as well regard G as connected at the outset.

We begin by taking any point of G; we call it u and colour it umber. Then we colour white all the points adjacent to u. Take each such point w and colour all points adjacent to it umber, and so forth. Since we are considering G as connected, all the points of G will certainly be reached in this way.

Now consider any cycle Z of G. By hypothesis of the if part, Z has even length. Therefore one does not need a third colour for any point of Z, as the colours can alternate in going around this cycle: umber, white, umber, white, etc. Since this is true for every cycle, all the points of G which lie on at least one cycle are 2-coloured in this way. Finally, any point of G which does not lie on a cycle offers no confusion. □

Corollary 2.2a. All trees are bipartite.

Proof. Let T be a tree. By definition, T has no cycles. *A fortiori*, T has no odd cycles. Hence by the theorem, T is bipartite. □

Theorem 2.3. A signed graph S has a colouring if and only if S has no cycle with exactly one negative line.

Proof in Cartwright and Harary (1968).

Theorem 2.4. The following statements are equivalent for a weak digraph D:
(1) D is functional.
(2) There is an arc x on a cycle of D such that $D-x$ is a tree toward fx.
(3) D has exactly one cycle Z, and after deleting its arcs each weak component of the resulting digraph consists of a tree toward a point of Z.

Proof in Harary, Norman, and Cartwright (1965).

Theorem 2.5. If a tournament T is strong, then it contains a cycle of each length $k = 3, 4, \ldots, p$.

Proof in Harary *et al.* (1965) and Harary and Moser (1966).

Theorem 3.1. A graph G with $p \geqslant 3$ is a block if and only if every two points of G lie on a common cycle.

Proof. As is so often the case in mathematical proofs of logical equivalence (if and only if proofs), the proof is quite easy in one direction but less easy in the other. The proof of the harder, only if, part is given in Harary (1969). We now present the proof of the easier half which does not appear there.

Proof of the if part. Given that every two points of a graph G lie on a common cycle we must prove that G is a block.

Take any two points u and v of G. As they lie on a common cycle by hypothesis, there must be a path joining them, so G is connected. Knowing that G is connected, we assume that G is not a block and derive a contradiction. By the assumption, G must contain a cutpoint w. By the definition of a cutpoint, there are two points u and v of G such that every path joining them goes through w. But by the hypothesis of the if part, there is a cycle of G containing both u and v. This cycle can be regarded as consisting of two paths joining u and v, only one of which can contain w. Therefore the other u, v path does not contain w, contradicting the assumption that G is not a block.□

Corollary 3.1a. A graph G with $p \geqslant 3$ is a block if and only if for every two distinct points u and v of G, there exist two disjoint u–v paths.

Theorem 4.1. Given the adjacency matrix A of a digraph D, the i, j entry of its rth power A^r is the number of walks of length r from v_i to v_j.

The proof for $r = 2$ already tells the entire method of proof by mathematical induction for r in general. Hence we now include this proof only. In this case we will find that the diagonal i, i entry for A^2 is the number of symmetric pairs which contain v_i, and the i, j entry when $i \neq j$ is the number of 2-step paths from v_i to v_j.

Proof of Theorem 4.1 for r = 2. This is demonstrated by looking into the meaning of each term in equation (4.5).

(4.5) $a_{ij}^{(2)} = a_{i1}\,a_{1j} + a_{i2}\,a_{2j} + \ldots + a_{in}\,a_{nj}$.

Consider the first term $a_{i1}\,a_{1j}$. As A is a binary matrix, $a_{i1} = 0$ or 1 and $a_{1j} = 0$ or 1. Therefore the only time that the numerical value of the first term is not zero is when $a_{i1} = a_{1j} = 1$. But this means that in D there is an arc v_iv_1 and another arc $v_1\,v_j$. The presence of both of these arcs establishes a 2-step path from v_i to v_j through v_1. In this argument, it is not stipulated that $v_i \neq v_j$. When $v_i = v_j$, these two arcs form a symmetric pair between v_i and v_1 showing that the i,i entry of A_2 is the number of symmetric pairs containing v_i.

As this reasoning for the first term involving v_1 also holds for each of the n terms of (4.5), we deduce that the exact value of $a_{ij}^{(2)}$ is the number of 2-step walks from v_i to v_j.□

Corollary 4.1a. The i, j entry $a_{ij}^{(n)}\#$ *of* $A^n\#$ is 1 if and only if there is at least one walk of length n in D from v_i to v_j.

Theorem 4.2. For every positive integer n,

(4.6) $R_n = (I + A + A^2 + \ldots + A^n)\# = (I + A)^n\#$, and

(4.7) $R = (I + A + A^2 + \ldots + A^{p-1})\# = (I + A)^{p-1}\#$.

Proof in Harary *et al.* (1965).

Theorem 4.3. Let $N = [d_{ij}]$ be the distance matrix of a given digraph D or graph G. Then,

(1) Every diagonal entry $d_{ii} = 0$,

(2) $d_{ij} = \infty$ if $r_{ij} = 0$, and

(3) Otherwise, d_{ij} is the smallest power n to which A must be raised so that $a_{ij}^{(n)} > 0$, that is, so that the i, j entry of $A^n\#$ is 1.

Proof in Harary *et al.* (1965).

Theorem 4.4. The graphs K_5 and $K_{3,3}$ are non-planar, and so is any graph which contains a subdivision of either of these as subgraphs.

Proof in Kuratowski (1930) and in Harary (1969).

Theorem 4.5. Let v_i and v_j be joined by a path in a graph H, that is, v_i and v_j are in the same connected component of H. The geodetic subgraph H_g of v_i and v_j consists of all the points v_k such that $d_{ik} + d_{kj} = d_{ij}$, and all the lines $v_r\,v_s$ of H such that both $d_{ir} + 1 = d_{is}$ and $d_{rj} = d_{sj} + 1$.

The proof in Harary *et al.* (1965) for digraphs applies at once to graphs.

Theorem 4.6. Let D_1 and D_2 be two digraphs having the same set of labelled points, with adjacency matrices A_1 and A_2. Then the adjacency matrices of the intersection, union, and symmetric difference digraphs are as follows:

$$A(D_1 \cap D_2) = A_1 \times A_2,$$
$$A(D_1 \cup D_2) = (A_1 + A_2)\#,$$
$$A(D_1 \oplus D_2) = (A_1 + A_2)\# - (A_1 \times A_2).$$

Proof. The matrix $A_1 \times A_2$ has an i,j entry of 1 if and only if both A_1 and A_2 have; hence it is the adjacency matrix of $D_1 \cap D_2$. Similarly, $(A_1 + A_2)\#$ has an entry of 1 when A_1 or A_2 has; hence $(A_1 + A_2)\# = A(D_1 \cup D_2)$. Since the symmetric difference is the union minus the intersection, it follows that

$$A(D_1 \oplus D_2) = A(D_1 \cup D_2) - A(D_1 \cap D_2) = (A_1 + A_2)\# - (A_1 \times A_2).\square$$

Theorem 5.1. Let M be the transition matrix of a given chain. Then in M^n the i,j entry is the nth transition probability from v_i to v_j.

Proof in Harary *et al.* (1965). Note the similarity in the statements of Theorems 4.1 and 5.1. The proof of Theorem 4.1 was given with the details provided for the case $n = 2$. The proof of Theorem 5.1 uses the same equation which defines matrix multiplication, but interprets its terms as the product of probabilities instead of the presence or absence of arcs.

Theorem 5.2. For a regular chain, the powers of its transition matrix M approach a limit. This limit has all rows alike, and each is equal to the unique probability vector P for which $PM = P$.

Proof in Harary *et al.* (1965).

Theorem 6.1. The following conditions are equivalent for a signed graph S.
(1) S is balanced: every cycle is positive.
(2) For each pair of points, u, v, of S all paths joining u and v have the same sign.
(3) There exists a partition of the points of S into two subsets (one of which may be empty) such that every positive line joins two points of the same subset and every negative line joins two points from different subsets.

Proof in Harary (1953).

Theorem 7.1. A graph is planar if and only if it does not have a subgraph contractible to K_5 or $K_{3,3}$.

Proof in K. Wagner (1937).

Theorem 7.2. Let $G = (V, E)$ be a graph. Let $\Gamma(G) = \{i, \alpha, \beta, \gamma, \ldots\}$ be the collection of all the automorphisms of G. Then $\Gamma(G)$ is a group under the binary operation 'is followed by'.

Proof. The set $\Gamma(G)$ is closed under the operation, because one automorphism followed by another must also be an automorphism, since the preservation of adjacency is a transitive relation.

The second axiom, which says that the associative law

$$\alpha(\beta\gamma) = (\alpha\beta)\gamma$$

holds for all automorphisms α, β, and γ, is true, because each side of the above equation is the result of α followed by β followed by γ.

The third axiom for a group says that there is an identity element. Of course there is, since the mapping which leaves all points fixed is the identity automorphism.

Finally, every automorphism α has an inverse written α^{-1}, which is also an automorphism by definition. For each permutation α can be regarded as a digraph in which every weak component is a directed cycle, and then α^{-1} is determined by the converse of this digraph.

Since all four group axioms hold for the set of all automorphisms of a graph under the operation 'is followed by', they constitute a group.□

Theorem 8.1. A graph is a parity relation if and only if every component is complete.

Proof. Let graph G be a parity relation. We first prove that every component H of G is complete. By definition G is transitive, t, so H is also t. Let u and v be any two points of H. It remains to show that u and v are adjacent. As H is connected, there is a path in H joining u and v. For graphs, which are symmetric, the definition of transitivity asserts that every pair of points connected by a 2-step path must be adjacent. Repeated application of this observation verifies that any two points u, v joined by a longer path must also be adjacent.

Proving the converse is also easy. Without loss of generality let G be connected. Then by hypothesis, G is K_p. To prove that G is a parity relation, we need to verify $\bar{r}$, s and t. The first two of these conditions follow immediately since $\bar{r}$ and s hold for all graphs. To prove t for G, consider any three distinct points u, v, w. Then both uRv and vRw since G is complete; and also uRw, so G is transitive.□

Bibliography

ALKIRE, W. H., 1965, *Lamotrek Atoll and Inter-Island Socioeconomic Ties*, Urbana, University of Illinois Press.

—— 1970, 'Systems of Measurement on Woleai Atoll, Caroline Islands', *Anthropos*, 65. 1–73.

—— 1972, 'Concepts of Order in Southeast Asia and Micronesia', *Comparative Studies in Society and History*, 14. 484–93.

—— 1977, *An Introduction to the Peoples and Cultures of Micronesia*, Menlo Park, Calif., Cummings.

—— 1978, *Coral Islanders*, Arlington Heights, Ill., AHM Publishing.

—— 1980, 'Technical Knowledge and the Evolution of Political Systems in the Central and Western Caroline Islands of Micronesia', *Canadian Journal of Anthropology*, 1. 229–37.

—— 1984, 'Central Carolinian Oral Narratives: Indigenous migration theories and principles of order and rank', *Pacific Studies*, 7. 1–14.

APPEL, K., and HAKEN, W., 1976, 'Every Planar Map is Four Colorable', *Bulletin of the American Mathematical Society*, 82. 711–12.

ARMSTRONG, W. E., 1928, *Rossel Island*, Cambridge, Cambridge University Press.

BAILEY, F. G., 1969, *Stratagems and Spoils*, Oxford, Blackwell.

BARNES, J. A., 1962, 'African Models in the New Guinea Highlands', *Man*, 62. 5–9.

—— 1969, 'Graph Theory and Social Networks: A technical comment on connectedness and connectivity', *Sociology*, 3. 215–32.

—— 1971, *Three Styles in the Study of Kinship*, Berkeley, University of California Press.

—— 1972, *Social Networks*, Addison-Wesley Module in Anthropology.

—— 1974, 'Genitrix : genitor :: nature : culture?', in J. Goody (ed.), *The Character of Kinship*, Cambridge, Cambridge University Press.

BATESON, G., 1958, *Naven*, 2nd edn., Stanford, Stanford University Press.

BEAGLEHOLE, E., and BEAGLEHOLE, P., 1938, *Ethnology of Pukapuka*, Honolulu, B. P. Bishop Museum Bulletin 150.

BELL, F. L. S., 1935, 'Warfare among the Tanga', *Oceania*, 5. 253–79.

BELLWOOD, P., 1979, *Man's Conquest of the Pacific*, New York, Oxford University Press.

BELSHAW, C. S., 1955, *In Search of Wealth*, American Anthropological Association Memoir no. 80.

BERNDT, C. H., 1953, 'Socio-Cultural Change in the Eastern Central Highlands of New Guinea', *Southwestern Journal of Anthropology*, 9. 114–38.

BERNDT, R. M., 1954, 'Reaction to Contact in the Eastern Highlands of New Guinea', *Oceania*, 24. 190–228, 255–74.

BIERSACK, A., 1974, 'Matrilaterality in Patrilineal Systems: The Tongan case', unpublished Curl Bequest Prize Essay, Royal Anthropological Institute.

—— 1982, 'Tongan Exchange Structures: Beyond descent and alliance', *Journal of the Polynesian Society*, 91. 181–212.

—— 1984, 'Paiela "Women-Men": The reflexive foundations of gender ideology', *American Ethnologist*, 11. 118–38.

BLUST, R., 1981, 'Dual Divisions in Oceania: Innovation or retention?', *Oceania*, 52. 66–79.

—— 1984, comment on Marshall (1984), *Current Anthropology*, 25. 626–8.

BORUVKA, O., 1926a, 'O jistém problému minimálním', *Acta Societatis Scientiarum Naturalium Moravicae*, 3. 37–58.

—— 1926b, 'Příspěvek k řešení otázky ekonomické stavby elektrovodních sítí', *Elektrotechnický Obzor*, 15. 153–4.

BOTT, E., 1981, 'Power and Rank in the Kingdom of Tonga', *Journal of the Polynesian Society*, 90. 7–81.

BROOKFIELD, H. C., and HART, D., 1971, *Melanesia: A geographical interpretation of an island world*, London, Methuen.

BRUNTON, R., 1975, 'Why do the Trobriands have Chiefs?', *Man* (NS), 10. 544–58.

BRYAN, E. H., 1971, *Guide to Place Names in the Trust Territory of the Pacific Islands*, Honolulu, Pacific Science Information Center, B. P. Bishop Museum.

BUCHER, B., 1985, 'An Interview with Claude Lévi-Strauss, 30 June 1982', *American Ethnologist*, 12. 360–8.

BUCK, P. H., 1930, *Samoan Material Culture*, Honolulu, B. P. Bishop Museum Bulletin 75.

BUCKLEY, F., 1984, 'Equalities Involving Certain Graphical Distributions', *Springer Lecture Notes in Mathematics*, 1073. 179–92.

BURRIDGE, K. O. L., 1958, 'Marriage in Tangu', *Oceania*, 29. 44–61.

—— 1959, 'Siblings in Tangu', *Oceania*, 30. 128–54.

BURROWS, E. G., 1936, *Ethnology of Futuna*, Honolulu, B. P. Bishop Museum Bulletin 138.

—— and SPIRO, M. E., 1957, *An Atoll Culture*, New Haven, HRAF.

CAMPBELL, S. F., 1983, '*Kula* in Vakuta: The mechanics of *keda*', in Leach and Leach (1983).

CANTOVA, J. A., 1728, 'Lettre du P. Jean Cantova, missionaire de C. de J. au R. P. Guillaume Daubenton, Mar. 20 1722', in *Lettres édifiantes et curieuses, écrites des missiones étrangères, par quelques missionaires de la Compagnie de Jésus*, vol. 18, Paris, N. Le Clerc.

CARTWRIGHT, D., and HARARY, F., 1956, 'Structural Balance: A generalization of Heider's theory', *Psychological Review*, 63. 277–93.

CARTWRIGHT, D., and HARARY, F., 1968, 'On the Coloring of Signed Graphs', *Elemente der Mathematik*, 23. 85–9.

—— —— 1977, 'A Graph Theoretic Approach to the Investigation of System–Environment Relationships', *Journal of Mathematical Sociology*, 5. 87–111.

CHRISTALLER, W., 1966, *Central Places in Southern Germany*, transl. C. W. Baskin, Englewood Cliffs, NJ, Prentice-Hall.

CLAIN, P., 1700, 'Lettre écrite de Manille le 10 de Juin 1697 par le Père Paul Clain, de la Compagne de Jésus au Révérend Père Thyrse Gonzalez, Général de la mesme Compagnie', in C. Le Gobien, *Histoire des Iles Marianes*, Paris, Pepie.

CLAY, B. J., 1975, *Pinikindu: Maternal nurture, paternal substance*, Chicago, University of Chicago Press.

CODRINGTON, R. H., 1891, *The Melanesians*, Oxford, Clarendon Press.

COOMBS, C. H., 1964, *A Theory of Data*, New York, Wiley.

COPI, I. M., 1954, *Symbolic Logic*, New York, Macmillan.

CZEKANOWSKI, J., 1909, 'Zur Differentialdiagnose der Neandertalgruppe', *Korrespondenzblatt der deutschen Gesellschaft für Anthropologie, Ethnologie und Urgeschichte*, 40. 44–7.

—— 1911, 'Objektive Kriterien in der Ethnologie', *Ebenda*, 43. 71–5.

DAMAS, D., 1979, 'Double Descent in the Eastern Carolines', *Journal of the Polynesian Society*, 88. 177–98.

DAMM, H., and SARFERT, E., 1935, *Inseln um Truk*, Ergebnisse der Südsee-Expedition 1908–10, ed. G. Thilenius, II B6, Hamburg, Friederichsen, de Gruyter.

DAMON, F. H., 1980, 'The *Kula* and Generalized Exchange: Considering some unconsidered aspects of *The Elementary Structures of Kinship*', *Man* (NS), 15. 267–92.

DANIELSSON, B., 1956, *Work and Life on Raroia*, London, George Allen and Unwin.

DAVENPORT, W., 1960, 'Marshall Islands Navigational Charts', *Imago Mundi*, 15. 19–26.

—— 1962, 'Red Feather Money', *Scientific American*, 206(3). 95–104.

DEACON, A. B., 1927, 'The Regulation of Marriage in Ambrym', *Journal of the Royal Anthropological Institute*, 57. 325–42.

DOREIAN, P., 1974, 'On the Connectivity of Social Networks', *Journal of Mathematical Sociology*, 3. 245–58.

DOUGLAS, M., 1966, *Purity and Danger*, London, Routledge and Kegan Paul.

—— 1973, *Natural Symbols*, New York, Vintage Books.

—— 1975, 'Couvade and Menstruation', in *Implicit Meanings*, London, Routledge and Kegan Paul.

DUTTON, T., 1982, *The Hiri in History*, Canberra, Australian National University.

DYEN, I., 1965, *A Sketch of Trukese Grammar*, American Oriental Society, essay No. 4, New Haven.

EGGAN, F., 1949, 'The Hopi and the Lineage Principle', in M. Fortes (ed.), *Social Structure: Studies presented to A. R. Radcliffe-Brown*, Oxford, Clarendon Press.

EKEH, P. P., 1974, *Social Exchange Theory: The two traditions*, Cambridge, Harvard University Press.

ELKIN, A. P., 1937, review of 1936 edn. of Hocart (1970), *Oceania*, 8. 120–1.

EPLING, P. J., KIRK, J., and BOYD, J. P., 1973, 'Genetic Relations of Polynesian Sibling Terminologies', *American Anthropologist*, 75. 1596–625.

ERDLAND, A., 1914, *Die Marshall-Insulaner*, Münster, Ethnologische Anthropos Bibliothek.

EVANS-PRITCHARD, E. E., 1970, foreword to Hocart (1970).

EYDE, D. B., 1983, 'Recursive Dualism in the Admiralty Islands', *Journal de la Société des Océanistes*, 39. 3–12.

FAITHORN, E., 1976, 'Women as Persons: Aspects of female life and male–female relations among the Kafe', in P. Brown and G. Buchbinder (eds.), *Man and Woman in the New Guinea Highlands*, Special Publication 8 of the American Anthropological Association.

FEIL, D. K., 1978, 'Enga Women in the *Tee* Exchange', *Mankind*, 11. 220–30.

—— 1980, 'When a Group of Women Takes a Wife: Generalized exchange and restricted marriage in the New Guinea Highlands', *Mankind*, 12. 286–99.

FINNEY, B. R., 1985, 'Anomalous Westerlies, *El Niño*, and the Colonization of Polynesia', *American Anthropologist*, 87. 9–26.

FIRTH, R., 1939, *Primitive Polynesian Economy*, London, Routledge.

—— 1957, *We, the Tikopia*, London, Allen and Unwin.

—— 1981, 'A Comment on A. Hooper, *Why Tikopia has Four Clans*', Royal Anthropological Institute, Occasional Paper no. 38.

—— 1983, 'Magnitudes and Values in *Kula* exchange', in Leach and Leach (1983).

FISCHER, J. L., 1957, *The Eastern Carolines*, New Haven, HRAF.

FLAMENT, C., 1963, *Applications of Graph Theory to Group Structure*, transl. M. Pinard, R. Breton, and F. Fontaine, Englewood Cliffs, NJ, Prentice-Hall.

FORGE, A., 1972, 'The Golden Fleece', *Man* (NS), 7. 527–40.

FORTES, M., 1949, *The Web of Kinship among the Tallensi*, London, Oxford University Press.

FORTUNE, R. F., 1932, *Sorcerers of Dobu*, London, Routledge.

—— 1933, 'A Note on Some Forms of Kinship Structure', *Oceania*, 4. 1–9.

FORTUNE, R. F., 1935a, 'Incest', *The Encyclopaedia of the Social Sciences*, New York, Macmillan.

—— 1935b, *Manus Religion*, Philadelphia, American Philosophical Society.

—— 1942, *Arapesh*, New York, J. J. Augustin.

—— 1947, 'Law and Force in Papuan Societies', *American Anthropologist*, 49. 244–59.

FOSBERG, F. R., 1956, 'Military Geography of the Northern Marshalls', Intelligence Division, Headquarters US Armed Forces, Far East, Tokyo.

FOX, R., 1983 [1967], *Kinship and Marriage*, Cambridge, Cambridge University Press.

FREEDMAN, M. P., 1970, 'Social Organization of a Siassi Island Community', in T. G. Harding and B. J. Wallace (eds.), *Cultures of the Pacific*, New York, Free Press.

FREEMAN, L. C., 1977, 'A Set of Measures of Centrality Based on Betweenness', *Sociometry*, 40. 35–41.

—— 1979, 'Centrality in Social Networks, I. Conceptual clarification', *Social Networks*, 1. 215–39.

FRENCH, J. R. P., JR., 1956, 'A Formal Theory of Social Power', *Psychological Review*, 63. 181–94.

FRIEDMAN, J., 1981, 'Notes on Structure and History in Oceania', *Folk*, 23. 275–95.

GALANTER, E. H., 1956, 'An Axiomatic and Experimental Study of Sensory Order and Measure', *Psychological Review*, 63. 16–28.

GARBETT, G. K., 1980, 'Graph Theory and the Analysis of Multiplex and Manifold Relationships', in J. C. Mitchell (ed.), *Numerical Techniques in Social Anthropology*, Philadelphia, Institute for the Study of Human Issues.

GARDNER, M., 1980, 'The Capture of the Monster: A mathematical group with a ridiculous number of elements', *Scientific American*, 242 (6), 20–32.

GELL, A., 1975, *Metamorphosis of the Cassowaries*, New Jersey, Humanities Press.

GEWERTZ, D. B., 1983, *Sepik River Societies: A historical ethnography of the Chambri and their neighbors*, New Haven, Yale University Press.

GIFFORD, E. W., 1929, *Tongan Society*, Honolulu, B. P. Bishop Museum Bulletin 61.

GLADWIN, T., 1958, 'Canoe Travel in the Truk Area: Technology and its psychological correlates', *American Anthropologist*, 60. 893–9.

—— 1962, 'Culture and Logical Process', in W. H. Goodenough (ed.), *Explorations in Cultural Anthropology: Essays in honor of George Peter Murdock*, New York, McGraw-Hill.

—— 1970, *East is a Big Bird*, Cambridge, Harvard University Press.

—— n.d., unpublished field notes from Puluwat.

GLASSE, R. M., and MEGGITT, M. J., 1969, *Pigs, Pearlshells, and Women*, Englewood Cliffs, NJ, Prentice-Hall.

GODELIER, M., 1972, *Rationality and Irrationality in Economics*, transl. B. Pearce, New York, Monthly Review Press.

—— 1982, *La Production des grands hommes*, Paris, Fayard.

GOLDMAN, I., 1970, *Ancient Polynesian Society*, Chicago, University of Chicago Press.

GOODALE, J. C., 1981, 'Siblings as Spouses: The reproduction and replacement of Kaulong society', in Marshall (1981).

GOODENOUGH, W. H., 1953, *Native Astronomy in the Central Carolines*, Philadelphia, University of Pennsylvania Museum Monographs.

—— 1986, 'Sky World and This World: The place of Kachaw in Micronesian cosmology', *American Anthropologist*, 88. 551–68.

GOODY, J., 1977, *The Domestication of the Savage Mind*, Cambridge, Cambridge University Press.

GOUGH, E. K., 1955, 'Female Initiation Rites on the Malabar Coast', *Journal of the Royal Anthropological Institute*, 85. 45–80.

GRAHAM, R. L., and HELL, P., 1985, 'On the History of the Minimum Spanning Tree Problem', *Annals of the History of Computing*, 7. 43–57.

GREEN, R. C., 1979, 'Lapita', in J. D. Jennings (ed.), *The Prehistory of Polynesia*, Cambridge, Harvard University Press.

GREGORY, C. A., 1982, *Gifts and Commodities*, London, Academic Press.

GROFMAN, B., and LANDA, J. T., 1983, 'The Development of Trading Networks among Spatially Separated Traders as a Process of proto-Coalition Formation: The *kula* trade', *Social Networks*, 5. 347–65.

GUIART, J., 1963, *Structure de la chefferie en Mélanésie du sud*, Paris, Institut d'Ethnologie.

GUIDERI, R., 1975, 'Note sur le rapport mâle/femelle en Mélanésie', *L'Homme*, 15. 103–19.

HAGE, P., 1973, 'A Graph Theoretic Approach to the Analysis of Alliance Structure and Local Grouping in Highland New Guinea', *Anthropological Forum*, 3. 280–94.

—— 1976a, 'The Atom of Kinship as a Directed Graph', *Man* (NS), 11. 558–68.

—— 1976b, 'Structural Balance and Clustering in Bushmen Kinship Relations', *Behavioral Science*, 21. 36–47.

—— 1977, 'Centrality in the Kula Ring', *Journal of the Polynesian Society*, 86. 27–36.

—— 1978, 'Speculations on Puluwatese Mnemonic Structure', *Oceania*, 49. 81–95.

—— 1979a, 'A Further Application of Matrix Analysis to Communication Structure in Oceanic Anthropology', *Mathématiques et Sciences Humaines*, 17. 51–69.

HAGE, P., 1979b, 'Symbolic Culinary Mediation: A group model', *Man* (NS), 14. 81–92.

—— 1981, 'On Male Initiation and Dual Organization in New Guinea', *Man* (NS), 16. 268–75.

—— and HARARY, F., 1981a, 'Mediation and Power in Melanesia', *Oceania*, 52. 124–35.

—— —— 1981b, 'Pollution Beliefs in Highland New Guinea', *Man* (NS), 16. 367–75.

—— —— 1983a, *Structural Models in Anthropology*, Cambridge, Cambridge University Press.

—— —— 1983b, 'Arapesh Sexual Symbolism, Primitive Thought and Boolean Groups', *L'Homme*, 23. 57–77.

—— —— 1985, 'Graph Theory', in A. Kuper and J. Kuper (eds.), *The Social Science Encyclopedia*, London, Routledge and Kegan Paul.

—— —— 1986, 'Some Genuine Graph Theoretic Models in Anthropology', *Journal of Graph Theory*, special issue commemorating Leonhard Euler, 10. 353–61.

—— —— and JAMES, B., 1986, 'Wealth and Hierarchy in the *Kula* Ring', *American Anthropologist*, 88. 108–15.

HAGGETT, P., 1967, 'Network Models in Geography', in R. J. Chorley and P. Haggett (eds.), *Models in Geography*, London, Methuen.

—— and CHORLEY, R. J., 1969, *Network Analysis in Geography*, New York, St Martin's Press.

HAINLINE, J., 1964, 'Human Ecology in Micronesia: Determination of population size, dynamics and structure', Ph.D. dissertation, University of California.

—— 1965, 'Culture and Biological Adaptation', *American Anthropologist*, 67. 1174–97.

HAKIMI, S. L., 1964, 'Optimum Locations of Switching Centers and the Absolute Centers and Medians of a Graph', *Operations Research*, 12. 450–9.

HARARY, F., 1953, 'On the Notion of Balance of a Signed Graph', *Michigan Mathematical Journal*, 2. 143–6.

—— 1957, 'Structural Duality', *Behavioral Science*, 2. 255–65.

—— 1959a, 'The Number of Functional Digraphs', *Mathematische Annalen*, 138. 203–10.

—— 1959b, 'Status and Contrastatus', *Sociometry*, 22. 23–43.

—— 1961a, 'A Very Independent Axiom System', *American Mathematical Monthly*, 68. 159–62.

—— 1961b, 'A Parity Relation Partitions its Field Distinctly', *American Mathematical Monthly*, 68. 215–17.

—— 1964, 'A Graph Theoretic Approach to Similarity Relations', *Psychometrika*, 29. 143–51.

—— 1969, *Graph Theory*, Reading, Mass., Addison-Wesley.

—— and MOSER, L., 1966, 'The Theory of Round Robin Tournaments', *American Mathematical Monthly*, 73. 231–46.

—— NORMAN, R. Z., and CARTWRIGHT, D., 1965, *Structural Models: An introduction to the theory of directed graphs*, New York, Wiley.

—— and PETERSON, G. E., 1961, 'Foundations of Phonemic Theory': The structure of language and its mathematical aspects', *Proceedings of the Symposium of the American Mathematical Society*, 12. 139–65.

—— and READ, R., 1966, 'The Probability of a Given 1-Choice Structure', *Psychometrika*, 31. 271–8.

HARDING, T. G., 1967, *Voyagers of the Vitiaz Strait*, Seattle, University of Washington Press.

HARRIS, M., 1971, *Culture, Nature and Man*, New York, Crowell.

HARVEY, D., 1967, 'Models of the Evolution of Spatial Patterns in Human Geography', in R. J. Chorley and P. Haggett (eds.), *Models in Geography*, London, Methuen.

HECHT, J., 1977, 'The Culture of Gender in Pukapuka: Male, female and the *mayakitanga* "sacred maid" ', *Journal of the Polynesian Society*, 86. 183–206.

HEIDER, F., 1946, 'Attitudes and Cognitive Organization', *Journal of Psychology*, 21. 107–12.

HERDT, G. H., 1984, 'Semen Transactions in Sambia Culture', in G. H. Herdt (ed.), *Ritualized Homosexuality in Melanesia*, Berkeley, University of California Press.

HÉRITIER, F., 1981, *L'Exercice de la parenté*, Paris, Gallimard, Le Seuil.

—— 1982, 'The Symbolics of Incest and its Prohibition', in M. Izard and P. Smith (eds.), *Between Belief and Transgression*, transl. J. Leavitt, Chicago, University of Chicago Press.

HEUSCH, L. DE, 1958, *Essai sur le symbolisme de l'inceste royal en Afrique*, Brussels, Institut de Sociologie.

—— 1971, *Pourquoi l'épouser? et autres essais*, Paris, Gallimard.

HEZEL, F. X., 1983, *The First Taint of Civilization*, Honolulu, University of Hawaii Press.

HOCART, A. M., 1915, 'Chieftainship and the Sister's Son in the Pacific', *American Anthropologist*, 17. 631–46.

—— 1929, *Lau Islands, Fiji*, Honolulu, B. P. Bishop Museum Bulletin 62.

—— 1952, *The Northern States of Fiji*, Royal Anthropological Institute, Occasional Publication no. 11.

—— 1970 [1936], *Kings and Councillors* (ed. R. Needham), Chicago, University of Chicago Press.

HOCKETT, C. F., 1958, *A Course in Modern Linguistics*, New York, Macmillan.

HODSON, T. C., 1925, 'Notes on the Marriage of Cousins in India', *Man in India*, 5. 163–74.

HOGBIN, H. I., 1947, 'Native Trade around the Huon Gulf, North-Eastern New Guinea', *Journal of the Polynesian Society*, 56. 242–55.

—— 1970, *The Island of Menstruating Men*, Scranton, Chandler.

HOMANS, G. C., and SCHNEIDER, D. M., 1955, *Marriage, Authority, and Final Causes*, Glencoe, Ill., The Free Press.

HOOPER, A., 1981, *Why Tikopia has Four Clans*, Royal Anthropological Institute, Occasional Paper no. 38.

HUNTSMAN, J., 1981, 'Complementary and Similar Kinsmen in Tokelau', in Marshall (1981).

IRWIN, G. J., 1974, 'The Emergence of a Central Place in Coastal Papuan Prehistory: A theoretical approach', *Mankind*, 9. 268–72.

—— 1978, 'Pots and Entrepôts: A study of settlement, trade and the development of economic specialization in Papuan prehistory', *World Archaeology*, 9. 299–319.

—— 1983, 'Chieftainship, *Kula* and Trade in Massim Prehistory', in Leach and Leach (1983).

JOHANSEN, J. P., 1954, *The Maori and his Religion*, Copenhagen, Munksgaard.

JOSSELIN DE JONG, J. P. B. DE, 1952, *Lévi-Strauss's Theory on Kinship and Marriage*, Leiden, Brill.

KAEPPLER, A. L., 1971, 'Rank in Tonga', *Ethnology*, 10. 174–93.

—— 1978, 'Exchange Patterns in Goods and Spouses: Fiji, Tonga and Samoa', *Mankind*, 11. 246–52.

KAPFERER, B., 1969, 'Norms and the Manipulation of Relationships in a Work Context', in J. C. Mitchell (ed.), *Social Networks in Urban Situations*, Manchester, Manchester University Press.

KELLY, R. C., 1968, 'L'Échange généralisé à Dobu', *L'Homme*, 8. 54–61.

—— 1974, *Etoro Social Structure*, Ann Arbor, University of Michigan Press.

—— 1976, 'Witchcraft and Sexual Relations: An exploration in the social and semantic implications of the structure of belief', in P. Brown and G. Buchbinder (eds.), *Man and Woman in the New Guinea Highlands*, Special Publication 8 of the American Anthropological Association.

KIRCH, P. V., 1984, *The Evolution of the Polynesian Chiefdoms*, Cambridge, Cambridge University Press.

KIRCHOFF, P., 1932, 'Verwandtschaftsbezeichnungen und Verwandtenheirat', *Zeitschrift für Ethnologie*, 64. 41–71.

KOMATSU, K., 1985, 'Two Tales from the Sacred Island of [the] Central Carolines', in E. Ishikawa (ed.), *The 1983–84 Cultural Anthropological Expedition to Micronesia: An interim report*, Tokyo, Tokyo Metropolitan University.

KÖNIG, D., 1936, *Theorie der endlichen und unendlichen Graphen*, Leipzig, Akademische Verlagsgesellschaft M.B.H.; reprinted by Chelsea, New York, 1950.

KOTZEBUE, O. VON, 1821, *A Voyage of Discovery into the South Sea and Beering's Straits ... Undertaken in the Years 1815–18 ... in the Ship Rurick*, transl. H. E. Lloyd, London, Longman, Hurst, Rees, Orme and Brown.

KRÄMER, A., 1935, *Inseln um Truk*, Ergebnisse der Südsee-Expedition 1908–10, ed. G. Thilenius, II B6, Hamburg, Friederichsen, de Gruyter.

—— 1937, *Zentralkarolinen*, Ergebnisse der Südsee-Expedition 1908–10, ed. G. Thilenius, II B10, Hamburg, Friederichsen, de Gruyter.

KRUSKAL, J. B., 1956, 'On the Shortest Spanning Tree of a Graph and the Traveling Salesman Problem', *Proceedings of the American Mathematical Society*, 7. 48–50.

KURATOWSKI, K., 1930, 'Sur le problème des courbes gauches en topologie', *Fundamenta Mathematicae*, 15. 271–83.

LABBY, D., 1976, *The Demystification of Yap*, Chicago, University of Chicago Press.

LANDA, J. T., 1983, 'The Enigma of the *Kula* Ring: Gift exchanges and primitive law and order', *International Review of Law and Economics*, 3. 137–60.

LANGNESS, L. L., 1967, 'Sexual Antagonism in the New Guinea Highlands: A Bena Bena example', *Oceania*, 37. 161–77.

—— 1977, 'Ritual, Power, and Male Dominance in the New Guinea Highlands', in R. D. Fogelson and R. N. Adams (eds.), *The Anthropology of Power*, New York, Academic Press.

LAUER, P. K., 1970, 'Amphlett Islands' Pottery Trade and the *Kula*', *Mankind*, 7. 165–76.

LAUTERBACH, C., 1898, 'Die geographischen Ergebnisse der Kaiser-Wilhelmsland Expedition', *Zeitschrift der Gesellschaft für Erdkunde zu Berlin*, 33. 141–75.

LAWRENCE, P., 1955, *Land Tenure among the Garia*, Canberra, Australian National University.

—— 1980, 'Obituary: Reo Franklin Fortune', *Oceania*, 51. 2–3.

LAYARD, J., 1942, *Stone Men of Malekula*, London, Chatto and Windus.

LEACH, E. R., 1951, 'The Structural Implications of Matrilateral Cross-Cousin Marriage', *Journal of the Royal Anthropological Institute*, 81. 23–55. Reprinted in Leach (1961).

—— 1957, 'On Asymmetrical Marriage Systems', *American Anthropologist*, 59. 343–4.

—— 1961, *Rethinking Anthropology*, London, Athlone Press.

—— 1972, 'The Structure of Symbolism', in J. S. La Fontaine (ed.), *The Interpretation of Ritual*, London, Tavistock.

—— 1983, 'The *Kula*: An alternative view', in Leach and Leach (1983).

LEACH, J. W., 1983, introduction to Leach and Leach (1983).

—— and LEACH, E. R. (eds.), 1983, *The Kula: New perspectives on Massim exchange*, Cambridge, Cambridge University Press.

LEENHARDT, M., 1937, *Gens de la Grande Terre: Nouvelle Calédonie*, Paris, Gallimard.
—— 1980 [1930], *Notes d'ethnologie néo-calédonienne*, Paris, Institut d'Ethnologie.
LEMAÎTRE, Y., 1970, 'Les Relations inter-insulaires traditionelles en Océanie: Tonga', *Journal de la Société des Océanistes*, 26. 93–105.
LESSA, W. A., 1950, 'Ulithi and the Outer Native World', *American Anthropologist*, 52. 27–52.
—— 1956, 'Myth and Blackmail in the Western Carolines', *Journal of the Polynesian Society*, 65. 66–74.
—— 1959, 'Divining from Knots in the Carolines', *Journal of the Polynesian Society*, 68. 188–204.
—— 1962, 'An Evaluation of Early Descriptions of Carolinian Culture', *Ethnohistory*, 9. 313–402.
—— 1966, *Ulithi: A Micronesian design for living*, New York, Holt, Rinehart, and Winston.
—— 1975, *Drake's Island of Thieves*, Honolulu, University of Hawaii Press.
—— 1983, 'The Mapia Islands and their Affinities', in N. Gunson (ed.), *The Changing Pacific: Essays in honor of H. E. Maude*, Oxford, Oxford University Press.
LÉVI-STRAUSS, C., 1944, 'Reciprocity and Hierarchy', *American Anthropologist*, 46. 266–8.
—— 1949, *Les Structures élémentaires de la parenté*, Paris, Presses Universitaires de France.
—— 1950, 'Introduction à l'œuvre de Marcel Mauss', in M. Mauss, *Sociologie et anthropologie*, Paris, Presses Universitaires de France.
—— 1963, *Structural Anthropology*, transl. C. Jacobson and B. G. Schoepf, New York, Basic Books.
—— 1963a, 'Social Structure', ibid.
—— 1963b, 'Do Dual Organizations Exist?', ibid.
—— 1963c, 'Structural Analysis in Linguistics and in Anthropology', ibid.
—— 1963d, 'Linguistics and Anthropology', ibid.
—— 1966, 'The Future of Kinship Studies: The Huxley Memorial Lecture 1965', *Proceedings of the Royal Anthropological Institute for 1965*, pp. 13–22.
—— 1969, *The Elementary Structures of Kinship*, revised edn., transl. J. H. Bell and J. R. von Sturmer, Boston, Beacon Press.
—— 1970, *The Raw and the Cooked*, transl. J. and D. Weightman, New York, Harper and Row.
—— 1973, *From Honey to Ashes*, transl. J. and D. Weightman, New York, Harper and Row.
—— 1976, 'Reflections on the Atom of Kinship', in *Structural Anthropology*, vol. 2, transl. M. Layton, New York, Basic Books.

—— 1978, *The Origin of Table Manners*, transl. J. and D. Weightman, New York, Harper and Row.

—— 1981, *The Naked Man*, transl. J. and D. Weightman, New York, Harper and Row.

—— 1984, *Paroles données*, Paris, Plon.

—— 1985, 'From Mythical Possibility to Social Existence', in *The View from Afar*, transl. J. Neugroschel and P. Hoss, New York, Basic Books.

LEWIS, D., 1972, *We, the Navigators*, Honolulu, University of Hawaii Press.

—— 1978, *The Voyaging Stars*, Sydney, Fontana Collins.

LINDENBAUM, S., 1972, 'Sorcerers, Ghosts, and Polluting Women: An analysis of religious belief and population control', *Ethnology*, 11. 241–53.

LINGENFELTER, S. G., 1975, *Yap: Political leadership and culture change in an island society*, Honolulu, University of Hawaii Press.

LIVINGSTONE, F. B., 1969, 'The Application of Structural Models to Marriage Systems in Anthropology', in I. R. Buchler and H. G. Nutini (eds.), *Game Theory in the Behavioral Sciences*, Pittsburgh, University of Pittsburgh Press.

LOEB, E. M., 1926, *History and Traditions of Niue*, Honolulu, B. P. Bishop Museum Bulletin 32.

LÖSCH, A., 1954, *The Economics of Location*, transl. W. F. Stolper, New Haven, Yale University Press.

LOWIE, R. H., 1928, 'A Note on Relationship Terminologies', *American Anthropologist*, 30. 263–7.

MABUCHI, T., 1960, 'The Two Types of Kinship Rituals among Malayo-Polynesian Peoples', *Proceedings of the Ninth International Congress for the History of Religions, 1958, Tokyo and Kyoto*, Tokyo, Maruzen.

—— 1964, 'Spiritual Predominance of the Sister', in A. H. Smith (ed.), *Ryūkyūan Culture and Society: A survey*, Tenth Pacific Science Congress Series, Honolulu, University of Hawaii Press.

MACARTHUR, R. H., and WILSON, E. O., 1967, *The Theory of Island Biogeography*, Princeton, Princeton University Press.

MCCOY, M., 1973, 'A Renaissance in Carolinian–Marianas Voyaging', *Journal of the Polynesian Society*, 82. 355–65.

—— 1974, 'Man and Turtle in the Central Carolines', *Micronesica*, 10. 207–21.

MCDOWELL, N., 1976, 'Kinship and Exchange: The *kaiman* relationship in a Yuat River village', *Oceania*, 47. 36–48.

—— 1980, 'It's not who you are but how you give that counts: The role of exchange in a Melanesian society', *American Ethnologist*, 7. 58–70.

MACINTYRE, M., 1983a, *The Kula: A bibliography*, Cambridge, Cambridge University Press.

MACINTYRE, M., 1983b, '*Kune* on Tubetube and in the Bwanabwana Region of the Southern Massim', in Leach and Leach (1983).

MALINOWSKI, B., 1920, '*Kula*: The circulating exchange of valuables in the archipelagoes of Eastern New Guinea', *Man*, 20. 97–105.

—— 1922, *Argonauts of the Western Pacific*, London, Routledge and Kegan Paul.

—— 1929, *The Sexual Life of Savages in North-Western Melanesia*, London, Routledge and Kegan Paul.

—— 1935, *Coral Gardens and their Magic*, 2 vols., London, Allen and Unwin.

MANCHESTER, C. A., 1951, 'The Caroline Islands', in O. W. Freeman (ed.), *Geography of the Pacific*, New York, Wiley.

MARSHALL, [K.] M., 1972, 'The Structure of Solidarity and Alliance on Namoluk Atoll', Ph.D. dissertation, University of Washington.

—— (ed.), 1981, *Siblingship in Oceania*, ASAO Monograph no. 8, Ann Arbor, University of Michigan Press.

—— 1981a, 'Introduction: Approaches to siblingship in Oceania', ibid.

—— 1981b, 'Sibling Sets as Building Blocks in Greater Trukese Society', ibid.

—— 1984, 'Structural Patterns of Sibling Classification in Island Oceania: Implications for culture history', *Current Anthropology*, 25. 597–637.

MASON, L., 1968, 'Suprafamilial Authority and Economic Process in Micronesian Atolls', in A. P. Vayda (ed.), *Peoples and Cultures of the Pacific*, New York, Natural History Press.

MAUSS, M., 1935, 'Les Techniques du corps', *Journal de Psychologie Normale et Pathologique*, 32. 271–93.

—— 1950, 'Essai sur le don: Forme et raison de l'échange dans les sociétés archaïques', in *Sociologie et anthropologie*, Paris, Presses Universitaires de France.

MAYBURY-LEWIS, D. H. P., 1965, 'Prescriptive Marriage Systems', *Southwestern Journal of Anthropology*. 21. 207–30.

MEAD, M., 1928, *Coming of Age in Samoa*, New York, Morrow.

—— 1934, *Kinship in the Admiralty Islands*, American Museum of Natural History, Anthropological Papers, vol. 34.

—— 1938, *The Mountain Arapesh I: An importing culture*, American Museum of Natural History, Anthropological Papers, vol. 36.

—— 1940, *The Mountain Arapesh II: Supernaturalism*, American Museum of Natural History, Anthropological Papers, vol. 37.

—— 1947, *The Mountain Arapesh III: Socioeconomic life*; *IV: Diary of events in Alitoa*, American Museum of Natural History, Anthropological Papers, vol. 40.

—— 1961, 'The Arapesh of New Guinea', in M. Mead (ed.), *Cooperation and Competition*, Boston, Beacon Press.

—— 1963 [1935], *Sex and Temperament in Three Primitive Societies*, New York, Morrow.

—— 1969 [1930], *Social Organization of Manu'a*, 2nd edn., Honolulu, B. P. Bishop Museum Bulletin 76.

—— 1972, *Blackberry Winter*, New York, Washington Square Press.

MEGGITT, M. J., 1964, 'Male–Female Relationships in the Highlands of Australian New Guinea', *American Anthropologist*, 66. 204–24.

—— 1974, ' "Pigs are our hearts!" The *te* exchange cycle among the Mae Enga of New Guinea', *Oceania*, 44. 165–203.

—— 1976, 'A Duplicity of Demons: Sexual and familial roles expressed in Western Enga stories', in P. Brown and G. Buchbinder (eds.), *Man and Woman in the New Guinea Highlands*, Special Publication 8 of the American Anthropological Association.

MEIGS, A. S., 1978, 'A Papuan Perspective on Pollution', *Man* (NS), 13. 304–18.

MITCHELL, J. C., 1969, 'The Concept and Use of Social Networks', in J. C. Mitchell (ed.), *Social Networks in Urban Situations*, Manchester, Manchester University Press.

MOSKO, M. S., 1985, *Quadripartite Structures*, Cambridge, Cambridge University Press.

MÜLLER-WISMAR, W., 1918, *Jap*, Ergebnisse der Südsee-Expedition 1908–10, ed. G. Thilenius, II B, Hamburg, Friederichsen, de Gruyter.

MURDOCK, G. P., 1949, *Social Structure*, New York, Free Press.

—— and GOODENOUGH, W. H., 1947, 'Social Organization of Truk', *Southwestern Journal of Anthropology*. 3, 331–43.

NASH, J., 1978, 'A Note on Groomprice', *American Anthropologist*, 80. 106–8.

NEEDHAM, R., 1958, 'The Formal Analysis of Prescriptive Patrilateral Cross-Cousin Marriage', *Southwestern Journal of Anthropology*, 14. 199–219.

—— 1968, translator's note to Wouden (1968).

—— 1970, editor's introduction to Hocart (1970).

—— 1971a, introduction to R. Needham (ed.), *Rethinking Kinship and Marriage*, London, Tavistock.

—— 1971b, 'Remarks on the Analysis of Kinship and Marriage', ibid.

—— 1983, 'Wittgenstein's Arrows', in *Against the Tranquility of Axioms*, Berkeley, University of California Press.

OLIVER, D. L., 1955, *A Solomon Island Society*, Cambridge, Harvard University Press.

—— 1974, *Ancient Tahitian Society*, 3 vols., Honolulu, University of Hawaii Press.

ORANS, M., 1966, 'Surplus', *Human Organization*, 25. 24–32.

ORE, O., 1962, *Theory of Graphs*, Providence, American Mathematical Society.

ORE, O., 1963, *Graphs and their Uses*, New York, Random House.

PEIRCE, C. S., 1931, 'Lessons from the History of Science', in *Collected Papers*, vol. 1, ed. C. Hartshorne and P. Weiss, Cambridge, Harvard University Press.

—— 1933, 'Nomenclature and Divisions of Dyadic Relations', in *Collected Papers*, vol. 3, ed. C. Hartshorne and P. Weiss, Cambridge, Harvard University Press.

PERSSON, J., 1983, 'Cyclical Change and Circular Exchange: A re-examination of the *kula* ring', *Oceania*, 54. 32–47.

PIAGET, J., 1971, *Structuralism*, transl. and ed. C. Maschler, New York, Harper and Row.

PIELOU, E. C., 1969, *An Introduction to Mathematical Ecology*, New York, Wiley.

PISARIK, S., 1975, 'Micronesian Atoll Populations: A path analysis', MA Thesis, Department of Anthropology, University of Iowa.

PITTS, F. R., 1965, 'A Graph Theoretic Approach to Historical Geography', *Professional Geographer*, 17. 15–20.

POLLOCK, N. J., 1975, 'The Risks of Dietary Change: A Pacific atoll example', in R. W. Casteel and G. I. Quimby (eds.), *Maritime Adaptations of the Pacific*, The Hague, Mouton.

POUILLON, J., 1970, 'L'Hôte disparu et les tiers incommodes', in J. Pouillon and P. Maranda (eds.), *Échanges et communications*, The Hague, Mouton.

POWDERMAKER, H., 1933, *Life in Lesu*, London, Williams and Norgate.

POWELL, H. A., 1969, 'Territory, Hierarchy and Kinship in Kiriwina', *Man* (NS), 4. 580–604.

PRIM, R. C., 1957, 'Shortest Connection Networks and some Generalizations', *Bell System Technical Journal*, 36. 1389–1401.

QUACKENBUSH, E. M., 1968, 'From Sonsorol to Truk: A dialect chain', Ph.D. Dissertation, University of Michigan.

QUAIN, B., 1948, *Fijian Village*, Chicago, University of Chicago Press.

RADCLIFFE-BROWN, A. R., 1913, 'Three Tribes of Western Australia', *Journal of the Royal Anthropological Institute*, 43. 143–94.

—— 1924, 'The Mother's Brother in South Africa', *South African Journal of Science*, 21. 542–55.

—— 1952, *Structure and Function in Primitive Society*, New York, Free Press.

RAPPAPORT, R. A., 1968, *Pigs for the Ancestors*, New Haven, Yale University Press.

—— 1979, 'Aspects of Man's Influence on Island Ecosystems: Alteration and control', in *Ecology, Meaning and Religion*, Berkeley, North Atlantic Books.

READ, K. E., 1952, '*Nama* cult of the Central Highlands, New Guinea', *Oceania*, 23. 1–25.

RESTLE, F., 1959, 'A Metric and an Ordering on Sets', *Psychometrika*, 24. 207–20.

RIESENBERG, S. H., 1968, *The Native Polity of Ponape*, Washington, Smithsonian Institution Press.

—— 1972, 'The Organisation of Navigational Knowledge on Puluwat', *Journal of the Polynesian Society*, 81. 19–56.

RIVERS, W. H. R., 1910, 'The Father's Sister in Oceania', *Folk-Lore*, 21. 42–59.

—— 1914, *The History of Melanesian Society*, 2 vols., Cambridge, Cambridge University Press.

ROBERTS, F. S., 1984, *Applied Combinatorics*, Englewood Cliffs, NJ, Prentice-Hall.

ROGERS, G., 1977, '"The father's sister is black": A consideration of female rank and power in Tonga', *Journal of the Polynesian Society*, 86. 157–82.

RÓHEIM, G., 1950, *Psychoanalysis and Anthropology*, New York, International Universities Press.

ROLETT, B., 1986, 'Turtles, Priests and the Afterworld: A study in the iconographic interpretation of Polynesian petroglyphs', in P. V. Kirch (ed.), *Island Societies*, Cambridge, Cambridge University Press.

ROMNEY, A. K., and EPLING, P. J., 1958, 'A Simplified Model of Kariera Kinship', *American Anthropologist*, 60. 59–74.

ROSMAN, A., and RUBEL, P. G., n.d., 'The Material Basis of Dual Organization'.

RUBEL, P. G., and ROSMAN, A., 1978, *Your own Pigs you may not Eat*, Chicago, University of Chicago Press.

RUSSELL, B., 1917, 'Mathematics and the Metaphysicians', in *Mysticism and Logic*, London, Allen and Unwin.

RYDER, J. W., and BLACKMAN, M. B., 1970, 'The Avunculate: A cross-cultural critique of Claude Lévi-Strauss', *Behavioral Science Notes*, 5. 97–115.

SAHLINS, M. D., 1958, *Social Stratification in Polynesia*, Seattle, University of Washington Press.

—— 1962, *Moala*, Ann Arbor, University of Michigan Press.

—— 1963, 'Poor Man, Rich Man, Big Man, Chief: Political types in Melanesia and Polynesia', *Comparative Studies in Society and History*, 5. 285–303.

—— 1972, 'On the Sociology of Primitive Exchange', in *Stone Age Economics*, Chicago, Aldine.

—— 1976, *Culture and Practical Reason*, Chicago, University of Chicago Press.

—— 1985, *Islands of History*, Chicago, University of Chicago Press.

SALISBURY, R. F., 1956, 'Asymmetrical Marriage Systems', *American Anthropologist*, 58. 639–55.

SCHEFFLER, H. W., 1978, *Australian Kin Classification*, Cambridge, Cambridge University Press.

SCHNEIDER, D. M., 1962, 'Double Descent on Yap', *Journal of the Polynesian Society*, 71. 1–24.

—— 1981, 'Conclusions', in Marshall (1981).

—— 1984, *A Critique of the Study of Kinship*, Ann Arbor, University of Michigan Press.

SCHWARTZ, T., 1963, 'Systems of Areal Integration: Some considerations based on the Admiralty Islands of northern Melanesia', *Anthropological Forum*, 2. 56–97.

SCHWIMMER, E., 1973, *Exchange in the Social Structure of the Orokaiva*, New York, St Martins Press.

—— 1977, 'F. E. Williams as Ancestor and Rain-Maker', introduction to Williams (1977).

SELIGMAN, C. G., 1910, *The Melanesians of British New Guinea*, Cambridge, Cambridge University Press.

SENFFT, A., 1904, 'Bericht über den Besuch einiger Inselgruppen der West-Karolinen', *Mitteilungen von Forschungsreisenden und Gelehrten aus den deutschen Schutzgebieten*, 17. 192–7.

SERPENTI, L. M., 1965, *Cultivators in the Swamps*, Assen, Netherlands, Van Gorcum.

SHAPIRO, W., 1976, review of P. O. Nsugbe, 'Ohaffia, A Matrilineal Ibo People', *Mankind*, 10. 285–6.

—— 1982, 'The Place of Cognitive Extensionism in the History of Anthropological Thought', *Journal of the Polynesian Society*, 91. 257–97.

SHARP, A., 1964, *Ancient Voyagers in Polynesia*, Berkeley, University of California Press.

SHORE, B., 1976, 'Incest Prohibitions and the Logic of Power in Samoa', *Journal of the Polynesian Society*, 85. 275–96.

—— 1982, *Sala'ilua: A Samoan mystery*, New York, Columbia University Press.

SMITH, C. A., 1976, *Regional Analysis*, New York, Academic Press.

SMITH, D. R., 1981, 'Palauan Siblingship: a study in structural complementarity', in Marshall (1981).

—— 1983, *Palauan Social Structure*, New Brunswick, Rutgers University Press.

SOHN, H. M., and TAWERILMANG, A. F., 1976, *Woleaian–English Dictionary*, Honolulu, University of Hawaii Press.

SPERBER, D., 1968, *Le Structuralisme en anthropologie*, Paris, Le Seuil.

STRATHERN, A. J., 1969, 'Descent and Alliance in the New Guinea Highlands: Some problems of comparison', *Proceedings of the Royal Anthropological Institute for 1968*, 37–52.

—— 1970, 'The Female and Male Spirit Cults in Mount Hagen', *Man* (NS), 5. 571–85.

—— 1971, *The Rope of Moka*, Cambridge, Cambridge University Press.

—— 1977, 'Melpa Food-Names as an Expression of Ideas on Identity and Substance', *Journal of the Polynesian Society*, 86. 503–11.

—— 1979, 'Men's House, Women's House: The efficacy of opposition, reversal and pairing in the Melpa *amb kor* cult', *Journal of the Polynesian Society*, 88. 37–51.

STRATHERN, M., 1972, *Women in Between*, London, Seminar Press.

—— 1984, 'Marriage Exchanges: A Melanesian comment', *Annual Review of Anthropology*, 13. 41–73.

SWADESH, M., 1954, 'Perspectives and Problems of Amerindian Comparative Linguistics', *Word*, 10. 306–32.

TAAFFE, E., and GAUTHIER, H. C., 1973, *Geography of Transportation*, Englewood Cliffs, NJ, Prentice-Hall.

TAMBIAH, S. J., 1983, 'On Flying Witches and Flying Canoes: The coding of male and female values', in Leach and Leach (1983).

TERRELL, J., 1977, 'Human Biogeography in the Solomon Islands', *Fieldiana: Anthropology*, 68. 1–47.

—— 1986, *Prehistory in the Pacific Islands*, Cambridge, Cambridge University Press.

THOMSON, B., 1908, *The Fijians*, London, Heinemann.

THURNWALD, H., 1938, 'Ehe und Mutterschaft in Buin', *Archiv für Anthropologie und Völkerforschung*, 24. 214–46.

THURNWALD, R., 1916, *Bánaro Society: Social organization and kinship system of a tribe in the interior of New Guinea*, American Anthropological Association Memoirs, vol. 3.

TURNER, G. A., 1861, *Nineteen Years in Polynesia*, London, Snow.

—— 1884, *Samoa, a Hundred Years Ago and Long Before*, London, Macmillan.

TUZIN, D. F., 1976, *The Ilahita Arapesh*, Berkeley, University of California Press.

UBEROI, J. P. S., 1971, *Politics of the Kula Ring*, Manchester, Manchester University Press.

UNITED STATES DEPARTMENT OF COMMERCE, 1968, *World Weather, 1951–1960*, vol. 6: *Antarctica, Australia, Oceanic Islands, and Ocean Weather Stations*, Environmental Services Administration, Environmental Data Service, Washington, DC.

VALERI, V., 1972, 'Le Fonctionnement du système des rangs à Hawaii', *L'Homme*, 12. 29–66.

—— 1985, *Kinship and Sacrifice*, transl. P. Wissing, Chicago, University of Chicago Press.

VAN DER LEEDEN, A. C., 1960, 'Social Structure in New Guinea', *Bijdragen tot de Taal-, Land-, en Volkenkunde*, 116. 119–49.

WAGNER, K., 1937, 'Ueber eine Eigenschaft der ebenen Komplexe', *Mathematische Annalen*, 114. 570–90.

WAGNER, R., 1967, *The Curse of Souw*, Chicago, University of Chicago Press.

—— 1972, 'Mathematical Prediction of Polygyny Rates among the Daribi of Karimui Patrol Post, Territory of Papua and New Guinea', *Oceania*, 42. 205–22.

WARNER, W. L., 1931, 'Morphology and Functions of the Australian Murngin Type of Kinship (Part 2)', *American Anthropologist*, 33. 172–98.

WATSON, J. B., 1964, 'Introduction: Anthropology in the New Guinea Highlands', in *American Anthropologist* Special Publication, vol. 66, no. 4, pt. 2, J. B. Watson (ed.), *New Guinea, the Central Highlands*.

WEINER, A. B., 1976, *Women of Value, Men of Renown*, Austin, University of Texas Press.

WEINER, J. F., 1982, 'Substance, Siblingship and Exchange: Aspects of social structure in New Guinea', *Social Analysis*, 11. 3–34.

WIENS, H. J., 1962, *Atoll Environment and Ecology*, New Haven, Yale University Press.

WILLIAMS, F. E., 1928, *Orokaiva Magic*, London, Oxford University Press.

—— 1930, *Orokaiva Society*, London, Oxford University Press.

—— 1932, 'Sex Affiliation and its Implications', *Journal of the Royal Anthropological Institute*, 62. 51–81.

—— 1934, 'Exchange Marriage and Exogamy: Summary of a communication', *Man*, 34. 110.

—— 1936, *Papuans of the Trans-Fly*, Oxford, Clarendon Press.

—— 1940, *Drama of Orokolo*, Oxford, Clarendon Press.

—— 1977, *'The Vailala Madness' and Other Essays*, ed. E. Schwimmer, Honolulu, University of Hawaii Press.

WILLIAMSON, I., and SABATH, M. D., 1982, 'Island Population, Land Area, and Climate: A case study of the Marshall Islands', *Human Ecology*, 10. 71–84.

—— —— 1984, 'Small Population Instability and Island Settlement Patterns', *Human Ecology*, 12. 21–34.

WOUDEN, F. A. E. VAN, 1968 [1935], *Types of Social Structure in Eastern Indonesia*, transl. R. Needham, The Hague, Martinus Nijhoff.

YOUNG, M. W., 1971, *Fighting with Food*, Cambridge, Cambridge University Press.

INDEX